R. Mahrwald (Ed.)

Modern Aldol Reactions

Vol. 2: Metal Catalysis

Also of Interest

Grubbs, R. H. (Ed.)

Handbook of Metathesis

3 Volumes

2003
ISBN 3-527-30616-1

Nicolaou, K. C., Snyder, S. A.

**Classics in Total Synthesis II
More Targets, Strategies, Methods**

2003
ISBN 3-527-30685-4 (Hardcover with CD-Rom)
ISBN 3-527-30684-6 (Softcover)

De Meijere, A., Diederich, F. (Eds.)

Metal-Catalyzed Cross-Coupling Reactions

Second, Completely Revised and Extended Edition

2 Volumes

2004
ISBN 3-527-30518-1

Krause, N., Hashmi, A. S. K. (Eds.)

Modern Allene Chemistry

2 Volumes

2004
ISBN 3-527-30671-4

Rainer Mahrwald (Ed.)

Modern Aldol Reactions

Vol. 2: Metal Catalysis

WITHDRAWN

WILEY-VCH Verlag GmbH & Co. KGaA

PD Dr. Rainer Mahrwald
Department of Organic Chemistry
Humboldt University
Brook-Taylor-Str. 2
12489 Berlin
Germany

■ This book was carefully produced. Nevertheless, editor, authors and publisher do not warrant the information contained therein to be free of errors. Readers are advised to keep in mind that statements, data, illustrations, procedural details or other items may inadvertently be inaccurate.

Library of Congress Card No.: Applied for
British Library Cataloguing-in-Publication Data: A catalogue record for this book is available from the British Library.
Bibliographic information published by Die Deutsche Bibliothek
Die Deutsche Bibliothek lists this publication in the Deutsche Nationalbibliografie; detailed bibliographic data is available in the Internet at http://dnb.ddb.de

Printed in the Federal Republic of Germany.
Printed on acid-free paper.

Typesetting   Asco Typesetters, Hong Kong
Printing   Strauss Gmbh, Mörlenbach
Bookbinding   Litges & Dopf Buchbinderei GmbH, Heppenheim

ISBN 3-527-30714-1

Foreword

Historically, the stimulus for the development of a particular reaction has been interconnected with a class of natural products whose synthesis would be greatly facilitated by the use of that particular bond construction. For example, the steroid synthesis challenges proved instrumental in the development of the Diels–Alder reaction. So too the synthesis challenges associated with the macrolide antibiotics have provided the motivation for the development of the full potential of the aldol addition reaction. R. B. Woodward's 1956 quote on the "hopelessly complex" architecture of the erythromycins was probably stimulated, in part, by the fact that the aldol reaction existed in a completely underdeveloped state five decades ago.

The erythromycin-A structure, as viewed by Woodward in the '50s

"Erythromycin, with all of our advantages, looks at present quite hopelessly complex, particularly in view of its plethora of asymmetric centers."

R. B. Woodward in *Perspectives in Organic Chemistry*; Todd, A. Ed.; Wiley-Interscience, New York, 1956, page 160.

The challenges associated with the development of this reaction are also embodied in the more general goals of acyclic stereocontrol that have been under active investigation for nearly twenty-five years. In these studies, the goal of understanding pi-face selectivity at trigonal carbon centers for a multitude of organic transformations has been the ultimate objective. From these research activities, a host of stereochemical models have evolved, such as the Felkin–Anh model for carbonyl addition and the Zimmermann–Traxler aldol stereochemical model for aldol diastereoselection.

The development of modern aldol reaction methods has evolved through a succession of pivotal discoveries that have advanced the whole field of stereoselective synthesis:

A. Development of enolization strategies for the formation of (E) and (Z) enolates.
B. Development of kinetic diastereoselective aldol addition variants through the discovery of optimal metal architectures [B(III), Ti(IV), Sn(II)].
C. Discovery of aldol reaction variants such as the Lewis acid catalyzed addition of enolsilanes to aldehydes (Mukaiyama aldol variant).
D. Development of chiral enolates exhibiting exceptional pi-face selectivities.
E. Development of chiral metal complexes as Lewis acid aldol catalysts.

This two-volume series on aldol addition reaction methodology brings together an up-to-date discussion of all aspects of this versatile process. The reader will gain an appreciation for the role of metal enolate architecture in aldol diastereoselectivities (Vol. I; Chapters 1–3) and for the utility of chiral metal complexes in the catalysis of the Mukaiyama aldol reaction (Vol. II; Chapters 1–3, 5). In Vol. II; Chapter 6, enantioselective catalytic processes incorporating both enolization and addition are surveyed, as is the exciting progress being made in the use of chiral amines as aldol catalysts (Vol. I; Chapter 4). This highly active area of research will continue to develop ever more versatile chiral catalysts and stereochemical control concepts.

Students and researchers in the field of asymmetric synthesis will greatly profit from the contributions of this distinguished group of authors who have so insightfully reviewed this topic.

May 2004

David A. Evans
Harvard University

Contents

Volume 1

Preface

The aldol reaction was first described by Kane in 1848. Thus it is high time to provide a comprehensive overview of the different developments in aldol chemistry, especially those of the past few decades. Demands for this important method of C–C–bond formation came and continue to come from every field of synthetic chemistry, particularly from natural product synthesis. Here, challenging problems in regioselectivity, chemoselectivity, diastereoselectivity and enantioselectivity frequently arise, many of which are still awaiting a solution. Symptomatically the word "selectivity" in its various connotations occurs no fewer than 1,100 times in both volumes, i.e. an average of twice a page.

This book examines the enormous variety of aldol chemistry from the view of both organic as well as inorganic and bioorganic chemistry. It presents a wide range of potent syntheses based on the discoveries from enolate chemistry or the catalysis of Lewis acids and Lewis bases, for instance. The important role of metal catalysis, organocatalysis and direct aldol addition is described, along with enzymatic methods. However, it was not our intention to simply list all existing publications about aldol chemistry. Instead, we wanted to point out fundamental and at the same time efficient ways leading to defined configured aldol products. Two of these are depicted on the cover: the metal catalysis and the enzymatic method.

It is now my pleasure to express my profound gratitude to the 22 authors and co-authors, all belonging to the elite of aldol chemistry, for their outstanding contributions and their professional cooperation. Special thanks are due to Wiley-VCH, especially Elke Maase and Rainer Münz, for their fine work in turning the manuscript into the finished book. Finally, I am indebted to my wife and my son for countless hours of assistance.

Last but not least, this book is also a tribute to the works of Teruaki Mukaiyama, who has done tremendous work in the field of aldol reaction and now celebrates his 77th birthday.

Berlin, Germany   Rainer Mahrwald
May 2004

List of Contributors

Editor

PD Dr. Rainer Mahrwald
Institut für Organische und Bioorganische
Chemie
der Humboldt-Universität zu Berlin
Brook-Taylor-Str. 2
12489 Berlin
Germany

Authors

Prof. Dr. Carlos F. Barbas, III
The Skaggs Institute for Chemical Biology
and the Department of Molecular Biology
The Scripps Research Institute
10550 North Torrey Pines Road
La Jolla, CA 92037
USA

Prof. Dr. Manfred Braun
Institut für Organische Chemie und
Makromolekulare Chemie I
Heinrich-Heine-Universität Düsseldorf
Universitätsstr. 1
40225 Düsseldorf
Germany

Prof. Dr. Scott E. Denmark
245 Roger Adams Laboratory, Box 18
Department of Chemistry
University of Illinois
600 S. Mathews Avenue
Urbana, IL 61801
USA

Prof. Dr. Wolf-Dieter Fessner
TU Darmstadt
Department of Organic Chemistry and
Biochemistry
Petersenstr. 22
64287 Darmstadt
Germany

Shinji Fujimori
236 Roger Adams Laboratory, Box 91-5
Department of Chemistry
University of Illinois
600 S. Mathews Avenue
Urbana, IL 61801
USA

Prof. Dr. Arun K. Ghosh
Department of Chemistry
University of Illinois at Chicago
845 West Taylor Street
Chicago, IL 60607
USA

Prof. Dr. Kazuaki Ishihara
Graduate School of Engineering
Nagoya University
Chikusa
Nagoya, 464-8603
Japan

Prof. Dr. Jeffrey S. Johnson
Department of Chemistry
University of North Carolina at Chapel
Hill
Chapel Hill, NC 27599-3290
USA

Prof. Dr. Shū Kobayashi
Graduate School of Pharmaceutical
Sciences
The University of Tokyo
Hongo, Bunkyo-ku
Tokyo 113-0033
Japan

Naoya Kumagai
Graduate School of Pharmaceutical
Sciences
The University of Tokyo
Hongo 7-3-1, Bunkyo-ku
Tokyo, 113-0033
Japan

Prof. Dr. Benjamin List
Max-Planck-Institut für Kohlenforschung
Kaiser-Wilhelm-Platz 1
45470 Mülheim an der Ruhr
Germany

PD Dr. Rainer Mahrwald
Institut für Organische und Bioorganische
Chemie
der Humboldt-Universität zu Berlin
Brook-Taylor-Str. 2
12489 Berlin
Germany

Prof. Dr. Shigeki Matsunaga
Graduate School of Pharmaceutical
Sciences
The University of Tokyo
Hongo 7-3-1, Bunkyo-ku
Tokyo, 113-0033
Japan

Dr. Jun-ichi Matsuo
The Kitasato Institute
Center for Basic Research
(TCI) 6-15-5 Toshima
Kita-ku, Tokyo 114-003
Japan

Prof. Dr. Teruaki Mukaiyama
The Kitasato Institute
Center for Basic Research
(TCI) 6-15-5 Toshima
Kita-ku, Tokyo 114-003
Japan

David A. Nicewicz
Department of Chemistry
University of North Carolina at Chapel
Hill
Chapel Hill, NC 27599-3290
USA

Prof. Dr. Dieter Schinzer
Otto-von-Guericke-Universität Magdeburg
Chemisches Institut
Universitätsplatz 2
39106 Magdeburg
Germany

Michael Shevlin
Department of Chemistry
University of Illinois at Chicago
845 West Taylor Street
Chicago, IL 60607
USA

Prof. Dr. Masakatsu Shibasaki
Graduate School of Pharmaceutical
Sciences
The University of Tokyo
Hongo 7-3-1, Bunkyo-ku
Tokyo, 113-0033
Japan

Prof. Dr. Isamu Shiina
Department of Applied Chemistry
Faculty of Science
Tokyo University of Science
Kagurazaka, Shinjuku-ku
Tokyo 162-8601
Japan

Prof. Dr. Fujie Tanaka
Department of Molecular Biology
The Scripps Research Institute
10550 North Torrey Pines Road
La Jolla, CA 92037
USA

Prof. Dr. Hisashi Yamamoto
Department of Chemistry
The University of Chicago
5735 S. Ellis Avenue
Chicago, IL 60637
USA

Dr. Yasuhiro Yamashita
Graduate School of Pharmaceutical
Sciences
The University of Tokyo
Hongo, Bunkyo-ku
Tokyo 113-0033
Japan

Prof. Dr. Akira Yanagisawa
Department of Chemistry
Faculty of Science
Chiba University
Inage, Chiba 263-8522
Japan

1
Silver, Gold, and Palladium Lewis Acids

Akira Yanagisawa

1.1
Introduction

Silver(I), gold(I), and palladium(II) salts have moderate Lewis acidity and
have been exploited as catalysts in organic reactions in recent years. Among
these salts, Pd(II) compounds are the most well-known reagents for catalyzing a variety of carbon–carbon bond-forming reactions such as allylic al-
kylations [1]. Ag(I) salts are also popular reagents for promoting transformations, including glycosylation, cycloadditions, and rearrangements, which
make use of their halophilicity or thiophilicity [2]. There are, however, few
examples of organic reactions employing Au(I) or Au(III) compounds as
Lewis acid catalysts. This chapter focuses on aldol reactions catalyzed by
silver(I), gold(I), or palladium(II) Lewis acids.

The Mukaiyama aldol reaction of silyl enol ethers or ketene silyl acetals
and related reactions using silver(I) and palladium(II) compounds are
reviewed in Section 1.2. The next section covers the diastereo- and enantioselective aldol-type reactions of activated isocyanides with aldehydes.

1.2
Mukaiyama Aldol Reaction and Related Reactions

Silver(I) compounds are known to promote the aldol condensation between
silyl enol ethers or ketene silyl acetals and aldehydes (the Mukaiyama aldol
reaction). For example, the adduct **3** is obtained in 72% yield when ketene
silyl acetal **1** is treated with α,β-unsaturated aldehyde **2** in the presence of a
catalytic amount of Ag(fod) (Scheme 1.1). Eu(fod)$_3$ or Yb(fod)$_3$ catalyzes a
hetero-Diels–Alder reaction of **1** and **2** [3]. A [2+2] cycloaddition followed
by a ring opening of the resulting oxetane is an alternative possible route to
the adduct **3**.

A BINAP-silver(I) complex is a superior asymmetric catalyst for allylation of aldehydes with allylic stannanes [4]. The chiral phosphine-silver(I)

Modern Aldol Reactions. Vol. 2: Metal Catalysis. Edited by Rainer Mahrwald
Copyright © 2004 WILEY-VCH Verlag GmbH & Co. KGaA, Weinheim
ISBN: 3-527-30714-1

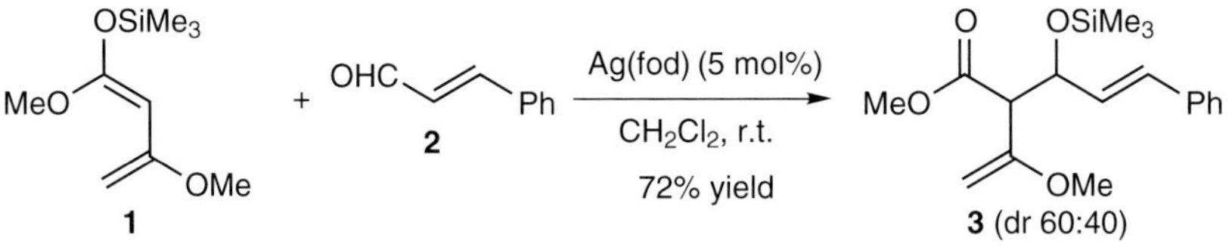

Scheme 1.1
Ag(fod)-catalyzed Mukaiyama aldol reaction of ketene silyl acetal.

catalyst is prepared simply by stirring a 1:1 mixture of BINAP and silver(I) compound in THF at room temperature. The BINAP-silver(I) complex can be also used as a chiral catalyst of asymmetric aldol reaction. Although a variety of beneficial methods have been developed for catalytic asymmetric aldol reaction, most of these are chiral Lewis acid-catalyzed Mukaiyama aldol reactions using ketene silyl acetals or silyl enol ethers [5] and there has been no example on enol stannanes. Yanagisawa, Yamamoto, and their colleagues first reported the enantioselective aldol addition of tributyltin enolates 4 to aldehydes catalyzed by a BINAP-silver(I) complex (Scheme 1.2) [6].

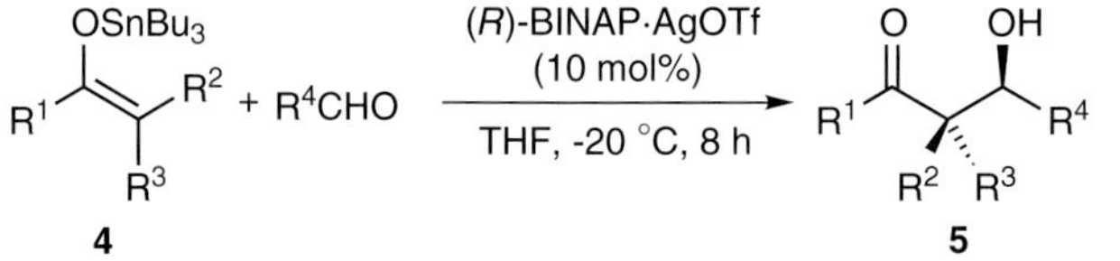

Scheme 1.2
Enantioselective aldol reaction of tributyltin enolates catalyzed by BINAP·silver(I) complex.

The tributyltin enolates 4 are easily generated from the corresponding enol acetates and tributyltin methoxide without any solvent [7]. The tin compounds thus prepared exist in the O-Sn form and/or the C-Sn form. Although the tin reagents themselves have sufficient reactivity toward aldehydes [7c], under the influence of the BINAP-silver(I) catalyst the reaction advances faster even at −20 °C. The results employing optimum conditions in the catalytic enantioselective aldol reaction of a variety of tributyltin enolates or α-tributylstannylketones with aromatic, α,β-unsaturated, and aliphatic aldehydes are summarized in Table 1.1. The characteristic features are: (i) all reactions occur to provide the corresponding aldol adducts 5 in moderate to high yield in the presence of 10 mol% (R)-BINAP-AgOTf complex at −20 °C, and no dehydrated aldol adduct is formed; (ii) with an α,β-unsaturated aldehyde, the 1,2-addition reaction is predominant (entry 3); (iii) use of a sterically hindered tin enolate results in an increase in the

Tab. 1.1
Diastereo- and enantioselective aldol addition of tin compounds to aldehydes in the presence of 10 mol% of (R)-BINAP·AgOTf complex in THF at −20 °C.

Entry	Tin Compound	Aldehyde	Product	Yield (%)[a]	anti:syn[b]	ee (%)[c]
1[d]	6	PhCHO	5a	73		77
2[d]	7	PhCHO	5b	78		95
3[d]		Ph‑CHO	5c	69		86
4[d]		Ph‑CHO	5d	75		94
5[e]	4a	PhCHO	5e	81	<1:99	95
6[e]		Ph‑CHO	5f	77	<1:99	95
7[f]	4b	PhCHO	5g-anti + 5g-syn	94	92:8	93[i]
8[f,h]		PhCHO		95	93:7	94[i]

[a] Isolated yield.
[b] Determined by ^{1}H NMR analysis.
[c] The value corresponds to the major diastereomer. Determined by HPLC analysis with chiral columns.
[d] O–Sn:C–Sn < 1:99.
[e] O–Sn:C–Sn > 99:1. The E:Z ratio for the O–Sn isomer was < 1:99.
[f] O–Sn:C–Sn > 99:1.
[g] The syn isomer: 25% ee.
[h] 1 mol% catalyst was used.
[i] The syn isomer: 33% ee.

enantioselectivity of the aldol reaction. For example, ee higher than 90% are observed when pinacolone and *tert*-butyl ethyl ketone-derived tin compounds **7** and **4a** are treated with aldehydes (entries 2 and 4–6); (iv) addition of the enol tributylstannane **4b** derived from cyclohexanone ((E)-enolate) to

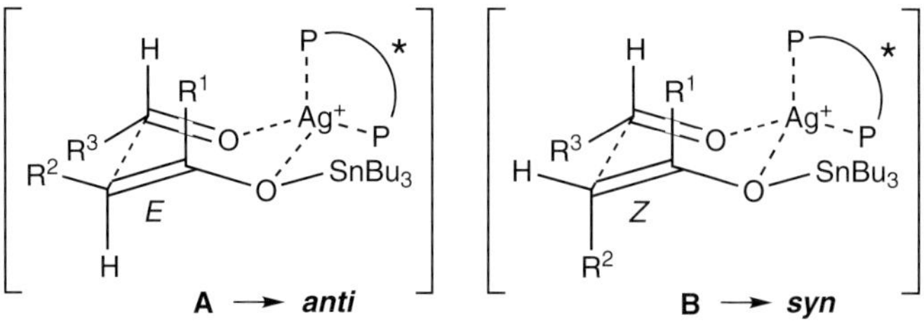

Fig. 1.1
Probable structures of cyclic transition states.

benzaldehyde in the presence of 10 mol% (*R*)-BINAP-AgOTf in THF at −20 °C yields the non-racemic *anti* aldol adduct **5g** selectively with an *anti:syn* ratio of 92:8, in contrast with the *syn* selectivity afforded by representative chiral Lewis acid catalysts [5]. The *anti* isomer indicates 93% ee (entry 7). The amount of catalyst can be reduced to 1 mol% without losing the isolated yield or diastereo- and enantioselectivity (entry 8). In contrast, the (*Z*) enolate generated from *tert*-butyl ethyl ketone **4a** produces the *syn*-aldol adducts **5e** and **5f** almost exclusively with 95% ee in the reaction with benzaldehyde and hydrocinnamaldehyde (entries 5 and 6).

These results reveal unambiguously that the diastereoselectivity relies on the geometry of tin enolate, and that cyclic transition-state structures (**A** and **B**, Figure 1.1) are plausible models. Accordingly, from the (*E*) enolate, the *anti*-aldol product forms via a model **A**, and another model **B** for the (*Z*) enolate leads to the *syn* product. Analogous six-membered cyclic models including a BINAP-coordinated silver atom in place of a tributylstannyl group are also probable substitutes when the transmetalation to silver enolate is sufficiently rapid.

Although the above-mentioned reaction is a superior asymmetric aldol process with regard to enantioselectivity and diastereoselectivity, it has a disadvantage of requiring the stoichiometric use of toxic trialkyltin compounds. The same group has shown that the amount of trialkyltin compounds can be reduced to a catalytic amount when an enol trichloroacetate is employed as a substrate for the reaction [8]. For example, treatment of benzaldehyde with the enol trichloroacetate of cyclohexanone **8** under the influence of (*R*)-BINAP-AgOTf complex (5 mol%), tributyltin methoxide (5 mol%), and MeOH (200 mol%) in dry THF at −20 °C to room temperature for 20 h provides a 92:8 mixture of non-racemic *anti* and *syn* aldol adduct, **5g**-*anti* and **5g**-*syn* respectively, in 82% yield (Scheme 1.3). The *anti* isomer **5g**-*anti* affords 95% ee, a grade of enantiomeric excess similar to that obtained from a BINAP-silver(I)-catalyzed aldol reaction with enol tributylstannanes [6].

A suggested catalytic cycle of this asymmetric aldol reaction is shown in Figure 1.2. To start with, Bu$_3$SnOMe reacts with enol trichloroacetate **9** to yield trialkyltin enolate **4** and methyl trichloroacetate. The tin enolate **4** then adds enantioselectively to an aldehyde under the influence of BINAP-AgOTf

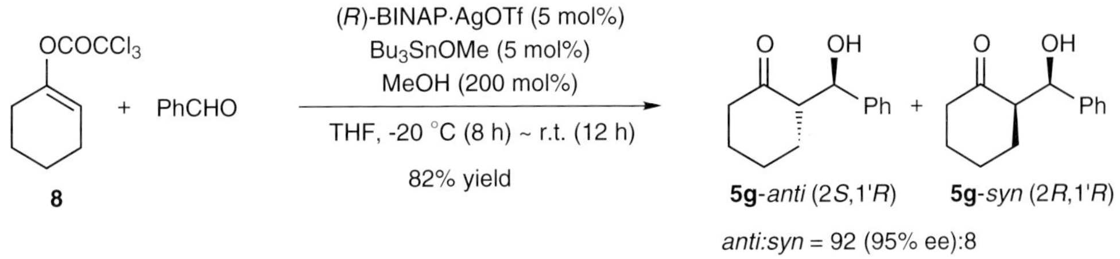

Scheme 1.3
Enantioselective aldol reaction catalyzed by tin methoxide and BINAP·silver(I) complex.

as an asymmetric catalyst to furnish the tin alkoxide of non-racemic aldol adduct **10**. Last, protonolysis of **10** by MeOH produces the optically active aldol product **5** and regenerates the tin methoxide. The rate of methanolysis is considered to be the key to success in the catalytic cycle.

The BINAP-Ag(I)-catalyzed asymmetric Mukaiyama aldol reaction using trimethylsilyl enol ethers was first developed by Yamagishi and co-workers, who found that the reaction was accelerated by BINAP-AgPF$_6$ in DMF containing a small amount of water, to give the aldol product with high enantioselectivity [9] (Scheme 1.4). In the reaction with BINAP-AgOAc, much higher catalytic activity and opposite absolute configuration of the aldol adduct were observed and ee was low [9].

Yanagisawa, Yamamoto, and their colleagues independently examined different combinations of BINAP-Ag(I) catalysts and silyl enol ethers and found that high enantioselectivity and chemical yields were obtained in the *p*-Tol-BINAP-AgF-catalyzed aldol reaction of trimethoxysilyl enol ethers in

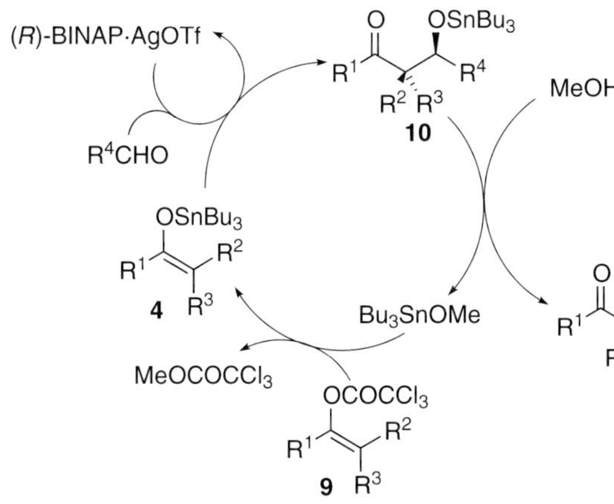

Fig. 1.2
A proposed catalytic mechanism for the asymmetric aldol reaction catalyzed by (*R*)-BINAP·AgOTf and tin methoxide.

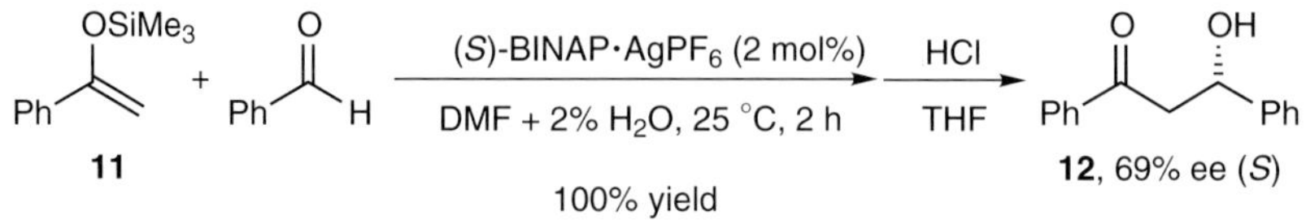

Scheme 1.4
BINAP·silver(I)-catalyzed asymmetric Mukaiyama aldol reaction.

methanol [10]. In addition, remarkable *syn* selectivity was observed for the reaction irrespective of the *E:Z* stereochemistry of the silyl enol ethers. For example, when the (*Z*)-trimethoxysilyl enol ether of *t*-butyl ethyl ketone **13** was treated with benzaldehyde the reaction proceeded smoothly at −78 to −20 °C and *syn*-aldol adduct **5e** was obtained almost exclusively with 97% ee (Scheme 1.5). In contrast, cyclohexanone-derived (*E*)-silyl enol ether gave the aldol adduct with an *anti:syn* ratio of 84:16 [10]. Use of a 1:1 mixture of MeOH and acetone as a solvent in the reaction of the trimethoxysilyl enol ethers resulted in higher enantioselectivity [10b].

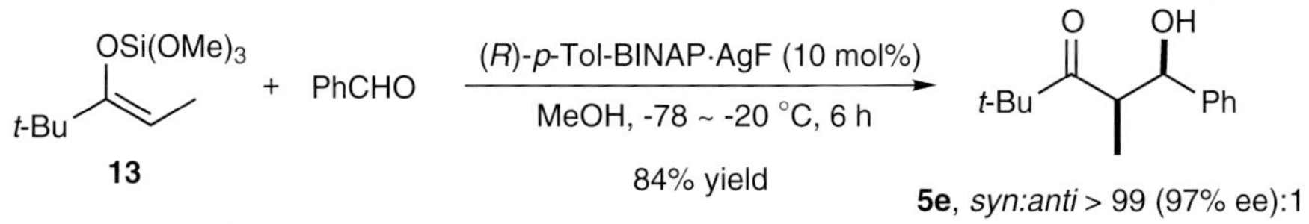

Scheme 1.5
Enantioselective aldol reaction of
trimethoxysilyl enol ether catalyzed by
p-Tol-BINAP·AgF complex.

The BINAP-silver(I) complex was further applied to asymmetric Mannich-type reactions by Lectka and coworkers [11]. Treatment of silyl enol ether **11** with a solution of α-imino ester **14** in the presence of 10 mol% (*R*)-BINAP-AgSbF$_6$ at −80 °C leads the corresponding α-amino acid derivative **15** in 95% yield with 90% ee (Scheme 1.6). They showed that (*R*)-BINAP-Pd(ClO$_4$)$_2$ was also an effective chiral Lewis acid for the reaction though it gave lower ee (80%).

Asymmetric Mukaiyama aldol reactions can also be catalyzed by cationic BINAP-Pd(II) complexes. In 1995 Sodeoka, Shibasaki et al. first reported

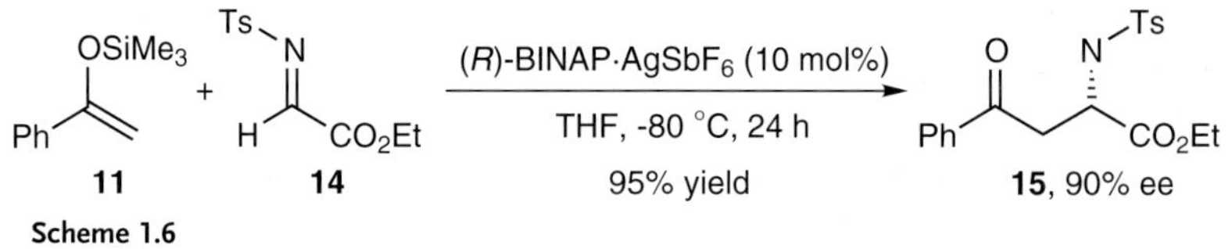

Scheme 1.6
Enantioselective Mannich-type reaction catalyzed by BINAP·silver(I) complex.

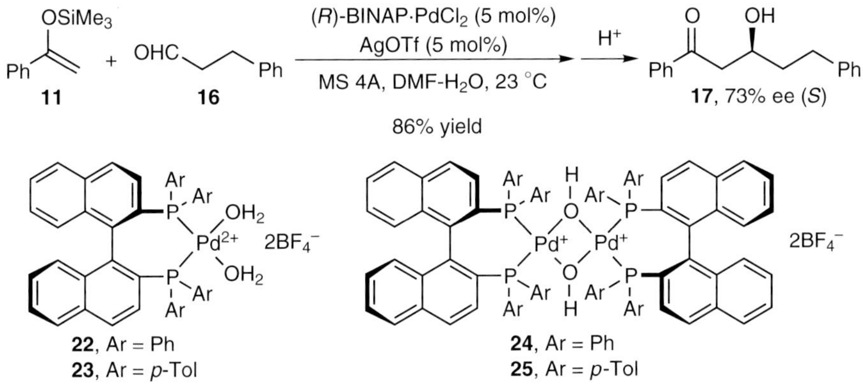

Scheme 1.7
Asymmetric Mukaiyama aldol reaction catalyzed by cationic BINAP·Pd(II) complexes.

that (*R*)-BINAP-PdCl$^+$, prepared from a 1:1 mixture of (*R*)-BINAP-PdCl$_2$ and AgOTf in wet DMF is an effective chiral catalyst for asymmetric aldol condensation of silyl enol ethers and aldehydes [12]. For instance, when hydrocinnamaldehyde (**16**) is treated with trimethylsilyl enol ether of aceto-phenone **11** under the influence of 5 mol% of this catalyst followed by de-silylation the desired aldol adduct **17** is obtained in 86% yield with 73% ee as shown in Scheme 1.7. Some examples of the aldol reaction are summarized in Table 1.2. The same group subsequently succeeded in generating chiral palladium diaquo complexes **22** and **23** from (*R*)-BINAP-PdCl$_2$ and (*R*)-*p*-Tol-BINAP-PdCl$_2$, respectively, by treatment with 2 equiv. AgBF$_4$ in wet acetone [13]. These complexes are inert toward air and moisture, and have similar reactivity and enantioselectivity in the aldol reaction of **11** with the aliphatic aldehyde **16**. Sodeoka et al. have further developed catalytic asym-metric Mannich-type reactions of silyl enol ethers with imines employing binuclear μ-hydroxo palladium(II) catalysts **24** and **25** generated from the diaquo complexes **22** and **23**, respectively [14]. In the reaction, a chiral palla-dium(II) enolate is assumed to be formed from the corresponding silyl enol ether. They later developed palladium complexes with polymer-supported BINAP ligands and showed that these reusable complexes are good catalysts for the asymmetric aldol reactions and Mannich-type reactions mentioned above [15].

Doucet and coworkers have shown that the complex dppe-Pd(OAc)$_2$ is an efficient catalyst for Mukaiyama aldol addition of ketene silyl acetals to aldehydes and ketones under neutral conditions [16]. The reaction proceeds smoothly, even in the presence of 0.1 mol% of the catalyst. [Bis(diphenylphosphino)alkane]bis(propenyl)ruthenium complexes also cat-alyze the aldol addition, furnishing a variety of 3-hydroxymethyl esters in good yields, as do other late transition metal complexes; platinum(II) cati-onic complexes are known to act as Lewis acids. Fujimura reported the first

Tab. 1.2
Catalytic asymmetric aldol reaction of silyl enol ethers with benzaldehyde in the presence of 5 mol% (*R*)-BINAP·PdCl$_2$–AgOTf in wet DMF.

Entry	Silyl Enol Ether	Product	Yield (%)[a]	ee (%)[b]
1[c]	OSiMe$_3$ Ph **11**	O OSiMe$_3$ Ph Ph **18**	87[d]	71
2	OSiMe$_3$ **19**	O OH Ph **20**	80	73
3	OSiMe$_3$ **21**	O OH Ph **5g**	58[e]	72[f]

[a] Isolated yield.
[b] Determined by HPLC analysis with chiral columns.
[c] Desilylation with acid was not done.
[d] Desilylated product was formed in 9% yield with 73% ee.
[e] The *syn:anti* ratio was 74:26.
[f] The value corresponds to the major diastereomer.

example of platinum-catalyzed enantioselective aldol reaction of ketene silyl acetals with aldehydes [17]. The chiral catalysts are prepared by treatment of chiral bisphosphine-Pt acyl complexes with triflic acid. In the aldol reaction, a C-bound platinum enolate is assumed to be an intermediate, on the basis of on ^{31}P NMR and IR studies.

1.3
Asymmetric Aldol Reactions of α-Isocyanocarboxylates

In 1986, Ito, Sawamura, and Hayashi showed that chiral ferrocenyl-phosphine **26**-gold(I) complexes catalyzed the aldol-type reaction of iso-cyanoacetate with aldehydes to provide optically active 5-alkyl-2-oxazoline-4-carboxylate (Scheme 1.8) [18]. Since then, they have extensively studied the chiral gold(I)-catalyzed reaction [19] as have Pastor and Togni [20]. The gold complexes can be generated in situ by mixing bis(cyclohexyl isocyanide)gold(I) tetrafluoroborate and (*R*)-*N*-methyl-*N*-[2-(dialkylamino)-ethyl]-1-[(*S*)-1′,2-bis(diphenylphosphino)ferrocenyl]ethylamine (**26**). Examples of the reaction of methyl isocyanoacetate (**27**) and different aldehydes in the presence of 1 mol% of **26c**-Au(I) complex are summarized in

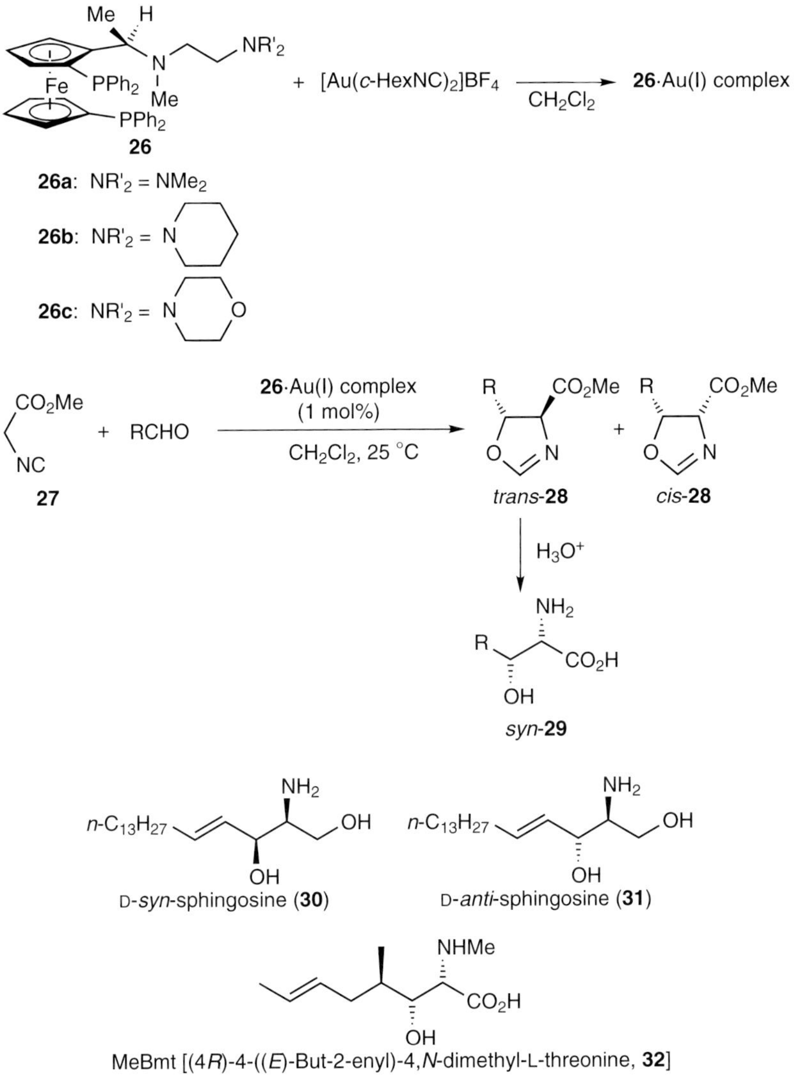

26a: NR'$_2$ = NMe$_2$

26b: NR'$_2$ = N (piperidine)

26c: NR'$_2$ = N—O (morpholine)

Scheme 1.8
Asymmetric aldol reaction of methyl
isocyanoacetate with aldehydes catalyzed by
chiral ferrocenylphosphine **26**–gold(I)
complexes.

Table 1.3. Benzaldehyde and substituted aromatic aldehydes, except 4-nitrobenzaldehyde, are transformed into the corresponding *trans*-oxazolines **28** with high enantio- and diastereoselectivity (entries 1–6). Secondary and tertiary alkyl aldehydes give *trans*-**28** nearly exclusively with high ee (entries 9 and 10). The *trans*-oxazolines **28** can be readily hydrolyzed to *threo*-β-

Tab. 1.3

Diastereo- and enantioselective aldol reaction of methyl isocyanoacetate (**27**) with aldehydes catalyzed by chiral ferrocenylphosphine **26c**·gold(I) complex.

Entry	Aldehyde	Product	Yield, %[a]	trans:cis[b]	% ee[c]
1	PhCHO	*trans-***28a** + *cis-***28a**	93	95:5	95
2	2-MeC$_6$H$_4$CHO	*trans-***28b** + *cis-***28b**	98	96:4	95
3	2-MeOC$_6$H$_4$CHO	*trans-***28c** + *cis-***28c**	98	92:8	92
4	4-ClC$_6$H$_4$CHO	*trans-***28d** + *cis-***28d**	97	94:6	94
5	4-O$_2$NC$_6$H$_4$CHO	*trans-***28e** + *cis-***28e**	80	83:17	86
6	piperonal CHO	*trans-***28f** + *cis-***28f**	86	95:5	96
7	*n*-Pr—CH=CH—CHO	*trans-***28g** + *cis-***28g**	85	87:13	92
8[d]	MeCHO	*trans-***28h** + *cis-***28h**	99	89:11	89
9	*i*-BuCHO	*trans-***28i** + *cis-***28i**	99	96:4	87

Tab. 1.3
(continued)

Entry	Aldehyde	Product	Yield, %[a]	trans:cis[b]	% ee[c]
10	*t*-BuCHO	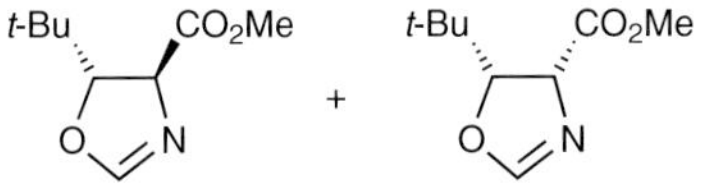 t-Bu CO$_2$Me *trans*-**28j** + t-Bu CO$_2$Me *cis*-**28j**	94	>99:1	97

[a] Isolated yield.
[b] Determined by ^{1}H NMR analysis.
[c] Determined by ^{1}H NMR analysis with chiral shift reagent Eu(dcm)$_3$.
[d] 0.2 mol% of the catalyst was used.

hydroxy α-amino acids **29**. The gold-catalyzed aldol reaction has been applied to the asymmetric synthesis of the biologically important compounds D-*threo*-sphingosine (**30**) [21], D-*erythro*-sphingosine (**31**) [21], and MeBmt (**32**) [22]. The enantioselective synthesis of (−)-α-kainic acid has also been achieved using this aldol reaction [23].

A proposed transition-state model for the reaction is shown in Figure 1.3. The presence of the 2-(dialkylamino)ethylamino group in **26** is necessary to obtain high selectivity [24]. The terminal amino group abstracts one of the α-protons of isocyanoacetate coordinated with gold, and the resulting ion pair causes advantageous arrangement of the enolate and aldehyde around the gold. In contrast, Togni and Pastor proposed an alternative acyclic transition-state model [20d].

The chiral ferrocenylphosphine-gold(I)-catalyzed aldol reaction of α-alkyl α-isocyanocarboxylates **33** with paraformaldehyde gives optically active 4-alkyl-2-oxazoline-4-carboxylates **34** with moderate to good enantioselectivity [25]. The absolute configuration (*S*) of the product indicates that the reaction proceeds selectively at the *si* face of the enolate, as illustrated in Figure 1.3. These oxazolines **34** can be converted into α-alkylserine derivatives **35** (Scheme 1.9).

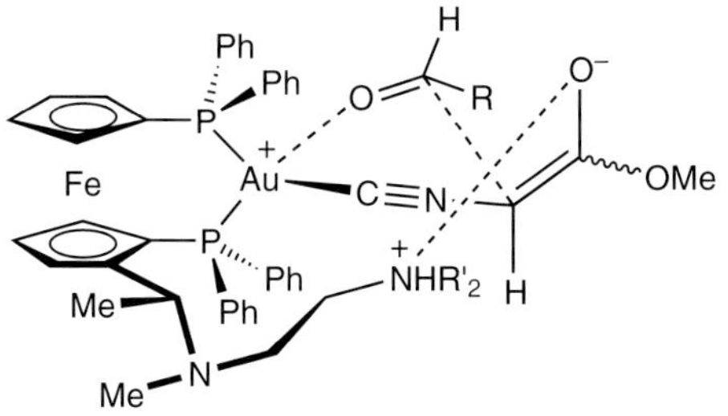

Fig. 1.3
Transition-state assembly in the gold-catalyzed asymmetric aldol reaction.

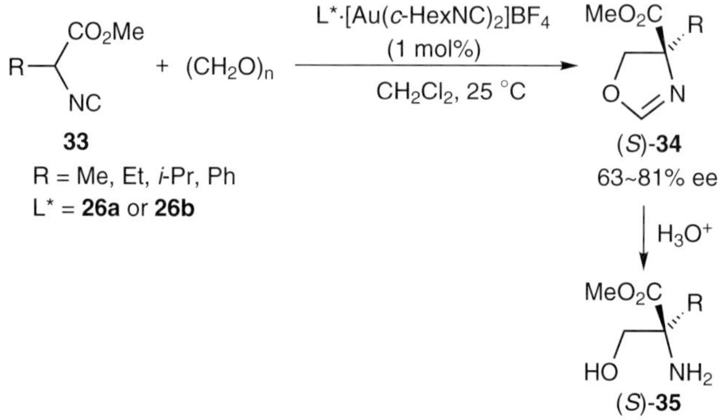

Scheme 1.9
Catalytic asymmetric synthesis of α-alkylserines.

This enantioselective aldol reaction using isocyanoacetate **27** is quite effective for aromatic aldehydes or tertiary alkyl aldehydes, but not for sterically less hindered aliphatic aldehydes, as described above. Ito and coworkers found that very high enantioselectivity is obtained even for acetaldehyde (R = Me) in the aldol reaction with *N,N*-dimethyl-α-isocyanoacetamide (**36**) (Scheme 1.10) [26]. Use of α-keto esters in place of aldehydes also results in moderate to high enantioselectivity of up to 90% ee [27].

The same group also developed an asymmetric aldol reaction of *N*-methoxy-*N*-methyl-α-isocyanoacetamide (α-isocyano Weinreb amide) with aldehydes (Scheme 1.10). For instance, reaction of the Weinreb amide **37** with acetaldehyde in the presence of **26c**-Au(I) catalyst gives the optically active *trans*-oxazoline **39** (E = CON(Me)OMe; R = Me) with high diastereo- and enantioselectivity similar to those of **36** [28]. The oxazoline can be transformed into *N,O*-protected β-hydroxy-α-amino aldehydes or ketones.

(Isocyanomethyl)phosphonate **38** is also a beneficial pronucleophile that leads to optically active (1-aminoalkyl)phosphonic acids, which are

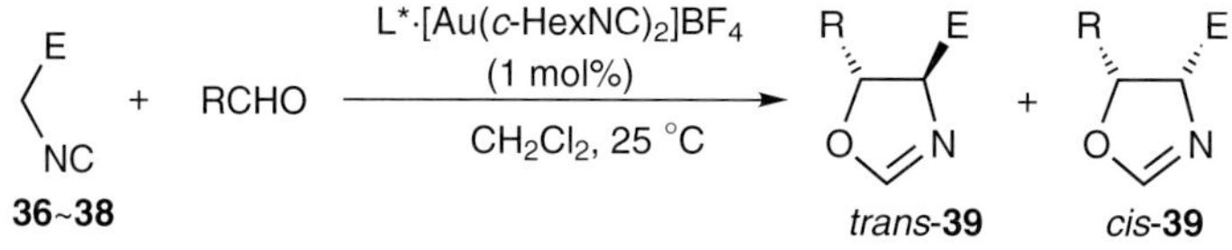

E		R	L*	*trans* (% ee) : *cis*
CONMe₂ (**36**)		Me	**26b**	91 (99) : 9
CON(Me)OMe (**37**)		Me	**26c**	95 (97) : 5
PO(OPh)₂ (**38**)		Ph	**26b**	>98 (96) : 2

Scheme 1.10
Gold(I)-catalyzed asymmetric aldol reaction of isocyanoacetamides and (isocyanomethyl)-phosphonate.

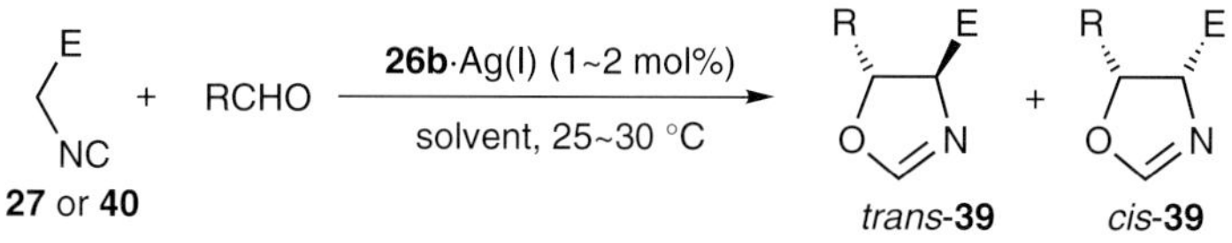

E	R	Ag(I)	Solvent	trans (% ee) : cis
CO$_2$Me (**27**)[a]	Ph	AgOTf	ClCH$_2$CH$_2$Cl	96 (80) : 4
CO$_2$Me (**27**)[a]	i-Pr	AgClO$_4$	ClCH$_2$CH$_2$Cl	99 (90) : 1
SO$_2$(p-Tol) (**40**)	Ph	AgOTf	CH$_2$Cl$_2$	>99 (77) : 1
SO$_2$(p-Tol) (**40**)	i-Pr	AgOTf	CH$_2$Cl$_2$	>99 (86) : 1

a) slow addition of **27** over 1 h.

Scheme 1.11
Asymmetric aldol reaction of methyl
isocyanoacetate and tosylmethyliso-
cyanide catalyzed by chiral
ferrocenylphosphine·silver(I) complex.

phosphonic acid analogs of α-amino acids, via *trans*-5-alkyl-2-oxazoline-4-phosphonates **39** (E = PO(OPh)$_2$, Scheme 1.10) [29].

Ito and coworkers found that chiral ferrocenylphosphine-silver(I) complexes also catalyze the asymmetric aldol reaction of isocyanoacetate with aldehydes (Scheme 1.11) [30]. It is essential to keep isocyanoacetate at a low concentration to obtain a product with high optical purity. They performed IR studies on the structures of gold(I) and silver(I) complexes with chiral ferrocenylphosphine **26a** in the presence of methyl isocyanoacetate (**27**) and found a significant difference between the coordination numbers of the isocyanoacetate to the metal in these metal complexes (Scheme 1.12). The

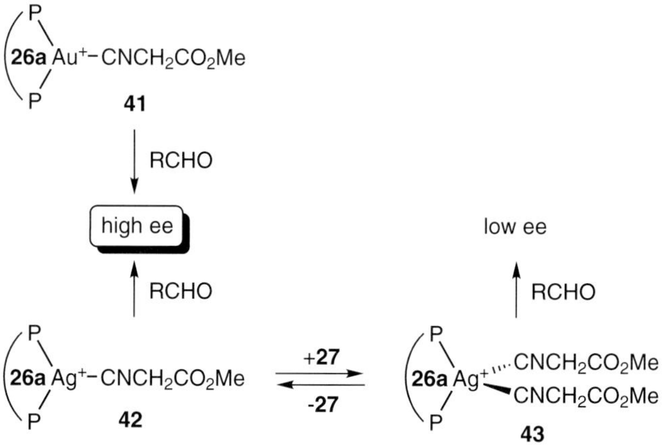

Scheme 1.12
A difference in the coordination number of
methyl isocyanoacetate to metal between
gold(I) and silver(I) complexes.

gold(I) complex has a tricoordinated structure **41**, which results in high ee, whereas the silver(I) complex is in equilibrium between tricoordinated structure **42** and tetracoordinated structure **43**, which results in low enantioselectivity. Slow addition of isocyanoacetate **27** to a solution of the silver(I) catalyst and aldehyde effectively reduces the undesirable tetracoordinated species and results in high enantioselectivity.

The asymmetric aldol-type addition of tosylmethyl isocyanide (**40**) to aldehydes can also be catalyzed by the chiral silver(I) complex giving, almost exclusively, *trans*-5-alkyl-4-tosyl-2-oxazolines **39** [E = SO$_2$(*p*-Tol)] with up to 86% ee, as shown in Scheme 1.11 [31]. The slow addition method described above is not necessary for this reaction system.

Soloshonok and Hayashi used chiral ferrocenylphosphine-gold(I) complexes in asymmetric aldol-type reactions of fluorinated benzaldehydes with methyl isocyanoacetate (**27**) and *N*,*N*-dimethyl-α-isocyanoacetamide (**36**). Interestingly, successive substitution of hydrogen atoms by fluorine in the phenyl ring of benzaldehyde causes a gradual increase in both the *cis* selectivity and the ee of *cis* oxazolines [32].

Cationic chiral palladium complexes are known to catalyze the aldol reaction of methyl isocyanoacetate (**27**) and aldehydes. For example, Richards et al. prepared cationic 2,6-bis(2-oxazolinyl)phenylpalladium(II) complex **44** from the corresponding bromopalladium(II) complex and AgSbF$_6$ in wet CH$_2$Cl$_2$ and showed that an increase in rate was observed for the aldol reaction of **27** with benzaldehyde in the presence of 1 mol% **44** and 10 mol% Hünigs base [33]. Zhang and coworkers developed a palladium(II) complex of PCP-type chiral ligand **46** (PCP is the monoanionic "pincer" ligand [C$_6$H$_3$(CHMePPh$_2$)$_2$-2,6]$^-$). Removal of the chloride with AgOTf produces an active cationic chiral Pd(II) catalyst for the asymmetric aldol reaction of aldehydes (Scheme 1.13) [34]. Several examples of the reaction under the influence of 1 mol% of catalyst **46** are summarized in Table 1.4. When the effects of solvent on enantioselectivity were examined in the reaction with benzaldehyde, THF was found to be solvent of choice (*trans*-**47a**: 24% ee, *cis*-**47a**: 67% ee, entry 1). The *trans* isomers were usually obtained as major products though with lower ee. In the reaction with aromatic aldehydes the enantioselectivity is almost constant (entries 1–5) and the trisubstituted aromatic aldehyde gives the highest ee (entry 5). It is noteworthy that higher enantioselectivity is observed with aliphatic aldehydes than with aromatic aldehydes with regard to their *cis* product (entries 6 and 7). L-Valine-derived NCN-type Pd(II) complex **48** (NCN is the monoanionic, *para*-functionalized "pincer" ligand [C$_6$H$_2$(CH$_2$NMe$_2$)$_2$-2,6]$^-$) synthesized by van Koten and coworkers is also an active catalyst for the aldol reaction after conversion into the corresponding cationic complex by treatment with AgBF$_4$ in wet acetone [35]. Motoyama and Nishiyama have shown that excellent *trans* diastereoselectivity (>99% *trans*) and moderate enantioselectivity (57% ee) was obtained in the asymmetric aldol-type condensation of tosylmethyl isocyanide with benzaldehyde employing cationic Pd(II) aqua complex **45** [36].

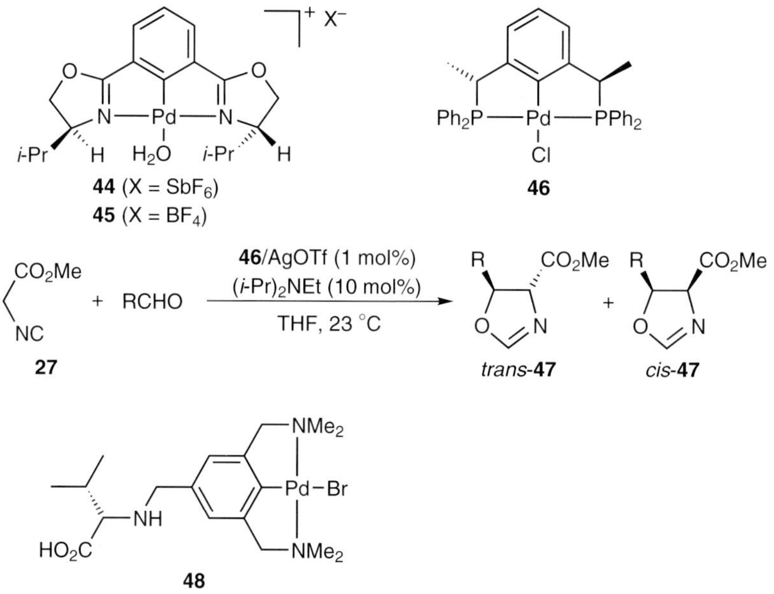

Scheme 1.13

Asymmetric aldol reaction of methyl isocyanoacetate with aldehydes catalyzed by cationic chiral palladium(II) complexes.

Other notable examples of the aldol-type reaction using a variety of palladium complexes have also appeared [37–41].

1.4
Summary and Conclusions

Described herein are examples of aldol reactions using silver(I), gold(I), or palladium(II) Lewis acids. The BINAP-silver(I) catalyst has made possible the aldol reaction of silyl enol ethers or trialkyltin enolates with high enantio- and diastereoselectivity. This silver catalyst is also effective in Mannich-type reactions of silyl enol ethers with α-imino esters. The remarkable affinity of the silver ion for halides is useful for accelerating chiral palladium-catalyzed asymmetric Mukaiyama aldol reactions. Isolated chiral palladium diaquo complexes and binuclear μ-hydroxo palladium(II) complexes can catalyze asymmetric Mannich-type reactions and the aldol reaction. The chiral ferrocenylphosphine gold(I)-catalyzed asymmetric aldol reaction results in high stereoselectivity, although the substrates are restricted to α-isocyanocarboxylates and their derivatives, and has proven to be an excellent method for synthesizing optically active α-amino acid derivatives and

Tab. 1.4
Asymmetric aldol reaction of methyl isocyanoacetate (**27**) with aldehydes catalyzed by a cationic Pd(II) complex generated from **46** and AgOTf.

Entry	Aldehyde	Product	Yield, %[a]	trans:cis[b]	% ee[c] trans	% ee[c] cis
1	PhCHO	trans-47a + cis-47a	85	78:22	24	67
2		trans-47b + cis-47b	81	81:19	23	66
3		trans-47c + cis-47c	80	82:18	11	61
4		trans-47d + cis-47d	60	74:26	21	57

5	CHO (2,4,6-trimethylbenzaldehyde)	2,4,6-Me$_3$C$_6$H$_2$ ···CO$_2$Me (N, O) *trans*-47e + 2,4,6-Me$_3$C$_6$H$_2$ CO$_2$Me (N, O) *cis*-47e	84	86:14	26	71
6	CHO (cyclohexanecarbaldehyde)	c-C$_6$H$_{11}$ ···CO$_2$Me (N, O) *trans*-47f + c-C$_6$H$_{11}$ CO$_2$Me (N, O) *cis*-47f	97	72:28	11	74
7	EtCHO	Et ···CO$_2$Me (N, O) *trans*-47g + Et CO$_2$Me (N, O) *cis*-47g	91	91:9	30	70

[a] Isolated yield.
[b] Determined by ^{1}H NMR analysis.
[c] Determined by GC analysis.

amino alcohols. The examples given here unambiguously indicate that silver(I), gold(I), and palladium(II) compounds in combination with chiral ligands are chiral Lewis acid catalysts of great promise for asymmetric synthesis.

1.5
Experimental Procedures

Typical Procedure for Asymmetric Aldol Reaction of Benzaldehyde with 3,3-Dimethyl-1-tributylstannyl-2-butanone (7) Catalyzed by BINAP-AgOTf Complex. Synthesis of (R)-4,4-Dimethyl-1-hydroxy-1-phenyl-3-pentanone (5b, entry 2 in Table 1.1) [6]. A mixture of AgOTf (26.7 mg, 0.104 mmol) and (R)-BINAP (64.0 mg, 0.103 mmol) was dissolved in dry THF (3 mL) under argon atmosphere and with direct light excluded and stirred at 20 °C for 10 min. To the resulting solution was added dropwise a THF solution (3 mL) of benzaldehyde (100 μL, 0.98 mmol), and then 3,3-dimethyl-1-tributylstannyl-2-butanone (**7**, 428.1 mg, 1.10 mmol) was added over a period of 4 h with a syringe pump at −20 °C. The mixture was stirred for 4 h at this temperature and treated with MeOH (2 mL). After warming to room temperature, the mixture was treated with brine (2 mL) and solid KF (ca. 1 g). The resulting precipitate was removed by filtration and the filtrate was dried over Na_2SO_4 and concentrated in vacuo. The crude product was purified by column chromatography on silica gel to afford the aldol adduct **5b** (161.7 mg, 78% yield as a colorless oil). TLC R_F 0.22 (1:5 ethyl acetate–hexane); IR (neat) 3625–3130, 3063, 3033, 2971, 2907, 2872, 1701, 1605, 1495, 1478, 1455, 1395, 1368, 1073, 1057, 1011, 984, 914, 878, 760, 747, 700 cm^{-1}; ^1H NMR ($CDCl_3$) δ (ppm) 1.14 (s, 9 H, 3 CH_3), 2.89 (d, 2 H, $J = 5.7$ Hz, CH_2), 3.59 (d, 1 H, $J = 3.0$ Hz, OH), 5.13 (m, 1 H, CH), 7.29–7.39 (m, 5 H, aromatic); $[\alpha]^{30}_D$ +61.5 (c 1.3, $CHCl_3$). The enantioselectivity was determined to be 95% ee by HPLC analysis using a chiral column (Chiralcel OD-H, Daicel Chemical Industries, hexane–i-PrOH, 20:1, flow rate 0.5 mL min^{-1}); $t_{minor} = 17.7$ min, $t_{major} = 20.2$ min.

Typical Procedure for Silver(I)-catalyzed Asymmetric Mukaiyama Aldol Reaction of Benzaldehyde with Acetophenone Silyl Enol Ether 11. Synthesis of (S)-1-Hydroxy-1,3-diphenyl-3-propanone (12, Scheme 1.4) [9]. Wet DMF containing 2% H_2O (206 μL) was added under nitrogen atmosphere to a mixture of $AgPF_6$ (4.0 mg, 0.013 mmol) and (S)-BINAP (8.4 mg, 0.013 mmol) and the solution was stirred at room temperature for 10 min. Benzaldehyde (69 μL, 0.67 mmol) was added and stirring was continued for 10 min. Acetophenone silyl enol ether **11** (276 μL, 1.35 mmol) was added and the mixture was stirred for 2 h at 25 °C. It was filtered through a short silica gel column, concentrated in vacuo, and clear oil was obtained. The oil was hydrolyzed with 1 M HCl THF/H$_2$O (1:1) solution, extracted with ether, dried over anhydrous sodium sulfate, concentrated in vacuo to afford a pale clear oil. This oil was purified by column chromatography on silica gel (1:1 ethyl acetate–hexane) to give the (*S*)-enriched aldol adduct **12** (100% yield, 69% ee). ^{1}H NMR (CDCl$_3$) δ (ppm) 3.38 (d, 2 H, *J* = 5.9 Hz, CH$_2$), 3.56 (d, 1 H, *J* = 3.0 Hz, OH), 5.35 (dt, 1 H, *J* = 3.0, 5.9 Hz, CH), 7.26–7.62 (m, 8 H, aromatic), 7.95 (d, 2 H, *J* = 7.3 Hz, aromatic); FAB-MS calcd. for C$_{15}$H$_{14}$O$_2$, 226 (M$^+$); found 227 ((M + 1)$^+$), 209 ((M-OH)$^+$); HPLC analysis using a chiral column (Chiralcel OB-H, Daicel Chemical Industries, hexane–*i*-PrOH, 9:1, flow rate 0.9 mL min^{-1}); t_{minor} = 22.5 min (*R*), t_{major} = 33.5 min (*S*).

Typical Procedure for Asymmetric Aldol Reaction of Trimethoxysilyl Enol Ether 13 with Benzaldehyde Catalyzed by (*R*)-*p*-Tol-BINAP-AgF Complex. Synthesis of 1-Hydroxy-2,4,4-trimethyl-1-phenyl-3-pentanone (5e, Scheme 1.5) [10]. A mixture of AgF (13.0 mg, 0.102 mmol) and (*R*)-*p*-Tol-BINAP (67.9 mg, 0.100 mmol) was dissolved in dry MeOH (6 mL) under an argon atmosphere and with direct light excluded, and stirred at 20 °C for 10 min. Benzaldehyde (100 μL, 0.98 mmol) and *t*-butyl ethyl ketone-derived trimethoxysilyl enol ether **13** (236.9 mg, 1.01 mmol) were successively added dropwise, at −78 °C, to the resulting solution. The mixture was stirred at this temperature for 2 h, then at −40 °C for 2 h, and finally at −20 °C for 2 h. It was then treated with brine (2 mL) and solid KF (ca. 1 g) at ambient temperature for 30 min. The resulting precipitate was removed by filtration through a glass filter funnel filled with Celite and silica gel. The filtrate was dried over Na$_2$SO$_4$ and concentrated in vacuo after filtration. The crude product was purified by column chromatography on silica gel (1:5 ethyl acetate–hexane as eluent) to afford a mixture of the aldol adduct **5e** (181.2 mg, 84% yield). The *syn/anti* ratio was determined to be >99/1 by ^{1}H NMR analysis. The enantioselectivity of the *syn* isomer was determined to be 97% ee by HPLC analysis using a chiral column (Chiralcel OD-H, Daicel Chemical Industries, hexane–*i*-PrOH, 40:1, flow rate 0.5 mL min^{-1}); $t_{syn\text{-}minor}$ = 17.1 min, $t_{syn\text{-}major}$ = 18.0 min. Specific rotation of the *syn* isomer (95% ee) $[\alpha]^{30}_D$ −67.1° (*c* 1.3, CHCl$_3$). Other spectral data (IR and ^{1}H NMR) of the *syn* isomer were in good agreement with reported data [6].

Typical Procedure for Asymmetric Mukaiyama Aldol Reaction of Silyl Enol Ethers with Aldehydes Catalyzed by Cationic BINAP-Pd(II) Complexes. Synthesis of (*R*)-1,3-Diphenyl-1-trimethylsiloxy-3-propanone (18, Scheme 1.7, and entry 1 in Table 1.2) [12, 13]. Wet DMF (8 mL dry DMF with 144 μL H$_2$O) was added to a mixture of (*R*)-BINAP-PdCl$_2$ (160 mg, 0.20 mmol), AgOTf (51 mg, 0.20 mmol), and MS 4 Å (powder, 1.2 g) and the suspension was stirred at 23 °C for 20 min. After cannula filtration, benzaldehyde (410 μL, 4.0 mmol) and acetophenone silyl enol ether **11** (1.23 mL, 6.0 mmol) were added to the resulting orange solution and the mixture was stirred for 13 h at 23 °C. Dilution of the reaction mixture with diethyl ether, filtration through a short silica gel column, and concentration afforded pale yellow oil. This crude product was purified by column chromatography on silica gel to give the (R)-enriched silylated aldol adduct **18** (1.04 g, 87% yield, 71% ee) and its desilylated product (82 mg, 9% yield, 73% ee). The enantio-selectivity of the silylated product **18** was determined by HPLC analysis using Chiralcel OJ (hexane–*i*-PrOH, 9:1) after conversion to the corresponding desilylated product (1 M HCl–THF, 1:2). Specific rotation of the desilylated aldol adduct (70% ee) $[\alpha]_D$ +32.4° (*c* 0.74, MeOH).

General Procedure for Asymmetric Aldol Reaction of Methyl Isocyanoacetate (27) with Aldehydes Catalyzed by Chiral Ferrocenylbisphosphine-Gold(I) Complexes (Scheme 1.8 and Table 1.3) [18]. Methyl isocyanoacetate (**27**, 5.0 mmol) was added to a solution of bis(cyclohexyl isocyanide)gold(I) tetra-fluoroborate (0.050 mmol), chiral ferrocenylbisphosphine **26** (0.050–0.055 mmol), and aldehyde (5.0–5.5 mmol) in CH_2Cl_2 (5 mL) and the mixture was stirred under nitrogen at 25 °C until **27** was not detected by silica gel TLC (hexane–ethyl acetate, 2:1) or IR. Evaporation of the solvent followed by bulb-to-bulb distillation gave oxazoline **28**. The *trans/cis* ratio was determined by ^1H NMR spectroscopy and the enantiomeric purity of *trans*-**28** and *cis*-**28**, readily separated by MPLC (hexane–ethyl acetate), were determined by ^1H NMR studies using Eu(dcm)$_3$. The OCH$_3$ singlet of the major enantiomer of *trans*-**28** always appeared at a higher field than that of the minor enantiomer.

General Procedure for Asymmetric Aldol Reaction of Tosylmethyl Isocyanide (40) with Aldehydes Catalyzed by Chiral Ferrocenylbisphosphine-Silver(I) Complexes (Scheme 1.11) [31]. Aldehyde (1.5 mmol) was added to a solution of silver(I) triflate (0.011 mmol), chiral ferrocenylbisphosphine **26** (0.010 mmol), and tosylmethyl isocyanide (**40**, 1.0 mmol) in dry CH_2Cl_2 (5 mL). The mixture was stirred under nitrogen at 25 °C for 2 h. The catalyst was removed by passing the mixture through a bed of Florisil (17 mm × 30 mm, EtOAc), and MPLC purification (silica gel, CH_2Cl_2–EtOAc, 15:1) gave oxazoline **39**. The *trans/cis* ratio was determined by ^1H NMR spectroscopy and the enantiomeric excess of *trans*-**39** was determined by HPLC analysis with a chiral stationary phase after conversion to the corresponding α-naphthylurea derivative of amino alcohol (1. LiAlH$_4$, 2. α-naphthyl isocyanate).

Experimental Procedure for Pd(II)-Catalyzed Asymmetric Aldol Reaction of Methyl Isocyanoacetate (27) with Aldehydes (Scheme 1.13 and Table 1.4) [34].

Preparation of [(1R,1′R)-2,6-bis[1-(diphenylphosphino)ethyl]phenyl]chloropalladium(II) (46). PdCl$_2$(PhCN)$_2$ (383 mg, 1.0 mmol) was added to a solution of (1R,1′R)-1,3-bis[1-(diphenylphosphino)ethyl]benzene (502 mg, 1.0 mmol) in CH_2Cl_2 (15 mL). The resulting orange solution was stirred at room temperature for 24 h. The reaction mixture was then reduced to one-third of its volume and absolute ethanol was added to precipitate the product. Filtration gave the desired product **46** as a yellow powder (418 mg, 85%), mp 249–252 °C; $[\alpha]_D$ −323.6 (*c* 1.0, CHCl$_3$). ^{1}H NMR (360 MHz, CDCl$_3$) δ (ppm) 1.19 (m, 6H, CH$_3$), 4.04 (m, 2H, CH), 7.10 (m, 3H), 7.34–7.45 (m, 12H), 7.70–7.74 (m, 4H), 7.95–7.97 (m, 4H); ^{13}C NMR (90 MHz, CDCl$_3$) δ (ppm) 22.1 (s, 2C, CH$_3$), 46.7 (m, 2C, CH), 122.5–156.8 (twelve different aromatic carbon atoms); ^{31}P NMR (145 MHz, CDCl$_3$) δ (ppm) 46.5. HRMS calculated for C$_{34}$H$_{31}$P$_2$PdCl (M$^+$) 642.0624; found 642.0634.

General Procedure for the Aldol Reaction.   A solution of Pd complex **46** (7 mg, 0.011 mmol, 1.0 mol%) and AgOTf (3 mg, 0.011 mmol) in CH$_2$Cl$_2$ (2 mL) was stirred for ca. 30 min at room temperature. The resulting cloudy solution was filtered through Celite and the solvent was removed under reduced pressure to give the active catalyst. The catalyst was then dissolved in EtOAc and passed through a plug of silica gel to remove excess AgOTf. After removal of EtOAc the catalyst was dissolved in THF (6 mL), and methyl isocyanoacetate (**27**, 110 µL, 1.1 mmol) was added followed by introduction of diisopropylethylamine (19 µL, 0.11 mmol) and aldehyde (1.1 mmol). The reaction was monitored by TLC (EtOAc–hexane, 1:1, visualized with KMnO$_4$). After removal of solvent, the pure product **47** was obtained by bulb-to-bulb distillation under reduced pressure (0.1 mmHg). The *trans/cis* ratio was determined by ^{1}H NMR spectroscopy by integration of the methyl ester protons and the enantiomeric excess for *trans*-**47** and *cis*-**47** were determined by GC analysis.

References

1 (a) B. M. Trost, in *Comprehensive Organometallic Chemistry, Vol. 8*, (ed.: G. Wilkinson), Pergamon Press, Oxford, **1982**, Chapter 57. (b) S. A. Godleski, in *Comprehensive Organic Synthesis, Vol. 4*, (eds.: B. M. Trost, I. Fleming), Pergamon Press, New York, **1991**, p. 585. (c) A. Pfaltz, M. Lautens, in *Comprehensive Asymmetric Catalysis, Vol. 2*, (eds.: E. N. Jacobsen, A. Pfaltz, H. Yamamoto), Springer, Heidelberg, **1999**, Chapter 24, p. 833. (d) B. M. Trost, C. B. Lee, in *Catalytic Asymmetric Synthesis, 2nd ed.*, (ed.: I. Ojima), Wiley–VCH, New York, **2000**, Chapter 8E, p. 593.

2 (a) D. R. Rae, in *Encyclopedia of Reagents for Organic Synthesis, Vol. 6*, (ed.: L. A. Paquette), John Wiley & Sons, Chichester, **1995**, p. 4461. (b) J. C. Lanter, in *Encyclopedia of Reagents for Organic Synthesis, Vol. 6*, (ed.: L. A. Paquette), John Wiley & Sons, Chichester, **1995**, p. 4469. (c) L.-G. Wistrand, in *Encyclopedia of Reagents for Organic Synthesis, Vol. 6*, (ed.: L. A. Paquette), John Wiley & Sons, Chichester, **1995**, p. 4472. (d) T. H. Black, in *Encyclopedia of Reagents for Organic Synthesis, Vol. 6*, (ed.: L. A. Paquette), John Wiley & Sons, Chichester, **1995**, p. 4476.

3 S. CASTELLINO, J. J. SIMS, *Tetrahedron Lett.* **1984**, *25*, 4059.

4 (a) A. YANAGISAWA, H. NAKASHIMA, A. ISHIBA, H. YAMAMOTO, *J. Am. Chem. Soc.* **1996**, *118*, 4723. See also: (b) C. BIANCHINI, L. GLENDENNING, *Chemtracts–Inorg. Chem.* **1997**, *10*, 339; (c) P. G. COZZI, E. TAGLIAVINI, A. UMANI-RONCHI, *Gazz. Chim. Ital.* **1997**, *127*, 247.

5 Reviews: (a) T. BACH, *Angew. Chem. Int. Ed. Engl.* **1994**, *33*, 417; (b) T. K. HOLLIS, B. BOSNICH, *J. Am. Chem. Soc.* **1995**, *117*, 4570; (c) M. BRAUN, in *Houben–Weyl: Methods of Organic Chemistry*, Vol. E 21, (eds.: G. HELMCHEN, R. W. HOFFMANN, J. MULZER, E. SCHAUMANN), Georg Thieme Verlag, Stuttgart, **1995**, p. 1730; (d) S. G. NELSON, *Tetrahedron: Asymmetry* **1998**, *9*, 357; (e) H. GRÖGER, E. M. VOGL, M. SHIBASAKI, *Chem. Eur. J.* **1998**, *4*, 1137.

6 A. YANAGISAWA, Y. MATSUMOTO, H. NAKASHIMA, K. ASAKAWA, H. YAMAMOTO, *J. Am. Chem. Soc.* **1997**, *119*, 9319.

7 (a) M. PEREYRE, B. BELLEGARDE, J. MENDELSOHN, J. VALADE, *J. Organomet. Chem.* **1968**, *11*, 97; (b) I. F. LUTSENKO, Y. I. BAUKOV, I. Y. BELAVIN, *J. Organomet. Chem.* **1970**, *24*, 359; (c) S. S. LABADIE, J. K. STILLE, *Tetrahedron* **1984**, *40*, 2329; (d) K. KOBAYASHI, M. KAWANISI, T. HITOMI, S. KOZIMA, *Chem. Lett.* **1984**, 497.

8 (a) A. YANAGISAWA, Y. MATSUMOTO, K. ASAKAWA, H. YAMAMOTO, *J. Am. Chem. Soc.* **1999**, *121*, 892. (b) A. YANAGISAWA, Y. MATSUMOTO, K. ASAKAWA, H. YAMAMOTO, *Tetrahedron* **2002**, *58*, 8331.

9 (a) M. OHKOUCHI, M. YAMAGUCHI, T. YAMAGISHI, *Enantiomer* **2000**, *5*, 71. (b) M. OHKOUCHI, D. MASUI, M. YAMAGUCHI, T. YAMAGISHI, *J. Mol. Catal. A: Chem.* **2001**, *170*, 1.

10 (a) A. YANAGISAWA, Y. NAKATSUKA, K. ASAKAWA, H. KAGEYAMA, H. YAMAMOTO, *Synlett* **2001**, 69. (b) A. YANAGISAWA, Y. NAKATSUKA, K. ASAKAWA, M. WADAMOTO, H. KAGEYAMA, H. YAMAMOTO, *Bull. Chem. Soc. Jpn.* **2001**, *74*, 1477.

11 D. FERRARIS, B. YOUNG, T. DUDDING, T. LECTKA, *J. Am. Chem. Soc.* **1998**, *120*, 4548.

12 (a) M. SODEOKA, K. OHRAI, M. SHIBASAKI, *J. Org. Chem.* **1995**, *60*, 2648. Review: (b) M. SODEOKA, M. SHIBASAKI, *Pure Appl. Chem.* **1998**, *70*, 411.

13 M. SODEOKA, R. TOKUNOH, F. MIYAZAKI, E. HAGIWARA, M. SHIBASAKI, *Synlett* **1997**, 463.

14 (a) E. HAGIWARA, A. FUJII, M. SODEOKA, *J. Am. Chem. Soc.* **1998**, *120*, 2474. (b) A. FUJII, E. HAGIWARA, M. SODEOKA, *J. Am. Chem. Soc.* **1999**, *121*, 5450.

15 A. FUJII, M. SODEOKA, *Tetrahedron Lett.* **1999**, *40*, 8011.

16 H. DOUCET, J.-L. PARRAIN, M. SANTELLI, *Synlett* **2000**, 871.

17 O. FUJIMURA, *J. Am. Chem. Soc.* **1998**, *120*, 10032.

18 (a) Y. ITO, M. SAWAMURA, T. HAYASHI, *J. Am. Chem. Soc.* **1986**, *108*, 6405; (b) Y. ITO, M. SAWAMURA, T. HAYASHI, *Tetrahedron Lett.* **1987**, *28*, 6215; (c) T. HAYASHI, M. SAWAMURA, Y. ITO, *Tetrahedron* **1992**, *48*, 1999.

19 Reviews: (a) M. SAWAMURA, Y. ITO, *Chem. Rev.* **1992**, *92*, 857; (b) M. SAWAMURA, Y. ITO, in *Catalytic Asymmetric Synthesis*, (ed.: I. OJIMA), VCH, New York, **1993**, p. 367.

20 (a) S. D. Pastor, *Tetrahedron* **1988**, *44*, 2883; (b) S. D. Pastor, A. Togni, *J. Am. Chem. Soc.* **1989**, *111*, 2333; (c) A. Togni, S. D. Pastor, *Helv. Chim. Acta* **1989**, *72*, 1038; (d) A. Togni, S. D. Pastor, *J. Org. Chem.* **1990**, *55*, 1649; (e) A. Togni, R. Häusel, *Synlett* **1990**, 633; (f) S. D. Pastor, A. Togni, *Tetrahedron Lett.* **1990**, *31*, 839; (g) A. Togni, S. D. Pastor, G. Rihs, *J. Organomet. Chem.* **1990**, *381*, C21; (h) S. D. Pastor, A. Togni, *Helv. Chim. Acta* **1991**, *74*, 905.

21 Y. Ito, M. Sawamura, T. Hayashi, *Tetrahedron Lett.* **1988**, *29*, 239.

22 A. Togni, S. D. Pastor, G. Rihs, *Helv. Chim. Acta* **1989**, *72*, 1471.

23 M. D. Bachi, A. Melman, *J. Org. Chem.* **1997**, *62*, 1896.

24 M. Sawamura, Y. Ito, T. Hayashi, *Tetrahedron Lett.* **1990**, *31*, 2723.

25 (a) Y. Ito, M. Sawamura, E. Shirakawa, K. Hayashizaki, T. Hayashi, *Tetrahedron Lett.* **1988**, *29*, 235. See also: (b) Y. Ito, M. Sawamura, E. Shirakawa, K. Hayashizaki, T. Hayashi, *Tetrahedron* **1988**, *44*, 5253.

26 Y. Ito, M. Sawamura, M. Kobayashi, T. Hayashi, *Tetrahedron Lett.* **1988**, *29*, 6321.

27 Y. Ito, M. Sawamura, H. Hamashima, T. Emura, T. Hayashi, *Tetrahedron Lett.* **1989**, *30*, 4681.

28 M. Sawamura, Y. Nakayama, T. Kato, Y. Ito, *J. Org. Chem.* **1995**, *60*, 1727.

29 (a) A. Togni, S. D. Pastor, *Tetrahedron Lett.* **1989**, *30*, 1071; (b) M. Sawamura, Y. Ito, T. Hayashi, *Tetrahedron Lett.* **1989**, *30*, 2247.

30 T. Hayashi, Y. Uozumi, A. Yamazaki, M. Sawamura, H. Hamashima, Y. Ito, *Tetrahedron Lett.* **1991**, *32*, 2799.

31 M. Sawamura, H. Hamashima, Y. Ito, *J. Org. Chem.* **1990**, *55*, 5935.

32 (a) V. A. Soloshonok, T. Hayashi, *Tetrahedron Lett.* **1994**, *35*, 2713; (b) V. A. Soloshonok, T. Hayashi, *Tetrahedron: Asymmetry* **1994**, *5*, 1091; (c) V. A. Soloshonok, A. D. Kacharov, T. Hayashi, *Tetrahedron* **1996**, *52*, 245.

33 M. A. Stark, C. J. Richards, *Tetrahedron Lett.* **1997**, *38*, 5881.

34 J. M. Longmire, X. Zhang, M. Shang, *Organometallics* **1998**, *17*, 4374.

35 G. Guillena, G. Rodríguez, G. van Koten, *Tetrahedron Lett.* **2002**, *43*, 3895.

36 Y. Motoyama, H. Kawakami, K. Shimozono, K. Aoki, H. Nishiyama, *Organometallics* **2002**, *21*, 3408.

37 R. Nesper, P. S. Pregosin, K. Püntener, M. Wörle, *Helv. Chim. Acta* **1993**, *76*, 2239.

38 C. Schlenk, A. W. Kleij, H. Frey, G. van Koten, *Angew. Chem. Int. Ed.* **2000**, *39*, 3445.

39 A. W. Kleij, R. J. M. Klein Gebbink, P. A. J. van den Nieuwenhuijzen, H. Kooijman, M. Lutz, A. L. Spek, G. van Koten, *Organometallics* **2001**, *20*, 634.

40 M. D. Meijer, N. Ronde, D. Vogt, G. P. M. van Klink, G. van Koten, *Organometallics* **2001**, *20*, 3993.

41 G. Rodriguez, M. Lutz, A. L. Spek, G. van Koten, *Chem. Eur. J.* **2002**, *8*, 45.

2
Boron and Silicon Lewis Acids for Mukaiyama Aldol Reactions

Kazuaki Ishihara and Hisashi Yamamoto

2.1
Achiral Boron Lewis Acids

2.1.1
Introduction

The classical boron Lewis acids, BX_3, RBX_2 and R_2BX (X = F, Cl, Br, I, OTf) are now popular tools in organic synthesis. B(III) can act as a Lewis acid because there is an empty *p*-orbital on the boron. Enthalpy values indicate that when pyridine is the reference base, the Lewis acidities of Group IIIB halides increase in the order $AlX_3 > BX_3 > GaX_3$. The Lewis acidity of BX_3 generally increases in the order fluoride < chloride < bromide < iodide, i.e. the exact reverse of the order expected on the basis of relative *σ*-donor strengths of the halide anions. The main reason for this anomaly is that in these BX_3 compounds the B–X bonds contain a *π*-component which is formed by overlap of a filled *p*-orbital on the halogen with the empty *p*-orbital on the boron. Because the latter orbital is used to form a *σ*-bond when BX_3 coordinates with a Lewis base, this *π*-component is completely destroyed by complex formation. The strength of the *π*-component now increases in the order iodide < bromide < chloride < fluoride, i.e. the amount of *π*-bond energy that is lost on complex formation increases as the atomic weight of the halogen decreases. Evidently, as far as the extent of complex formation is concerned, this is a more important factor than the corresponding decrease in the *σ*-donor strength of the halogen.

The BF_3 and BCl_3 complexes of diethyl ether are less stable than those of dimethyl ether, and the same order of stability is observed for complexes of diethyl and dimethyl sulfides. As expected, steric interaction decreases as the distance between the metal and ligand atom is increased. Thus, it decreases when the metal atom is changed from boron to aluminum, and when the ligand atom is changed from oxygen to sulfur.

The classical boron Lewis acids are used stoichiometrically in Mukaiyama

Modern Aldol Reactions. Vol. 2: Metal Catalysis. Edited by Rainer Mahrwald
Copyright © 2004 WILEY-VCH Verlag GmbH & Co. KGaA, Weinheim
ISBN: 3-527-30714-1

aldol reactions under anhydrous conditions, because the presence of even a small amount of water causes rapid decomposition or deactivation of the promoters. To obviate some of these inherent problems, the potential of aryl-boron compounds, $Ar_n B(OH)_{n-3}$ ($n = 1–3$), bearing electron-withdrawing aromatic groups as a new class of boron catalyst has recently been demonstrated. For example, tris(pentafluorophenyl)borane, $B(C_6F_5)_3$, is a convenient, commercially available Lewis acid of strength comparable with that of BF_3, but without the problems associated with reactive B–F bonds. Although its primary commercial application is as a co-catalyst in metallocene-mediated olefin polymerization, its potential as a Lewis acid catalyst for Mukaiyama aldol reactions is now recognized as being much more extensive. Diarylborinic acids and arylboronic acids bearing electron-withdrawing aromatic groups are also highly effective Lewis acid catalysts [1].

2.1.2
$BF_3.Et_2O$

Although $TiCl_4$ is a better Lewis acid at effecting aldol reactions of aldehydes, acetals, and silyl enol ethers, $BF_3.Et_2O$ is more effective for aldol reactions with anions generated from transition metal carbenes and with tetrasubstituted enol ethers such as (Z)- and (E)-3-methyl-2-(trimethylsilyloxy)-2-pentane [2, 3]. One exception is the preparation of substituted cyclopentanediones from acetals by aldol condensation of protected four-membered acyloin derivatives with $BF_3.Et_2O$ rather than $TiCl_4$ (Eq. (1)) [2, 4]. Use of the latter catalyst results in some loss of the silyl protecting group. The pinacol rearrangement is driven by the release of ring strain in the four-membered ring and is controlled by an acyl group adjacent to the diol moiety.

$$(1)$$

This reagent is the best promoter of the aldol reaction of 2-(trimethylsiloxy)acrylate esters, prepared by the silylation of pyruvate esters, to afford γ-alkoxy-α-keto esters (Eq. (2)) [5] These esters occur in a variety of important natural products.

$$
\begin{array}{c}
\text{[Structures: PhCH}_2\text{CH(OMe)}_2 \text{ (with OMe, OMe) + silyl enol ether (OSiMe}_3\text{, OEt, C=O)]}
\end{array}
$$

$$
\xrightarrow[\substack{\text{CH}_2\text{Cl}_2 \\ -78\,^\circ\text{C to } 0\,^\circ\text{C} \\ 86\% \text{ yield}}]{\text{BF}_3\cdot\text{Et}_2\text{O}}
$$

[Product structure: Ph chain with OMe, O, OEt, C=O]

$$(2)$$

$BF_3.Et_2O$ can improve or reverse aldehyde diastereofacial selectivity in the aldol reaction of silyl enol ethers with aldehydes, to give *syn* adducts. For example, Heathcock and Flippin have reported that the reaction of the silyl enol ether of pinacolone with 2-phenylpropanal using $BF_3.Et_2O$ results in enhanced Felkin selectivity (up to 36:1) compared with addition of the corresponding lithium enolate [6, 7]. When the α-substituents are more subtly differentiated, however, it is still difficult to achieve acceptable levels of selectivity. Davis et al. have reported that use of triisopropylsilyl enol ether and $i\text{-}Pr_3SiB(OTf)_4$ results in selectivity of ca. 100:1 with 2-phenylpropanal and a useful level of 7:1 with 2-benzylpropanal (Eq. (3)) [8]. Control experiments employing $BF_3.Et_2O$ catalysis and 2-benzylpropanal as substrate results in lower selectivity (ca. 3:1) that does not depend substantially on the bulk of the silyl group in the enolate (Eq. (3)). In contrast, both levels of 1,2-asymmetric induction in the $i\text{-}Pr_3SiB(OTf)_4$ (5 mol%)- and the $BF_3.Et_2O$ (1 equiv.)-promoted additions of silyl ketene thioacetals to α-asymmetric aldehydes are affected by the bulk of the silyl group (Eqs. (4) and (5)) [8].

[Scheme for Eq. (3): Bn–CHO + silyl enol ether (OTIPS, X) with X=Me, t-Bu, OMe, Ot-Bu, Lewis acid $i\text{-}Pr_3SiB(OTf)_4$, $BF_3\cdot Et_2O$, giving Cram and anti-Cram products (Bn, TIPSO, O, X)]

	Cram		anti-Cram
	7	:	1
	3	:	1

$$(3)$$

[Scheme for Eq. (4): Bn–CHO + silyl ketene thioacetal (OR, St-Bu), $i\text{-}Pr_3SiB(OTf)_4$ (5 mol%), giving Cram and anti-Cram products (Bn, RO, O, St-Bu)]

	Cram		anti-Cram
R=TIPS	5.5	:	1
R=TBDMS	3.6	:	1

$$(4)$$

$$
\begin{array}{c}
\text{Bn} \diagup \text{CHO} + \underset{St\text{-Bu}}{\overset{OR}{\diagup}} \xrightarrow[\text{(1 equiv)}]{BF_3 \cdot Et_2O} \text{Bn} \diagup \underset{HO \quad O}{\diagup} St\text{-Bu} + \text{Bn} \diagup \underset{HO \quad O}{\diagup} St\text{-Bu}
\end{array}
$$

	Cram		anti-Cram	
R=TIPS	13	:	1	
R=TBDMS	5.8		1	(5)

Addition of the tetrasubstituted selenoketene silyl acetal to β-benzyloxy aldehyde in the presence of Et_2BOTf leads to good yields and high ratios of products with 3,4-*anti* relative stereochemistry (Cram chelate model) (Eq. (6)). In contrast, reversed diastereoselectivity and a synthetically interesting ratio of 1:11 in favor of the 3,4-*syn* products is obtained when the bulky β-silyloxy aldehyde is used in the presence of the monodentate Lewis acid $BF_3 \cdot OEt_2$ (Felkin–Anh model) (Eq. (7)) [9].

$$
\begin{array}{c}
\underset{BnO \quad O}{\diagup} H + \underset{SePh}{\overset{OSiMe_3}{\diagup}} OMe \xrightarrow[\text{CH}_2\text{Cl}_2, -78\,°C]{\begin{array}{c}Et_2BOTf\\(1.2 \text{ equiv})\end{array}} \underset{BnO \quad OH \quad O}{\diagup} \underset{SePh}{OMe}
\end{array}
$$

86% yield, *syn:anti*=<1:20

$$(6)$$

$$
\begin{array}{c}
\underset{TBDPSO \quad O}{\diagup} H + \underset{SePh}{\overset{OSiMe_3}{\diagup}} OMe \xrightarrow[\text{CH}_2\text{Cl}_2, -78\,°C]{\begin{array}{c}BF_3 \cdot OEt_2\\(1.5 \text{ equiv})\end{array}} \underset{TBDPSO \quad OH \quad O}{\diagup} \underset{SePh}{OMe}
\end{array}
$$

84% yield, 2,3-(*syn:anti*)=11:1

$$
\left[\begin{array}{c} BnO\cdots\text{-}BEt_2OTf \\ O \\ Me \diagup \underset{H}{\overset{H}{|}} \end{array} \text{enol ether} \right]
\qquad
\left[\text{enol ether} \begin{array}{c} \cdots BF_3 \\ Me \quad O \\ \underset{H \ H}{|} OTBDPS \end{array} \right]
$$

Cram chelate   Felkin-Anh   (7)

The subsequent free-radical-based hydrogen transfer reaction of the products obtained from the Mukaiyama reactions occurs under the control of the exocyclic effect. Excellent diastereoselectivity favoring the 2,3-*anti* relative stereochemistry is obtained from the reduction with Bu_3SnH in the presence of Et_2BOTf and *N,N*-diisopropylethylamine (Eq. (8)) [9].

The initial product of the aldol condensation is a β-hydroxy carbonyl compound, which is often transformed into the corresponding α,β-unsaturated derivative. Interestingly, gaseous BF_3, which is a much stronger Lewis acid than $BF_3 \cdot Et_2O$ promotes the fragmentation reaction from β-aryl-

Exocyclic effect

$$(8)$$

β-hydroxyketones to (E)-arylalkenes and carboxylic acids in good yields (Eq. (9)) [10]. The combination of powerful Lewis acid and non-nucleophilic solvents, for example CCl_4, are keys to this unexpected behavior. BCl_3 and BBr_3 are less effective in this reaction.

$$(9)$$

2.1.3
$B(C_6F_5)_3$

$B(C_6F_5)_3$ is an air-stable, water-tolerant Lewis acid catalyst, which can be readily prepared as a white solid by reacting BCl_3 with C_6F_5Li [11, 12]. This compound does not react with pure oxygen [12]. It is very thermally stable, even at 270 °C, and is soluble in many organic solvents [12]. Although $B(C_6F_5)_3$ catalyzes reactions most effectively under anhydrous conditions, $B(C_6F_5)_3$ exposed to air is also available (not anhydrous grade).

Mukaiyama aldol reactions of a variety of silyl enol ethers or ketene silyl acetals with aldehydes or other electrophiles proceed smoothly in the presence of 2 mol% $B(C_6F_5)_3$ [13a,c]. The following characteristic features can be noted:

- the products can be isolated as β-trimethylsilyloxy ketones when crude adducts are worked-up without exposure to acid;

- the reaction can be conducted in aqueous media, so reaction of the silyl enol ether derived from propiophenone with a commercial aqueous solution of formaldehyde does not present any problems;
- the rate of an aldol reaction is markedly increased by use of an anhydrous solution of $B(C_6F_5)_3$ in toluene under an argon atmosphere; and
- silyl enol ethers can be reacted with chloromethyl methyl ether or trimethylorthoformate; hydroxymethyl, methoxymethyl, or dimethoxymethyl C1 groups can be introduced at the position α to the carbonyl group.

These aldol-type reactions do not proceed when triphenylborane is used (Eq. (10)).

(10)

2.1.4
Ar_2BOH

Diarylborinic acids, Ar_2BOH, bearing electron-withdrawing aromatic groups are effective catalysts for Mukaiyama aldol condensation and subsequent selective dehydration of β-hydroxy carbonyl compounds [14]. The catalytic activity of diarylborinic acids $(C_6F_5)_2BOH$ and $[3,5\text{-}(CF_3)_2C_6H_3]_2BOH$ in Mukaiyama aldol reactions is much higher than that of the corresponding arylboronic acids. It is worthy of note that small amounts of E-isomeric dehydrated product has been isolated in reactions catalyzed by diarylborinic acids $(C_6F_5)_2BOH$ and $[3,5\text{-}(CF_3)_2C_6H_3]_2BOH$. In contrast, no dehydrated products have been isolated in the presence of $(C_6F_5)_3B$, despite its extremely high catalytic activity (Eq. (11)).

Significant features of these active borinic acid catalysts are that they are strong Lewis acids and have a hydroxy group on the boron atom. Dehydration is strongly favored in THF. The reaction usually proceeds smoothly, and α,β-enones are obtained in high yields as E isomers. In reactions of α-substituted β-hydroxy carbonyl compounds, α,β-enones are preferentially

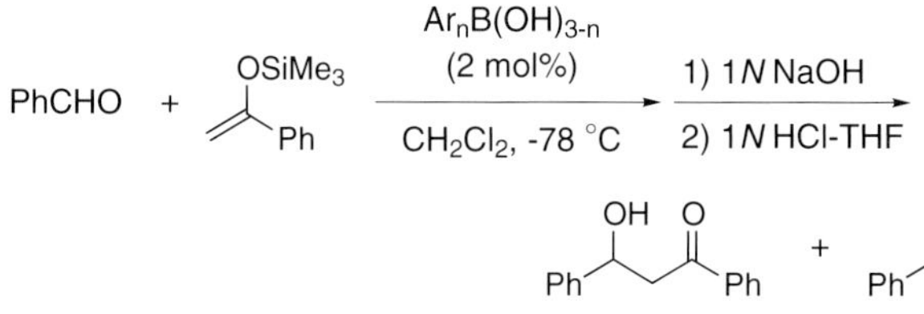

$$PhCHO \quad + \quad \overset{OSiMe_3}{\underset{Ph}{\diagup\!\!\!\diagdown}} \quad \xrightarrow[\substack{CH_2Cl_2,\ -78\ °C}]{\substack{Ar_nB(OH)_{3-n}\\(2\ mol\%)}} \quad \xrightarrow{\substack{1)\ 1N\ NaOH\\2)\ 1N\ HCl\text{-}THF}}$$

$$\underset{Ph}{\overset{OH\ \ O}{\diagup\!\!\diagdown\!\!\diagup\!\!\diagdown}}Ph \quad + \quad \underset{Ph}{\diagup\!\!\diagdown\!\!\diagup}\overset{O}{\diagdown}Ph \quad (11)$$

$C_6F_5B(OH)_2$	0%	0%
$(C_6F_5)_2BOH$	89%	7%
$[3,5\text{-}(CF_3)_2C_6H_3]_2BOH$	89%	10%
$B(C_6F_5)_3$	98%	0%

obtained from *anti* aldols whereas most of the *syn* aldols are recovered. This dehydration is, therefore, a useful and convenient method for isolating pure *syn* aldols from *syn/anti* isomeric mixtures (Eq. (12)). Reaction of the β-hydroxy function with the diarylborinic acid leads to a cyclic intermediate, which should be susceptible to dehydration. Subsequent transformation to α,β-enones occurs via an enolate intermediate resulting from selective abstraction of a pseudo-axial α-proton perpendicular to the carbonyl face. A cyclic intermediate formed from a *syn* aldol and a diarylborinic acid would be thermodynamically less stable than the cyclic intermediate. Thus, dehydration to (E)-α,β-enones occurs selectively for *anti*-aldols.

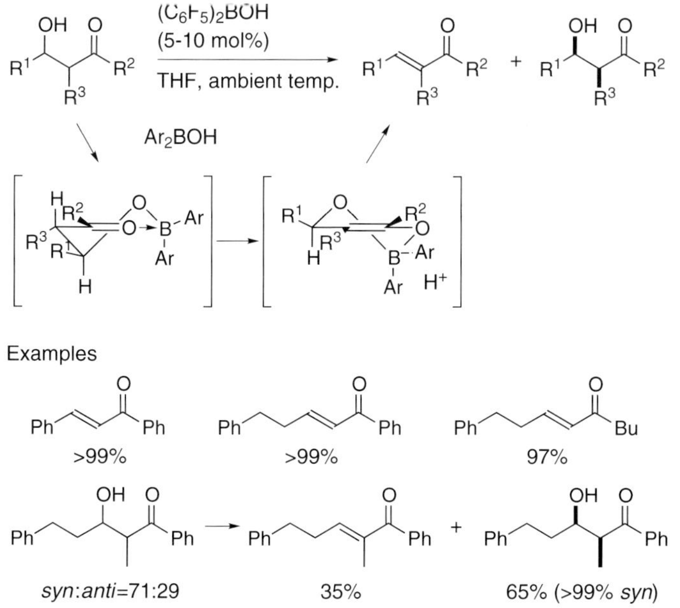

(12)

Examples

Ph—CH=CH—CO—Ph >99%

Ph—CH₂—CH=CH—CO—Ph >99%

Ph—CH₂—CH=CH—CO—Bu 97%

syn:anti=71:29   35%   65% (>99% *syn*)

Diphenylborinic acid (10 mol%), which is stable in water, is an effective catalyst for the Mukaiyama aldol reaction in the presence of benzoic acid (1 mol%) as a co-catalyst and sodium dodecyl sulfate (10 mol%) as a surfactant (Eq. (13)) [15]. Use of water as solvent is essential in this reaction. The reaction proceeds sluggishly in organic solvents such as dichloromethane and diethyl ether. Much lower yield than in water is obtained under neat conditions. Not only aromatic aldehydes but also α,β-unsaturated and aliphatic aldehydes give high *syn* selectivity. Although lower diastereoselectivity is observed when *E* enolates are used, reverse diastereoselectivity is observed when both stereoisomers of the silyl enolate derived from *tert*-butyl thiopropionate are used.

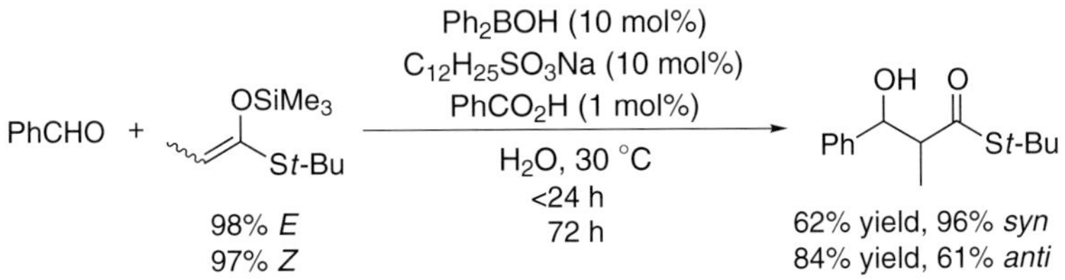
Ph2BOH (10 mol%)
C12H25SO3Na (10 mol%)
PhCO2H (1 mol%)
H2O, 30 °C
<24 h
72 h
PhCHO +
OSiMe3
St-Bu
98% E
97% Z
OH O
Ph
St-Bu
62% yield, 96% syn
84% yield, 61% anti

(13)

This reaction can be explained by a mechanism via the boron enolate as reaction intermediate generated by Si–B exchange (Figure 2.1). That the diastereoselectivity is reversed by using the stereoisomers of the silyl enolate supports the hypothesis because this type of reversal has also been observed in the traditional boron enolate mechanism which involves a chair-like six-membered transition-state. Furthermore, the trend that *anti* selectivity is poorer than *syn* selectivity in the reactions is also found in the traditional boron enolate-mediated aldol reactions. The mechanism is based on the hypothesis that Ph_2BOH can react with a silyl enolate to form the corresponding boron enolate under these conditions. When a *Z* enolate is used, an aldehyde and the boron enolate react via a chair-like six-membered tran-

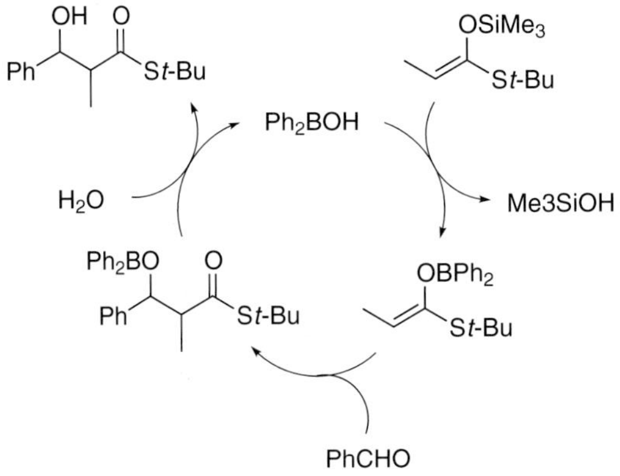
OH O
Ph
St-Bu
OSiMe3
St-Bu
Ph2BOH
H2O
Me3SiOH
Ph2BO O
Ph
St-Bu
OBPh2
St-Bu
PhCHO

Fig. 2.1
Proposed boron enolate mechanism.

sition state to give the *syn* aldol product. The B–O bond of the initial aldol product is presumed to be easily cleaved by hydrolysis, and Ph$_2$BOH can be regenerated. In this mechanism benzoic acid might accelerate the Si–B exchange step, which is thought to be rate-determining.

2.2
Chiral Boron Lewis Acids

2.2.1
Introduction

Asymmetric aldol synthesis has recently been the focus of intense interest. Especially worthy of note is the development of homogeneous catalytic enantioselective Mukaiyama aldol reactions in which a small amount of chiral ligand can induce asymmetry for a given reaction. The possible applications depend on the selectivity of the homogeneous catalysts, which are of great interest because they provide simple methods for synthesizing complex molecules for which enantiocontrol is needed. This section addresses chirally modified boron Lewis acid complexes, in which there has been increased interest because of their capacity to induce chirality. They have been successfully used for Mukaiyama aldol reactions.

2.2.2
Chiral Boron Lewis Acids as Stoichiometric Reagents

This reaction of silyl ketene acetals with aldehydes using **1** as a stoichiometric chiral reagent was originally reported by Reetz et al. (Eq. (14)) [16]. The aldol addition of 1-(trimethylsiloxy)-1-methoxy-2-methyl-1-propene and 3-methylbutanal provides the aldol in only 57% yield, but with 90% ee.

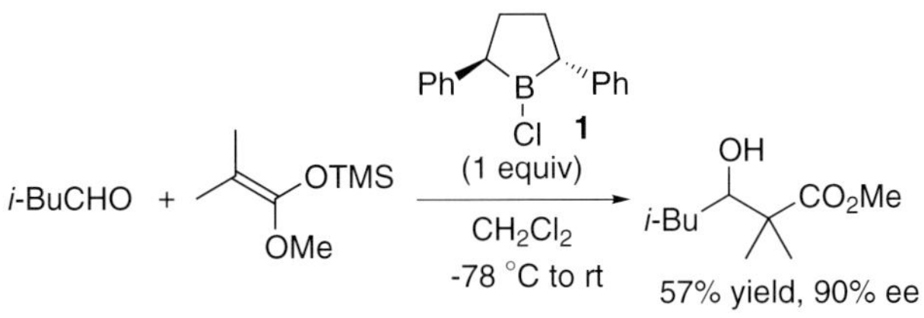

The use of chiral acyloxyborane (CAB) as a chiral reagent seems to be more effective for this reaction, which proceeds faster and with higher yields and enantiomeric excess. Kiyooka and his colleagues first described the use of a variety of chiral oxaborolidines, derived from sulfonamides of α-amino acids and borane, in the course of the selective aldol reaction between silyl ketene acetals and aldehydes (Eq. (15)) [17a]. This reaction gives β-hydroxy esters in high enantioselectivity and yields.

$$R^1CHO \quad + \quad \overset{OTMS}{\underset{OEt}{\diagup}} \quad \xrightarrow[\substack{CH_2Cl_2 \\ -78\ °C\ to\ rt}]{\substack{\textbf{2a} \\ (1\ equiv)}} \quad R^1 \overset{OH}{\underset{*}{\diagup}} CO_2Et \tag{15}$$

77-87% yield, 83-93% ee

The role played by the trialkylsilyl group is unclear. Changing the trimethylsilyl group increases not only the selectivity but also the product of the reaction – β-hydroxy acetals are now obtained instead of β-hydroxy esters. They investigated the course of the reaction of a variety of *tert*-butyldimethylsilyl ketene acetals and aldehydes with **2a** (Eq. (16)). The acetal is probably formed by hydride transfer to an intermediate ester. The *tert*-butyl group apparently stabilizes the second intermediate and consequently changes the course of the reaction. It should be noted that the first cyclic intermediate is stabilized by coordination of the borane with the oxygen of the carbonyl. The results are outstanding when $R^2 = Me$ (92–98%). Selectivity and yield decrease, however, when $R^2 = H$ (45–62%) (Eq. (16)).

$$R^1CHO + \overset{R^2}{\underset{OEt}{\underset{R^2}{\diagup}}}OTBDMS \xrightarrow[CH_2Cl_2,\ -78\ °C]{\textbf{2a}\ (1\ equiv)} \tag{16}$$

$$\xrightarrow{} \quad R^1 \overset{OH\quad OTBDMS}{\underset{R^2\ R^2}{\diagup}}OEt$$

R^2=H: 77-82% yield, 45-62% ee
R^2=Me: 79-85% yield, 92-98% ee

A stoichiometric amount of **2a** catalyzed the asymmetric aldol reaction of aldehydes with enol silyl ethers and subsequent asymmetric reduction in one pot to afford *syn* 1,3-diols with high enantioselectivity (Eq. (17)) [17b]. With a variety of aldehydes, 1,3-diols were obtained in moderate yields (53–70%) with high *syn* diastereoselectivity. The *syn* 1,3-diols prepared from aliphatic aldehydes in the reaction (EtCN) are almost enantiomerically pure

(96–99% ee). Propionitrile was the best solvent for reaction selectivity. The TBDMS substituent of enol silyl ether reduced the *syn* selectivity.

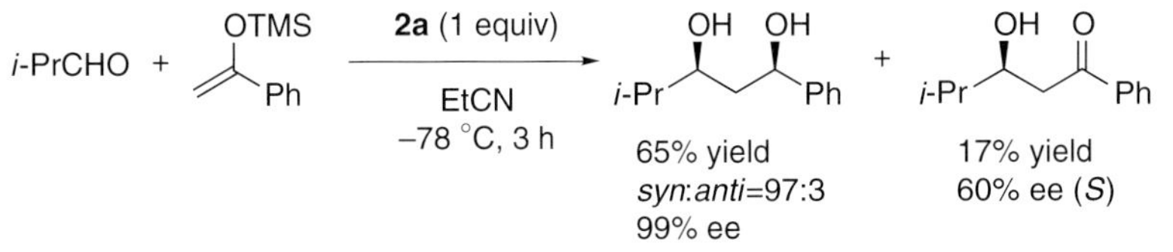

(17)

Thus, good to excellent diastereo- and enantioselectivity are achieved simultaneously with 1,3-diols whereas the enantioselectivity for β-hydroxy ketones is substantially lower. These observations on the selectivity of the products suggest that *syn*-selective reduction of the reaction intermediate occurs after enantioselective aldol addition.

On the basis of Kiyooka's working hypothesis for the aldol reaction mechanism, the reduction proceeds via intramolecular hydride transfer, which is accelerated by a matching mode between the promoter's chirality and that of the newly formed aldol (Eq. (18)). An alternative mechanism without chelation is also possible; this involves hydride delivery to the preferred *O*-silyl oxocarbenium ion conformer (Eq. (19)).

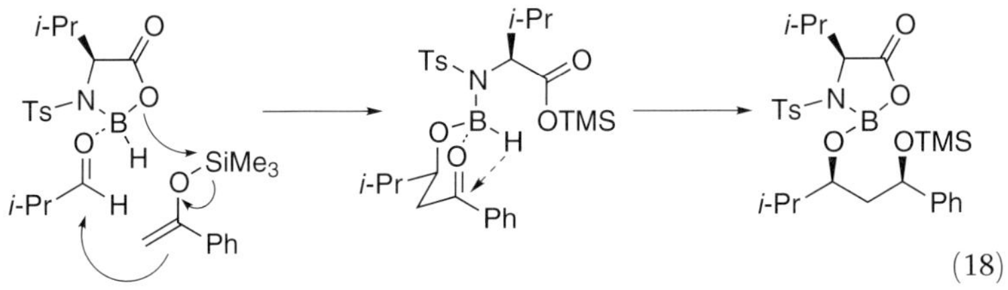

(18)

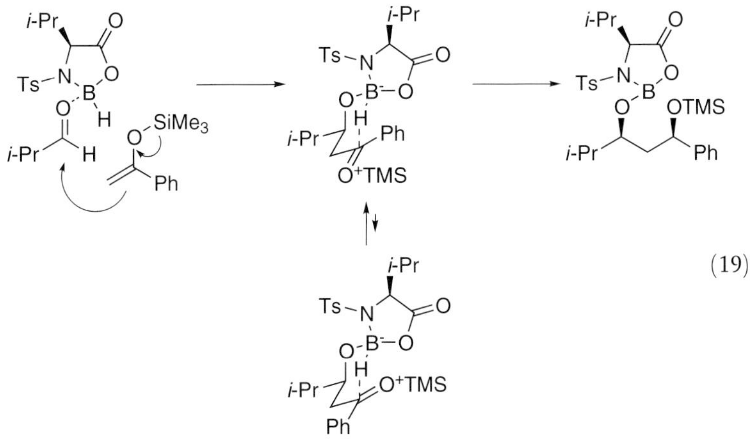

(19)

A very short asymmetric synthesis of an insect attractant, (1*S*,3*S*,5*R*)-1,3-dimethyl-2,9-dioxabicyclo[3.3.1]nonane, a host-specific substance for the ambrosia beetle that infests the bark of the Norway spruce, has been realized with a **2a**-mediated aldol reaction strategy; enantio- and diastereoselectivity were high (Eq. (20)) [17c].

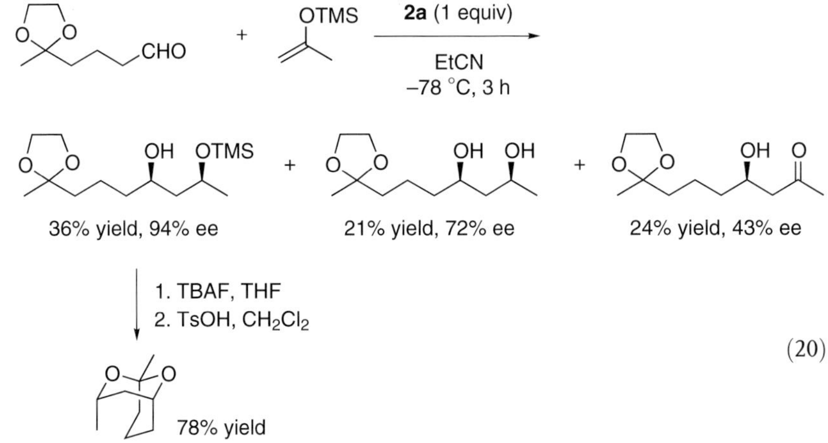

CHO
OTMS
2a (1 equiv)
EtCN
−78 °C, 3 h
OH OTMS
OH OH
OH O
36% yield, 94% ee
21% yield, 72% ee
24% yield, 43% ee
1. TBAF, THF
2. TsOH, CH2Cl2
78% yield

(20)

The chiral borane **2a**-mediated aldol reaction proceeds with α-chiral aldehydes in a reagent-controlled manner. Both enantiomers are obtained almost optically pure from one racemic aldehyde (Eqs. (21) and (22)) [17d].

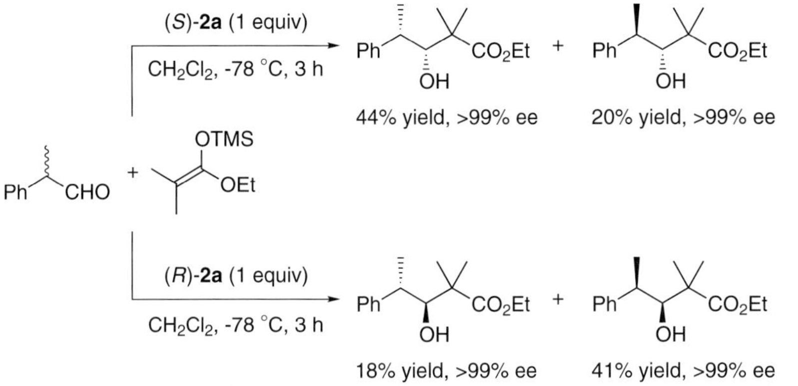

(S)-2a (1 equiv)
CH2Cl2, -78 °C, 3 h
Ph CO2Et
OH
44% yield, >99% ee
20% yield, >99% ee
Ph CHO
OTMS
OEt
(R)-2a (1 equiv)
CH2Cl2, -78 °C, 3 h
18% yield, >99% ee
41% yield, >99% ee

(21)

(22)

The reaction of β-chiral aldehydes with ketene silyl acetals gives both *syn* and *anti* aldols in similar yields without any Cram selectivity (Eq. (23)) [17d].

$$(23)$$

In Kiyooka's approach to acetate aldols using a stoichiometric amount of **2a**, a serious reduction (ca. 10–20%) in enantiomeric excess was observed in the reaction with silyl ketene acetals derived from α-unsubstituted acetates, compared with the high level of enantioselectivity ($> 98\%$ ee) in the reaction with 1-ethoxy-2-methyl-1-(trimethylsiloxy)-1-propene. Introduction of an eliminable substituent, e.g. a methylthio or bromo substituent, after aldol reaction at the α-position of chiral esters resolves this problem [17e]. Asymmetric synthesis of dithiolane aldols has been achieved in good yields by using the silyl ketene acetal derived from 1,3-dithiolane-2-carboxylate in the **2a**-promoted aldol reaction; desulfurization of the dithiolane aldols produces the acetate aldols in high enantiomeric purity (Eq. (24)).

$$(24)$$

A very short asymmetric synthesis of the bryostatin C_1–C_9 segment has been achieved using three sequential **2a**-promoted aldol reactions under reagent control [17f]. This synthetic methodology is based on the direct asymmetric incorporation of two acetate and one isobutyrate synthones into a framework (Scheme 2.1).

The **2a**-promoted asymmetric aldol reaction of a variety of aldehydes with a silyl nucleophile derived from phenyl propionate (E isomer 98%) results in moderate *anti* diastereoselectivity with relatively low enantioselectivity. On the other hand, with pivalaldehyde and the silyl nucleophile derived from ethyl propionate ($E/Z = 85{:}15$), the *syn* isomer is obtained as a major product (22:1) with 96% ee (Eq. (25)) [17g]. This unexpected switching of diastereoselectivity observed in the reaction of the bulky aldehyde can be explained by merging Corey's hydrogen-bond model between the aldehyde hydrogen and the catalyst borane-ring oxygen [18e] and Yamamoto's ex-

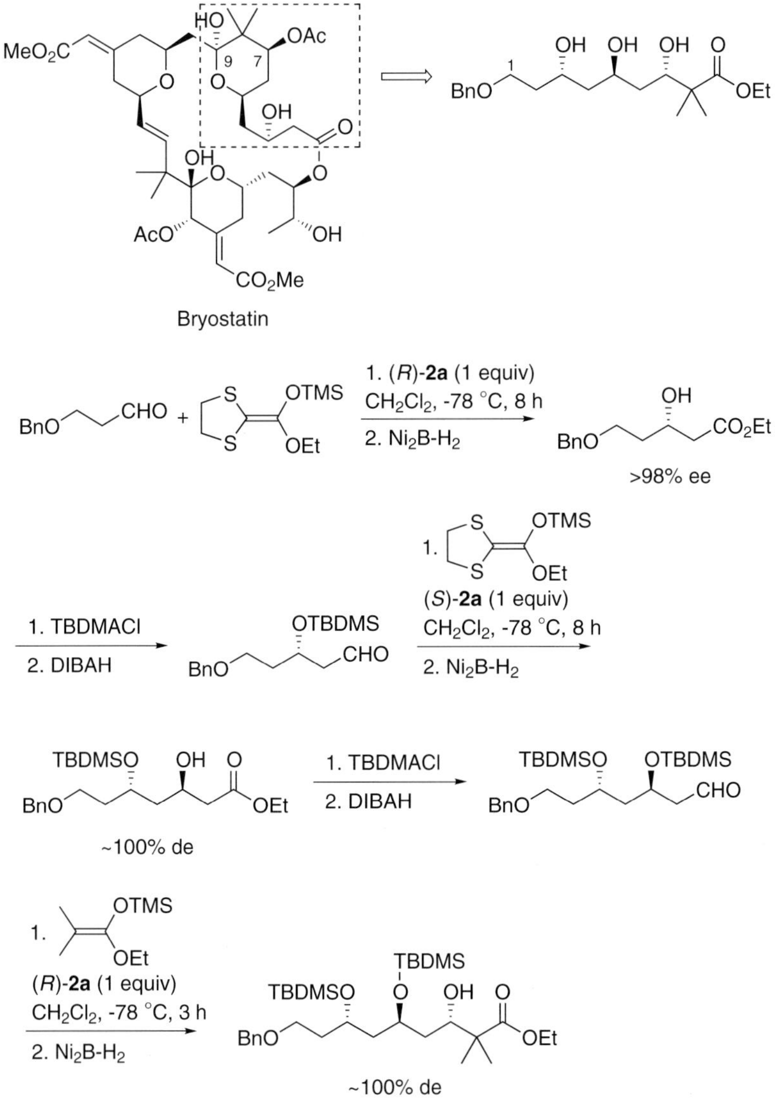

Bryostatin

Scheme 2.1

tended transition model **3** (see also Figure 2.4) [19] as depicted in Figure 2.2, where **4** is destabilized by gauche interaction between the methyl and *tert*-butyl groups.

Important limitations have been observed with regard to reagent control in reactions with highly sterically hindered aldehydes involving a chiral hydroxy function at the β-position (Eq. (26)) [17g]. When (*S*)-**2a** is used for **5**, the diastereo- and enantioselectivity are less satisfactory. When (*R*)-**2a**,

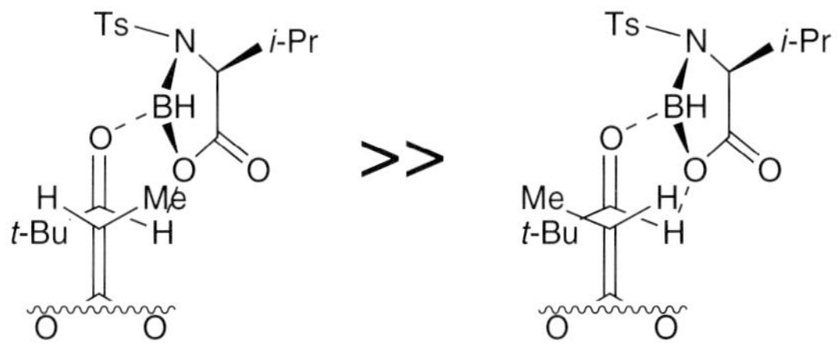

Fig. 2.2
Kiyooka's transition-state models.

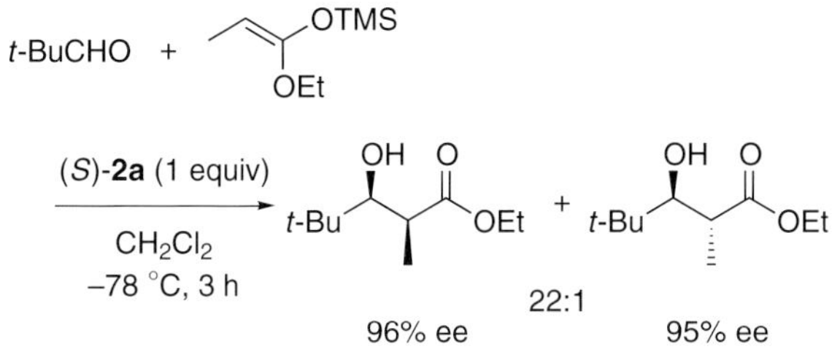

$$ (25) $$

is used, however, the reaction proceeds more smoothly to give the corresponding aldols with moderate *syn* selectivity in 87% yield. Each of the isomers obtained is almost enantiomerically pure. The spatial orientation of the siloxy group at C-3, which is presumably fixed by introduction of two methyl groups at C-2, affects the entire conformation of the aldehydes, and when the chiral borane coordinates to the aldehyde an adequate fit might be needed between the stereocenters of the reagent and the substrate (at C-3) for the stereochemical outcome expected from reagent control. Reaction with (*S*)-**2a** loses reagent control because of stereochemically mismatched interactions. Even in such complex circumstances, however, the reaction with (*R*)-**2a** gives products with stereochemistry at C-3 similar to that expected on the basis of reagent control. Effective approach of the silyl nucleophile might occur via a path similar to **11** in Figure 2.3.

Fig. 2.3
Kiyooka's transition-state models.

OTMS
2 CHO + OTMS   **2a** (1 equiv)
TBDMSO 3 1   OEt
5   CH₂Cl₂
−78 °C, 24 h

TMSO OH O   TMSO OH O
TBDMSO ⤳ OEt + TBDMSO ⤳ OEt
6   **7**

+ TMSO OH O   + TMSO OH O
TBDMSO ⤳ OEt   TBDMSO ⤳ OEt
8   **9**

(*S*)-**2a**   34% yield, *syn*(**6**+**8**):*anti*(**7**+**9**)=2:1, **6**:**8**=7:5, **7**:**9**=5:3

(*R*)-**2a**   87% yield, *syn*(**6**+**8**):*anti*(**7**+**9**)=4:1, **6**:**8**=>50:1, **7**:**9**=>50:1   (26)

2.2.3
Chiral Boron Lewis Acids as Catalytic Reagents

CAB **12**, R = H, derived from monoacyloxytartaric acid and diborane, is an excellent catalyst (20 mol%) for the Mukaiyama condensation of simple enol silyl ethers of achiral ketones with a variety of aldehydes. The reactivity of aldol-type reactions can be improved without reducing enantioselectivity by using 10–20 mol% **12**, R = 3,5-(CF₃)₂C₆H₃, prepared from 3,5-bis(tri-fluoromethyl)phenylboronic acid (**13**) and a chiral tartaric acid derivative. Enantioselectivity could also be improved, without reducing the chemical yield, by using 20 mol% **12**, R = *o*-PhOC₆H₄, prepared from *o*-phenoxy-phenylboronic acid and a chiral tartaric acid derivative. The **12**-catalyzed aldol process enables preparation of adducts highly diastereo- and enantio-selectively (up to 99% ee) under mild reaction conditions [19a,c]. These reactions are catalytic, and the chiral source is recoverable and reusable (Eq. (27)).

The relative stereochemistry of the major adducts is assigned to be *syn*, and the predominant *re*-face attack of enol ethers at the aldehyde carbonyl carbon has been confirmed when a natural tartaric acid derivative is used as Lewis acid ligand. The use of an unnatural form of tartaric acid as a chiral source gives the other enantiomer, as expected. Almost perfect asymmetric induction is achieved with the *syn* adducts, reaching 99% ee, although a slight reduction in both enantio- and diastereoselectivity is observed in re-actions with saturated aldehydes. Irrespective of the stereochemistry of the starting enol silyl ethers generated from ethyl ketone, *syn* aldols are ob-tained with high selectivity in these reactions. The observed high *syn* selec-tivity, and its lack of dependence on the stereoselectivity of the silyl enol ethers, in **12**-catalyzed reactions are fully consistent with Noyori's TMSOTf-

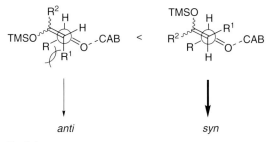

(27)

catalyzed aldol reactions of acetals, and thus might reflect the acyclic extended transition state mechanism postulated in the latter reactions (Figure 2.4). Judging from the product configurations, 12 (from natural tartaric acid) should effectively cover the *si* face of the carbonyl after its coordination, and selective approach of nucleophiles from the *re* face should result. This behavior is totally systematic and in good agreement with the results from previously described 12-catalyzed reactions for all of the aldehydes examined.

A catalytic enantioselective aldol-type reaction of ketene silyl acetals with achiral aldehydes also proceeds smoothly with 12, R = H; this can furnish erythro β-hydroxy esters with high optical purity (Eq. (28) [19b,c].

Fig. 2.4
Extended transition-state model.

1) **12** (R=H, 20 mol%)
EtCN, -78 °C

R^1CHO + R^2—OTMS/OR^3 → HO / O / R^1—OR^3 / R^2

2) TBAF

HO O
Ph—OPh
84% ee

HO O
Pr—OPh
76% ee

HO O
Ph—OPh
92% ee *syn*
syn:anti=79:21

(28)

HO O
Pr—OPh
88% ee *syn*
syn:anti=79:21

HO O
Pr—OPh
97% ee *syn*
syn:anti=96:4

HO O
—OPh
94% ee *syn*
syn:anti=95:5

A remarkable finding is the sensitivity of this reaction to the substituents of the starting silyl ketene acetals. The reactions of silyl ketene acetals derived from more common ethyl esters are totally stereorandom, and give a mixture of *syn* and *anti* isomers in even ratios with improved chemical yields. In sharp contrast, the use of silyl ketene acetals generated from phenyl esters leads to good diastereo- and enantioselectivity with excellent chemical yields. The reason for this finding is not clear, but a secondary interaction between electron-rich silyl ketene acetals derived from alkyl esters and Lewis acid might be responsible.

Analogous to the previous results with silyl enol ethers of ketones, nonsubstituted silyl ketene acetals lead to lower levels of stereoregulation. On the other hand, propionate-derived silyl ketene acetals lead to high asymmetric induction. Reactions with aliphatic aldehydes, however, result in a slight reduction in optical yields. With phenyl ester-derived silyl ketene acetals, *erythro* adducts predominate, but the selectivity is usually moderate compared with the reactions of silyl enol ethers. Exceptions are α,β-unsaturated aldehydes, for which diastereo-and enantioselectivity are excellent. The observed *erythro* selectivity and *re*-face attack of nucleophiles on the carbonyl carbon of aldehydes are consistent with the aforementioned aldol reactions of silyl enol ethers [19].

After the enantioselective aldol reaction using CAB **2a** under stoichiometric conditions has been reported by Kiyooka and his colleagues in 1991 [17], Masamune [20], Kiyooka [21a], and Corey [22] and their co-workers all independently developed CAB-catalyzed systems of enantioselective aldol reactions (Eq. (29)).

Masamune and colleagues examined several oxazaborolidines derived from a series of simple α-amino acid ligands derivatized as the corresponding *N-p*-toluenesulfonamides. A dramatic improvement in reaction enantio-

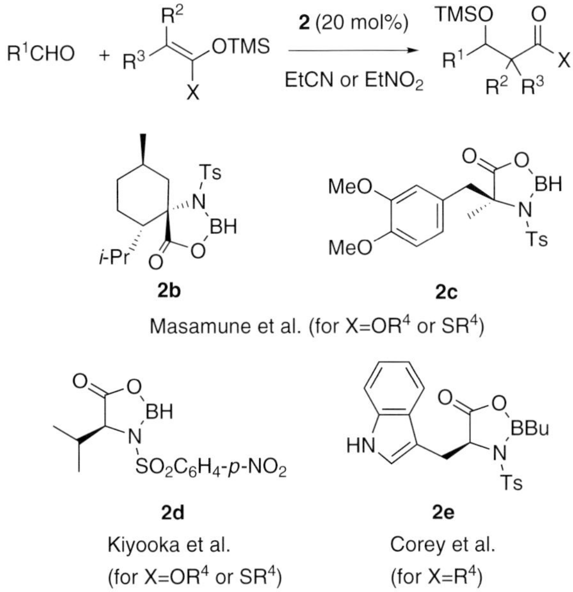
R1CHO
R2
R3
OTMS
X
2 (20 mol%)
EtCN or EtNO2
TMSO
O
R1
X
R2 R3
Ts
N
BH
O
i-Pr
O
2b
MeO
MeO
O
O
BH
N
Ts
2c
Masamune et al. (for X=OR4 or SR4)
O
O
BH
N
SO2C6H4-p-NO2
2d
Kiyooka et al.
(for X=OR4 or SR4)
HN
O
O
BBu
N
Ts
2e
Corey et al.
(for X=R4)

(29)

selectivity was observed when complexes prepared from α,α-disubstituted glycine arylsulfonamides were used. This suggests that the initial aldol adduct must undergo ring-closure to release the final product **15** and to regenerate the catalyst **2** (Figure 2.5) [20]. Slow addition of the aldehyde to the reaction mixture (making enough time available for **14** to undergo ring closure) has often been beneficial in improving the enantioselectivity of the reaction. Kiyooka and his colleagues have reported a straightforward improvement of this reaction to a catalytic version by using an *N-p*-nitro-

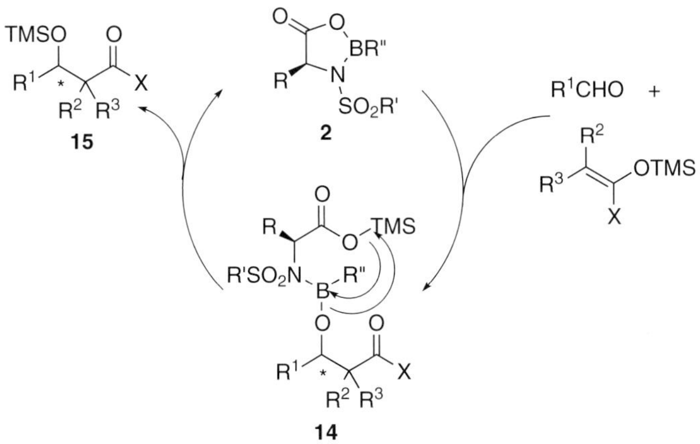
TMSO
O
R1
X
R2 R3
15
O
O
BR"
R
N
SO2R'
2
R1CHO +
R2
R3
OTMS
X
O
R
O
TMS
R'SO2N
B
R"
O
O
R1
X
R2 R3
14

Fig. 2.5
The proposed catalytic cycle.

benzenesulfonyl-derived ligand and nitroethane instead of dichloromethane as solvent [21a].

Product enantioselectivity has also been optimized as a function of substitution of the arylsulfonamide. (Eq. (30)) [20]. Thus, for complexes with the general structure **2c**, the enantiomeric excess of the benzaldehyde adduct varies along the series R′ = 3,5-bis(trifluoromethyl)phenyl (52% ee); mesityl (53% ee); 1-naphthyl (67% ee); 2-naphthyl (78% ee); 4-*tert*-butylphenyl (81% ee); phenyl (83% ee); 4-methoxyphenyl (86% ee); 4-acetamidophenyl (86% ee).

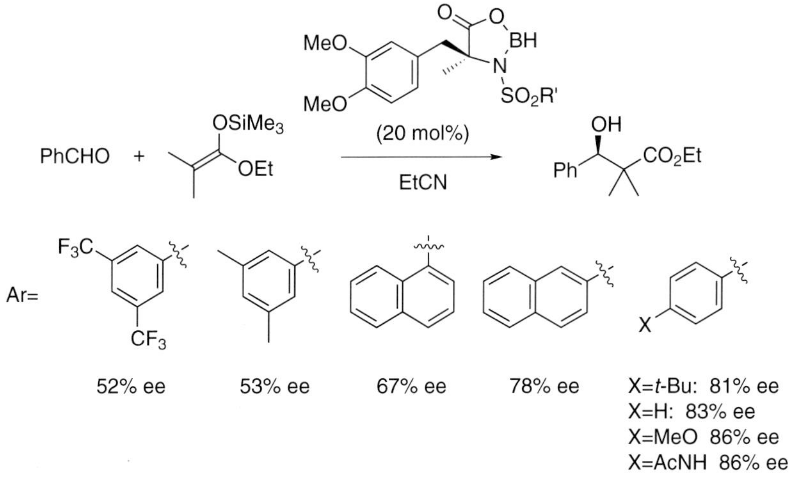

$$\text{(30)}$$

An AM1-optimized structure of the chiral borane complex was used as the centerpiece of a model proposed by Kiyooka and co-workers to account for the stereochemical outcome of the reaction (Figure 2.6) [21a]. It was suggested the aldehydes coordinate with the boron on the face opposite the isopropyl substituent, thereby minimizing steric interactions. The Kiyooka model places the formyl-H over the five-membered ring chelate subtending an obtuse H–B–O–C dihedral angle. Analogous modes of binding have

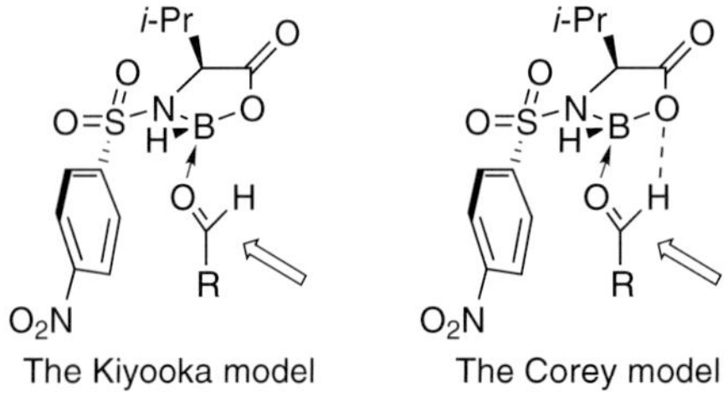

The Kiyooka model   The Corey model

Fig. 2.6
The proposed transition-state models.

been proposed in other chiral acid boron compounds that have been ingeniously used for Diels–Alder cycloaddition reactions [23]. The preference for this orientation might result from presence for a stabilizing anomeric interaction. Alternatively, the bound aldehyde might be locked in the conformation invoked by Kiyooka as a result of a formyl C–H hydrogen bond to the acyloxy donor, in accordance with the bonding model proposed by Corey [18e, 21a].

Kiyooka et al. reported that the **2d**-catalyzed aldol reaction of a silyl ketene acetal involving a dithiolane moiety with a β-siloxy aldehyde results in the production of *syn*- and *anti*-1,3-diols with complete stereoselectivity if the stereochemistry of the catalyst is chosen appropriately [21b]. This method has been applied to enantioselective synthesis of the optically pure lactone involving a *syn*-1,3-diol unit, which is known to be a mevinic acid lactone derivative of the HMG-CoA reductase inhibitors mevinolin and compactin (Scheme 2.2).

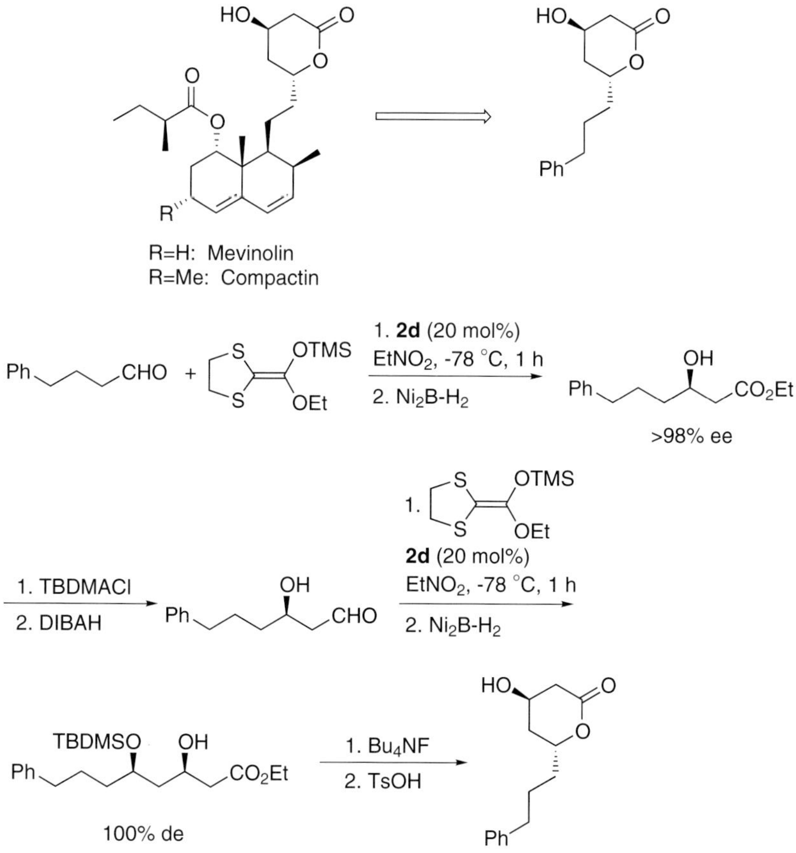

R=H: Mevinolin
R=Me: Compactin

Scheme 2.2

Corey et al. used **2e** in the conversion of aldehydes to 2-substituted 2,3-dihydro-4H-pyran-4-ones by reacting them with 1-methoxy-3-trimethylsilyloxy-1,3-butadiene in propionitrile at −78 °C for 14 h and then treating them with trifluoroacetic acid (Eq. (31)) [22].

$$(31)$$

Corey's tryptophan-derived chiral oxazaborolidine **2e** is highly effective for not only the Mukaiyama aldol reaction of aldehydes with silyl enol ethers [22] but also the Diels–Alder reaction of α-substituted α,β-enals with dienes [23], although more than 20 mol% **2e** is required for the former reaction. Other chiral oxazaborolidines that have been developed for enantioselective aldol reaction of aldehydes with relatively more reactive ketene silyl acetals also require large amounts (more than 20 mol%) to give aldol adducts in good yield [20, 21]. Yamamoto and his colleagues succeeded in enhancing the catalytic activity of CAB derived from 2,6-di(isopropoxy)benzoyltartaric acid and borane.THF by using **13** instead of borane.THF [19c]. In a similar manner they developed a new and extremely active Corey's catalyst, **2f**, using arylboron dichlorides bearing electron-withdrawing substituents as Lewis acid components [24].

A new chiral oxazaborolidine catalyst **2f** has been prepared by treating N-(p-toluenesulfonyl)-(S)-tryptophan with an equimolar amount of 3,5-bis(trifluoromethyl)phenylboron dichloride (**15**) in dichloromethane and subsequent removal of the resulting HCl and the solvent in vacuo (Scheme 2.3). Moisture-sensitive boron dichloride **15** and boron dibromide **16** are synthesized by dehydration of **13** to trimeric anhydride **14** and subsequent halogenation of **14** with 2 equiv. of BCl_3 and BBr_3, respectively [24]. The preparation of oxazaborolidines from arylboron dichlorides has been also reported by Reilly and Oh [25] and Harada and co-workers [26]. Although B-butyloxazaborolidine **2e** has been prepared from N-(p-toluenesulfonyl)-(S)-tryptophan and butylboronic acid by dehydration [22], B-aryloxazaborolidine cannot be prepared from arylboronic acid, as observed by Nevalainen et al. [27] and by Harada et al. [26b]. In contrast, CAB derived from 2,6-di(isopropoxy)benzoyltartaric acid in place of N-sulfonylamino acids has been easily prepared by adding an equimolar amount of the corresponding arylboronic acid at room temperature [19c].

According to Corey and co-workers [22], terminal trimethylsilyloxy (vinylidene) olefins seem to be more suitable substrates for enantioselective Mu-

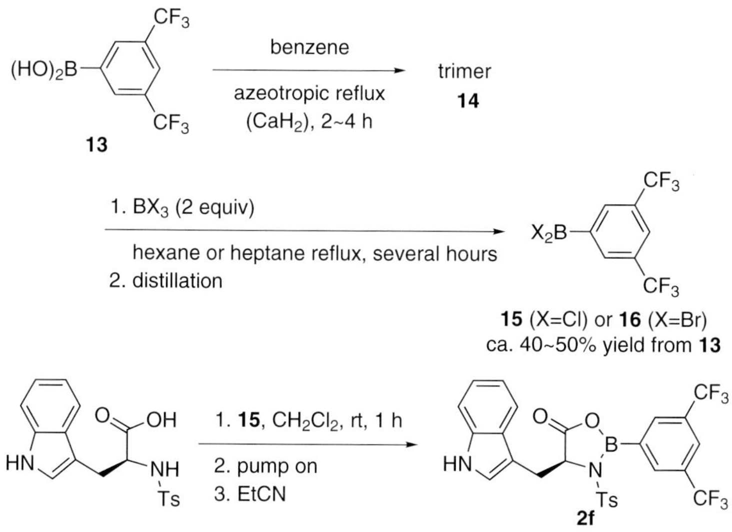

Scheme 2.3

kaiyama aldol coupling catalyzed by **2e** than more highly substituted olefins such as RCH=C(OSiMe₃)R' or R₂C=C(OSiMe₃)R'. In fact, reaction of the trimethylsilyl enol ether derived from cyclopentanone with benzaldehyde afforded the aldol products in only 71% yield even in the presence of 40 mol% **2e** [22].

According to Yamamoto et al. [24], reaction of benzaldehyde with the trimethylsilyl enol ether derived from acetophenone in the presence of 10 mol% **2e** gives the trimethylsilyl ether of aldol and the free aldol in yields of only 38% and 15%, respectively (Eq. (32)). When the *B*-3,5-bis(trifluoromethyl)phenyl analog **2f** is used, however, catalytic activity and enantioselectivity are increased to a turnover of 25 and 91–93% ee, respectively. The absolute configuration of the aldol adducts is uniformly *R*.

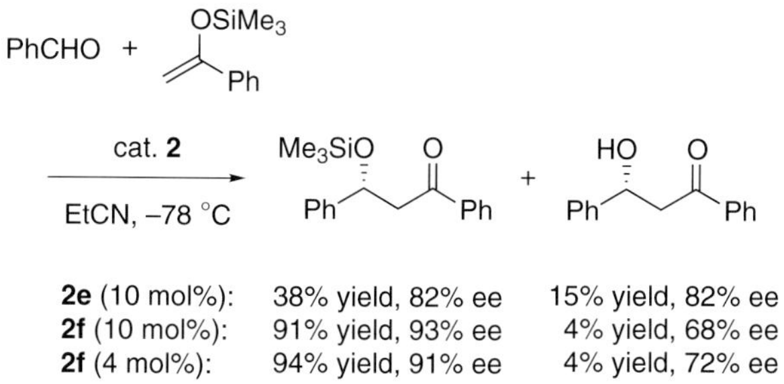

(32)

	Me₃SiO / O	HO / O
2e (10 mol%):	38% yield, 82% ee	15% yield, 82% ee
2f (10 mol%):	91% yield, 93% ee	4% yield, 68% ee
2f (4 mol%):	94% yield, 91% ee	4% yield, 72% ee

These results indicate that introduction of an electron-withdrawing substituent such as the 3,5-bis(trifluoromethyl)phenyl group to the *B* atom of chiral boron catalysts is an effective method for enhancing their catalytic activity. The method is especially attractive for large-scale synthesis (Eq. (33)).

$$\text{(33)}$$

CAB **2d** is effective for reaction not only with terminal trimethylsilyloxy olefins but also trisubstituted (*E*)- and (*Z*)-trimethylsilyl enol ethers (Table 2.1). In the reaction of aromatic aldehydes such as benzaldehyde with the trimethylsilyl enol ether of cyclohexanone, both substrates should be sequentially added to a solution of **2f** in propionitrile at −78 °C according to Corey's procedure (method A) [22]. The reaction proceeds quantitatively to give only the aldol products in a 78:22 *syn/anti* ratio, and the optical yield of the *syn* isomer **17** is 89% ee. Reaction of aliphatic aldehydes such as iso-butyraldehyde with the same silyl enol ether does not proceed well, however, probably because of decomposition of isobutyraldehyde in the presence of the strong Lewis acid **2f** before addition of the trimethylsilyl enol ether. On the other hand, sequential addition of silyl enol ethers and aldehydes to a solution of catalyst **2f** (method B) gives the aldol adducts in higher yield, but the enantioselectivity is relatively low. High enantioselectivity is also observed in the reaction with acyclic (*E*)- and (*Z*)-silyl enol ethers. Reaction with (*Z*)-trimethylsilyl enol ethers also gives *syn* aldol adducts as major diastereomers.

The *syn* preference and the absolute preference for carbonyl *re*-face attack observed in the reactions of aldehydes with (*E*)- and (*Z*)-trimethylsilyl enol ethers suggests that the reaction occurs via an extended-transition state assembly (Figure 2.7) [19, 22]. *Anti* preference has been observed in the reaction of aldehydes with (*E*)-ketene trimethylsilyl acetals catalyzed by other chiral oxazaborolidines [20, 21].

Harada and co-workers reported that arylboron complex **2g** derived from *N*-tosyl-(*αS,βR*)-*β*-methyltryptophan [23] and (*p*-chlorophenyl)dibromoborane is an excellent catalyst for enantioselective ring-cleavage reactions of 2-substituted 1,3-dioxolanes with enol silyl ethers [26c]. Interestingly, chiral boron complexes prepared by reacting the sulfonamide ligands with BH_3-THF do not have appreciable catalytic activity [26a,b]. Successful results have been obtained in the ring cleavage of 1,3-dioxolanes with aryl and alkenyl groups at the 2-position. Reaction of 2-alkyl derivatives is, however, very sluggish under these conditions. The 2-hydroxyethyl group in the ring-

Tab. 2.1

Mukaiyama aldol reaction of aldehydes with (*E*)- and (*Z*)-silyl enol ethers

$$R^1CHO \; + \quad \underset{R^3}{\overset{OSiMe_3}{\diagup\!\!\!\diagdown}}R^2 \quad \xrightarrow[\text{EtCN, } -78\,^\circ C, \text{ 12 h}]{\text{cat. } \mathbf{2f} \text{ (5 or 10 mol\%)}}$$

$$\underset{\mathbf{17}}{\overset{OH \quad O}{Ph\diagup\!\!\!\diagdown\underset{R^3}{\diagup}R^2}} \quad + \quad \underset{\mathbf{18}}{\overset{OH \quad O}{Ph\diagup\!\!\!\diagdown\underset{R^3}{\diagup}R^2}}$$

R¹	Silyl Enol Ether	Method[a]	Yield (%)[b]	17:18[c]	ee (%)[d] 17	ee (%)[d] 18
Ph	OSiMe₃ (cyclohexenyl)	A	>99	78:22	89	5
i-Pr	OSiMe₃ (cyclohexenyl)	A[e]	36	77:23	96	96
Pr	OSiMe₃[f] / Et	B	>99	48:52	95	93
Pr	Ph (>99:1)	A	23	>99:1	96	–
Pr	Ph (>99:1)	B	>99	>99:1	>99	–
Pr	Et (97:3)	B	>99	62:38	92	77
i-Pr	Ph (>99:1)	B	>99	97:3	98	–
i-Pr	Et (97:3)	B[e]	94	83:17	92	91
(*E*)-MeCH=CH	Ph (>99:1)	B	85	95:5	97	–
PhC≡C	Ph (>99:1)	B	92	89:11	90	58

[a] Method A: A solution of silyl enol ether (0.96 mmol) in propionitrile (0.32 mL) was added over 2 min to a mixed solution of **2f** (0.08 mmol) and an aldehyde (0.8 mmol) in propionitrile (0.65 mL). Method B: A solution of aldehyde (0.8 mmol) in propionitrile (0.32 mL) was added over 10 min to a mixed solution of the silyl enol ether (0.96 mmol) and **2f** (0.08 mmol) in propionitrile (0.65 mL). [b] Isolated yield. [c] Determined by ¹H NMR analysis. [d] Determined by HPLC. [e] 5 mol% of **2d** was used. [f] *E*:*Z* = 70:30.

cleavage products can be removed simply by conversion to the iodides then treatment with zinc powder (Eq. (34)).

Desymmetrization of *meso*-1,2-diols has been realized by chiral Lewis acid **2h**-mediated enantioselective ring-cleavage of dioxolane derivatives [26d]. Transacetalization of 3,3-diethoxy-1-phenylpropyne with *meso*-2,3-butanediol gives a 86:14 mixture of *syn*- and *anti*-**19** stereoselectively. Treatment of *syn*-**19** with 3 equiv. Me₂C=C(OTMS)OEt and 1.0 equiv. **2h** at −78 °C gives the ring-cleavage product **20** (> 20:1 diastereoselectivity) in 72% yield with 94%

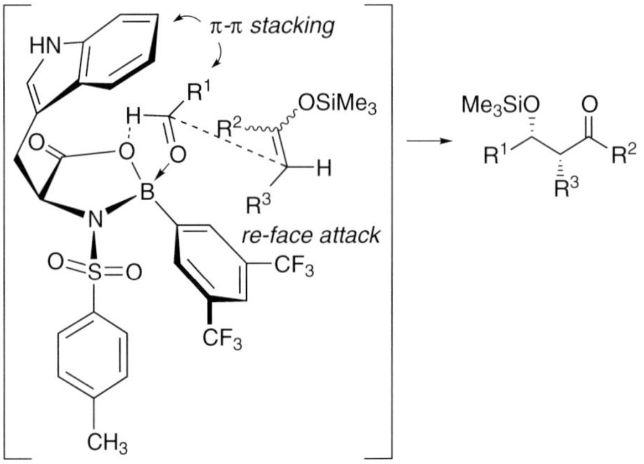

Fig. 2.7
Proposed extended-transition state assembly.

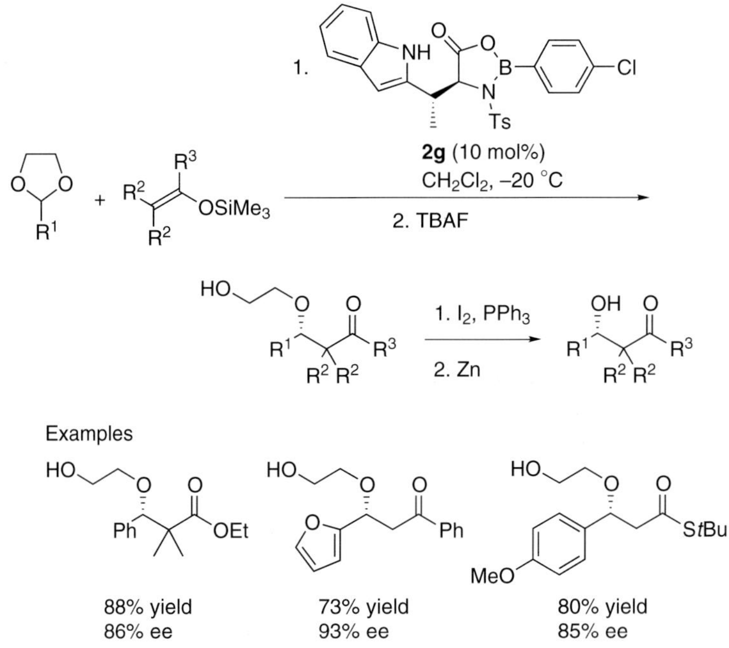

(34)

Examples

88% yield
86% ee

73% yield
93% ee

80% yield
85% ee

ee (Eq. (35)). A separate experiment using pure *anti*-**19** showed that it is unreactive under these conditions. Boron complex **2h** is also effective in the ring-cleavage of other dioxolanes that can be prepared stereoselectively (*syn:anti* > 20:1) from the diols under kinetically controlled conditions. The results obtained by using other catalysts, for example **2i** and **2j**, suggest that the structure of the *N*-sulfonyl moiety affects the enantioselectivity.

2h (1 equiv)

OEt / OTMS (3 equiv)

syn-**19**

20 (major) + ent-**20** (minor)

2h: R^1=tol → **20**: 94% ee
2i: R^1=Me → **20**: 48% ee
2j: R^1=CF$_3$ → **20**: 58% ee

(35)

Other examples

2h → 96% ee   **2h** → 93% ee   **2h** → 85% ee

Itsuno et al. have developed novel polyaddition reactions based on the Mukaiyama aldol reaction of silyl enol ethers with aldehydes. Bis(triethylsilyl enol ether) and bis(triethylsilyl ketene acetal) are prepared as stable and isolable monomers. In the presence of Lewis acid catalysts these monomers react smoothly with dialdehydes to afford the poly(β-hydroxy carbonyl) compounds. By asymmetric synthetic polymerization of such monomers with chiral modified Lewis acid it is possible to obtain optically active poly(β-hydroxycarbonyl) compounds with main-chain chirality [28].

For example, CAB **21**, which is highly efficient in the asymmetric aldol reaction of silyl enol ether with aldehyde, has been examined as a chiral catalyst for asymmetric aldol polymerization of **22** with **23**. Unfortunately, CAB **21** is not sufficiently active to polymerize these monomers at −78 °C. Increasing the temperature made it possible to obtain the chiral polymer in low yield, accompanied by partial decomposition of the catalyst. The polymer obtained is optically active, however (Eq. (36)) [28a].

Silyl ketene acetals also react enantioselectively with aldehydes in the presence of a chiral Lewis acid. Several useful chiral Lewis acids have recently been developed for this reaction. Itsuo et al. found that Kiyooka's catalyst **2a**.SMe$_2$ acts as a chiral catalyst of asymmetric aldol polymerization

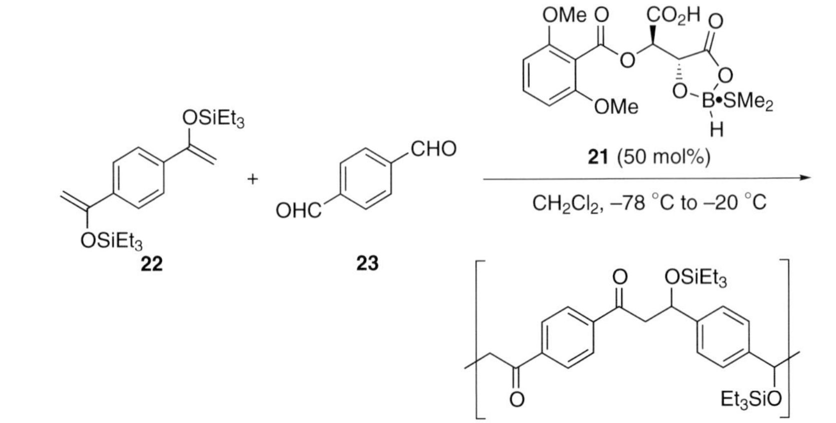

5% yield, Mn=1300, Mw/Mn=1.67, [Φ]=−64

(36)

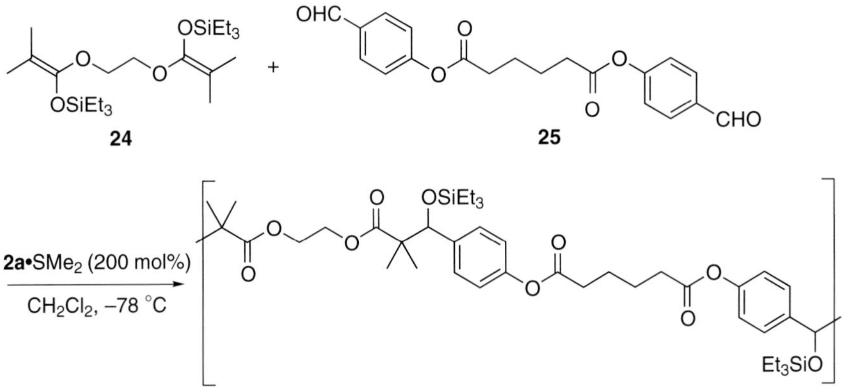

62% yield, Mn=1900, Mw/Mn=2.21, [Φ]=−49

(37)

of **24** with **25** even at −78 °C. The aldol polymer with optical activity is again obtained in 62% yield (Eq. (37)) [28a].

Itsuno and co-workers also reported that CAB **2d** is a more effective catalyst than other chiral oxazaborolidines **2** for asymmetric polymerization of bis(triethylsilyl enol ether)s and dialdehydes [28b]. The reactivity of dialdehydes containing ether linkages is quite low for formation of polymers, mainly because of the low solubility of dialdehyde monomers in propionitrile. Introduction of a silyl group into the monomeric structure of the dialdehyde dramatically improves the solubility. The asymmetric polymerization of silyl-containing dialdehyde **26** with **22** affords the chiral polymer in high yield with high molecular weight (Eq. (38)). This polymer is soluble in

$$71\% \text{ yield, } Mw=48200, Mw/Mn=10.3, [\Phi]_{435}=1670$$

common organic solvents such as THF, CH_2Cl_2, $CHCl_3$, DMF, and DMSO. All the chiral polymers obtained using **2d** as catalyst have positive optical rotation.

2.3
Silicon Lewis Acids

2.3.1
Introduction

Silicon Lewis acids have advantages over traditional metal-centered activators. For example, silicon Lewis acids are compatible with many synthetically valuable C-nucleophiles, such as silyl enol ethers. Unlike metal halides, silicon Lewis acids are not prone to aggregation, which substantially simplifies the analysis of the reaction mechanisms. Furthermore, the reactivity of silicon Lewis acids of R_3SiX structure can be finely controlled by varying the steric volume of alkyl substituents.

The most advantageous circumstance is the opportunity to realize the processes in the presence of catalytic amounts of silicon Lewis acids if silicon Lewis acids and silyl enol ethers have identical trialkylsilyl fragments. Thus, depending on the type of electrophile, two mechanistically different pathways can be considered (Scheme 2.4). For acetals and acetal-like compounds, silicon Lewis acids abstract the heteroatomic substituent, followed by reaction of electrophilic species formed with a nucleophile (left circle). When the substrates have a carbon–heteroatom double bond (e.g. carbonyl compounds, imines) silicon Lewis acids bind to their basic function leading, after carbon–carbon bond formation, to products containing the silyl group (right circle).

The approach generalized in Scheme 2.4 was first realized by Noyori and co-workers in the early eighties [29]. Subsequently silicon Lewis acid gained wide acceptance as mediators of a variety of transformations. This section

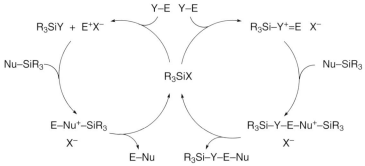

Scheme 2.4

surveys data on the behavior of silicon Lewis acids of general formula R_3SiX in Mukaiyama aldol reactions [30].

2.3.2
Lewis Acidity of Silicon Derivatives

In the last two decades, the problem of observation of trialkylsilyl cations R_3Si^+, apparently the strongest silicon Lewis acids, attracted considerable attention. According to the results of ab initio calculations [31] and experimental data [32] the equilibrium shown in Eq. (39) is substantially to the right.

$$R_3SiH + R_3C^+ \rightleftharpoons R_3Si^+ + R_3CH \tag{39}$$

Correspondingly, R_3Si^+ can be readily formed in the gas phase where they can be characterized and studied [33]. Observation of these cations in the condensed state (in solution or in the crystalline state) is, however, very difficult [34]. Nevertheless, Lambert demonstrated recently that silyl cations containing bulky substituents which hinder the approach of nucleophilic reagents to the silicon atom can be observed in solution. He succeeded in detecting $Mes_3Si^+B(C_6F_5)_4^-$ (^{29}Si NMR δ (ppm) = 225.5) [35], or $Dur_3Si^+B(C_6F_5)_4^-$ (^{29}Si NMR δ = 226.8) [36] (Mes = 2,4,6-trimethylphenyl, Dur = 2,3,5,6-tetramethylphenyl) in benzene, the chemical shifts being very close to the calculated value (δ (Mes_3Si^+, calcd.) = 230.1) [37].

Hence, covalent compounds of the type R_3SiX, where X is either the conjugated base of a strong acid (for example $CF_3SO_2^-$ or ClO_4^-) or a solvent molecule (for example, MeCN), generally serve as Lewis acids in carbon–carbon bond-forming reactions.

Several approaches have been proposed for estimation of the Lewis acidity of R_3SiX. One of these assumes that the positive charge on the silicon atom is proportional to the chemical shift in ^{29}Si NMR spectra. This scale can, however, only be used as a reliable indication of the relative reactivity of compounds in which the silicon atom is bound to the same heteroatom. Another procedure for estimating the Lewis acidity of Me_3SiX, suggested by Hergott and Simchen, is based on comparison of the silylation rate constants of cyclopentanone and diisopropyl ketone with these reagents in the presence of triethylamine in dichloroethane [38]. Bassindale et al. have proposed estimating the strength of silicon Lewis acids from their ability to form the *N,N*-bis(trimethylsilyl)imidazolium cation in reactions with *N*-trimethylsilylimidazole [39]. On the basis of results from studies of the kinetics and thermodynamics of this reaction, silicon Lewis acids were arranged $Me_3SiCl < Me_3SiBr < Me_3SiI < Me_3SiOTf < Me_3SiClO_4$ in order of silyl-donating capacity. Although quantitative data on Me_3SiNTf_2 and $Me_3SiN(SO_2F)_2$ derivatives are lacking, the results of comparative experiments provide evidence that these reagents are much more reactive than Me_3SiOTf.

The results obtained by different research groups make it possible to arrange the most commonly used neutral silicon Lewis acids in the quantitative activity series: $Me_3SiCl < Me_3SiOMs < Me_3SiOTs < Me_3SiBr \ll Me_3SiOTf \approx Me_3SiOSO_2F \leq Me_3SiI \leq Me_3SiClO_4 < Me_3SiN(SO_2F)_2 < Me_3SiNTf_2$. Positively charged species such as $MeCN-SiMe_3^+$ or complexes generated from neutral silicon Lewis acids and metal-centered Lewis acids might be even more reactive than Me_3SiNTf_2.

2.3.3
Silicon Lewis Acids as Catalytic Reagents

The reactions of carbonyl compounds with silyl enol ethers can be described by the mechanism shown in Scheme 2.5. Thus, the reaction of a substrate with a silicon Lewis acid initially affords a five-coordinate complex **27** which can exist in equilibrium with cation **28** [40]. Subsequent nucleophilic attack on the carbon atom of complexes **27** or **28** is accompanied by formation of the carbon–carbon bond to give intermediate **29** or **30**, respectively. The intermediate **30** is rapidly transformed into the final product **31**. The position of the equilibrium between $RCHO + Me_3SiX$, **27**, and **28** depends on the cation-stabilizing effect of the substituents R^1 and R^2 and on the nature of the leaving group X.

Attempts to observe complexes **27** or **28** generated from benzaldehyde and Me_3SiOTf by NMR spectroscopy have failed (only the starting component provided unambiguous evidence of a very small contribution of **27** and **28** to the equilibrium mixture). Hence, it can be tentatively assumed that cation **28** is not formed if the Si–X bond is sufficiently strong, e.g. an Si–O bond. Neutral complex **27**, in turn, is a much weaker electrophile than **28** or the oxocarbenium cation generated from acetals on elimination of the alkoxy group. Consequently, one would expect carbonyl compounds to be less reactive than the corresponding acetals in reactions with nucleophiles.

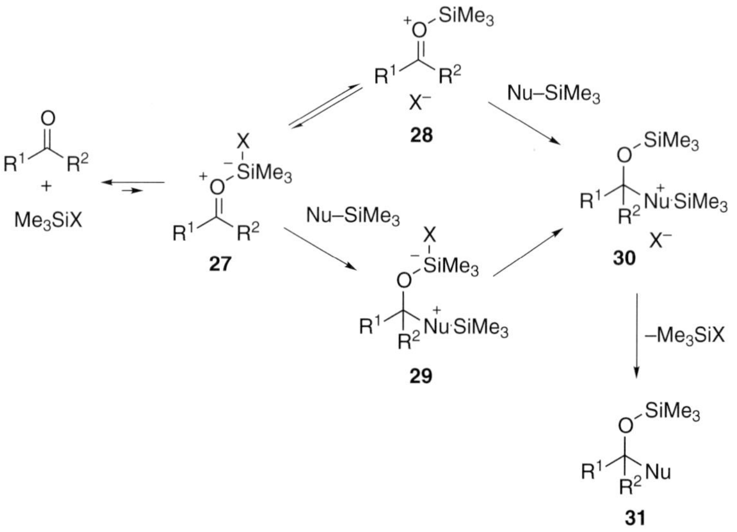

Scheme 2.5

The different reactivity of the acetal and carbonyl groups is demonstrated by the bifunctional substrate **32**, which reacts with 1-trimethylsiloxy-cyclohexene exclusively at the acetal fragment (Eq. (40)) [41].

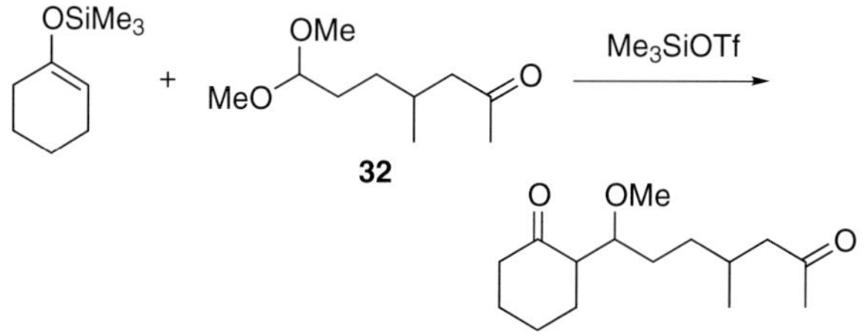

(40)

In the early 1980s, Noyori demonstrated that aldehydes and ketones do not react with 1-siloxycyclohexene in the presence of Me_3SiOTf in CH_2Cl_2 at $-78\ °C$ [30, 41, 42]. The reaction of benzaldehyde with 1-trime-thylsiloxycyclohexene catalyzed by Me_3SiOTf proceeds only at room temperature to give the target silyl ether of aldol (in toluene the yield was 60%, *syn:anti* = 49:51) or benzylidene-cyclohexanone (in CH_2Cl_2, 85%). Aliphatic aldehydes are not involved in this reaction. According to results from other studies benzaldehyde reacts smoothly with silyl enol ethers on catalysis by Me_3SiOTf (5 mol%) in CH_2Cl_2 at $-78\ °C$ to give the silyl ether of aldol in 89% yield in the ratio *syn:anti* = 63:37 [43]. The latter reaction is probably catalyzed by traces of TfOH rather than by Me_3SiOTf itself. Reaction of trimethylsiloxycyclohexene with benzaldehyde or isobutyraldehyde in the presence of 5 mol% TfOH in CH_2Cl_2 at $-78\ °C$ is complete in 30 min to give aldol products in 86% (*syn:anti* = 69:31) and 82% (*syn:anti* = 73:27) yields, respectively [41, 44].

Me$_3$SiNTf$_2$, however, a considerably stronger silyl donor than Me$_3$SiOTf, efficiently catalyzes addition of silyl enol ethers to aldehydes and ketones (Eq. (41)). The reaction is best performed in diethyl ether as solvent at $-78\,°C$ with as little as 0.5–1.0 mol% silicon Lewis acid, generated in situ from HNTf$_2$ and silyl enol ether.

R^1, R^2=H, Alkyl, Ph

R^3, R^4=H, Ph, (CH$_2$)$_4$

Examples (yield)

92%   87%   92% (*syn:anti*=70:30)

88% (*syn:anti*=76:24)   87%   92%
(silyl enol ether, 96% *cis*)

To minimize the formation of side products it is necessary to add the carbonyl compound slowly [45]. The presence of excess molar amounts of carbonyl compounds per desired adducts produced in the reaction concurrently promotes at least three reactions (Scheme 2.6): (1) cyclic trimerization of the aldehyde (path a), (2) dimerization of the desired adducts (path b), and (3) acetalization of the desired adducts (path c). Slow addition of the carbonyl compound to a mixed solution of silyl enol ether and Me$_3$SiNTf$_2$ is the best way to obtain the desired products selectively.

The following mechanism has been proposed for this aldol reaction pathway [45b]. Electrophilic attack of silyl-activated aldehyde species on the silyl enol ether produces cationic species **33** which subsequently acts as a source of Lewis acidic silyl group without regeneration of Me$_3$SiNTf$_2$ (Scheme 2.7).

In accord with such a mechanism is the observation that the silylated aldol initially formed by coupling of benzaldehyde with silyl enol ethers derived from acetophenone contains the silyl group derived from the nucleophile, and not from R$_3$SiNTf$_2$ (Eqs. (42) and (43) [45b]. In a similar experiment with Me$_3$SiOTf performed at $-78\,°C$ a mixture of **34** and **35** is obtained (Eq. (44)), suggesting that after carbon–carbon bond formation the silyl triflate with a silyl group originating from the enol ether is generated.

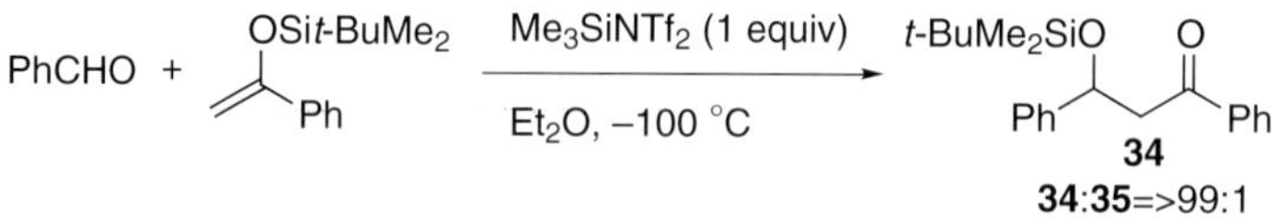

PhCHO + [OSi*t*-BuMe$_2$ / Ph enol ether] $\xrightarrow[\text{Et}_2\text{O, }-100\,°\text{C}]{\text{Me}_3\text{SiNTf}_2\ (1\ \text{equiv})}$ [*t*-BuMe$_2$SiO ... Ph ... O ... Ph] **34** (42)

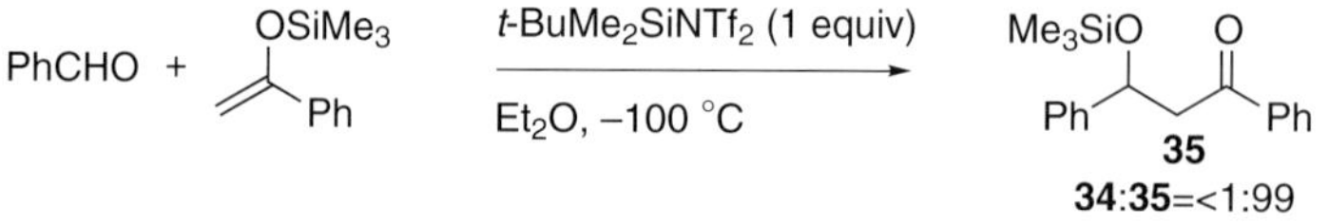

34:35=>99:1

PhCHO + [OSiMe$_3$ / Ph enol ether] $\xrightarrow[\text{Et}_2\text{O, }-100\,°\text{C}]{t\text{-BuMe}_2\text{SiNTf}_2\ (1\ \text{equiv})}$ [Me$_3$SiO ... Ph ... O ... Ph] **35** (43)

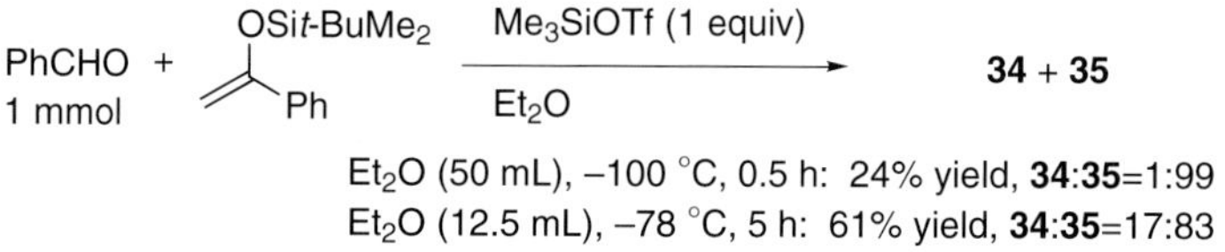

34:35=<1:99

PhCHO + [OSi*t*-BuMe$_2$ / Ph enol ether]
1 mmol $\xrightarrow[\text{Et}_2\text{O}]{\text{Me}_3\text{SiOTf}\ (1\ \text{equiv})}$ **34 + 35**

(44)

Et$_2$O (50 mL), −100 °C, 0.5 h: 24% yield, **34:35**=1:99
Et$_2$O (12.5 mL), −78 °C, 5 h: 61% yield, **34:35**=17:83

When, moreover, two enol ethers of different ketones bearing different silyl groups are used simultaneously scrambling of the silyl groups occurs [45b, 46].

These observations indicate that the ligand (X) of the silicon Lewis acid (R$_3$SiX) plays a crucial role in the Mukaiyama aldol reaction of trimethylsilyl enol ethers (Me$_3$SiNu). In the R$_3$SiOTf-induced reaction transfer of TfO$^-$ from siloxocarbenium ion **36** is expected to occur by electrophilic attack of

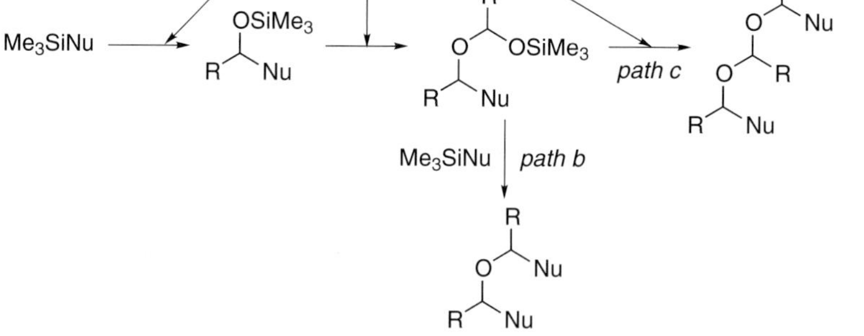

Scheme 2.6

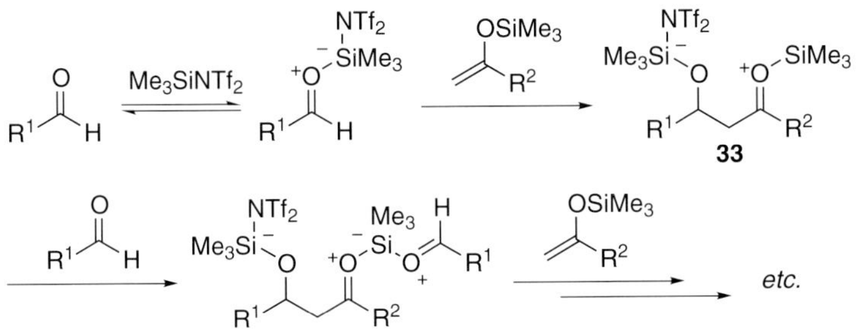

Scheme 2.7

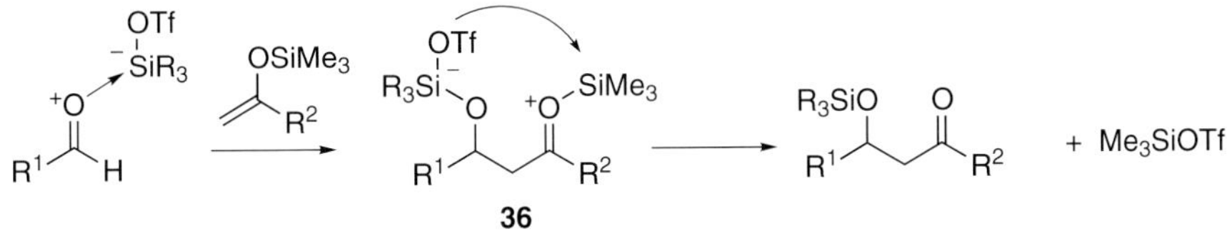

Scheme 2.8

the "Me$_3$Si–O$^+$ silicon" of **36** (Scheme 2.8). Me$_3$SiOTf would be generated by electrophilic attack of the "Me$_3$Si–O$^+$ silicon" on the "S=O oxygens" or the "S–O oxygen" of $^-$OTf. In the R$_3$SiNTf$_2$-induced reaction, in contrast, less nucleophilicity and/or more bulkiness of $^-$NTf$_2$ might suppress electrophilic attack of the "Me$_3$Si–O$^+$ silicon" on the nitrogen or oxygen atoms of $^-$NTf$_2$, and might increase the Lewis acidity of siloxocarbenium ion **33** (Scheme 2.7).

Unlike silyl enol ethers, silyl ketene acetals react with aldehydes and ketones on catalysis by Me$_3$SiOTf [47], and carbonyl compounds often seem to be more reactive in these reactions than their acetals (Eq. (45)) [47b]. Bis(silyl)methylacetoacetate is a synthetic equivalent of the corresponding dianion and its terminal carbon atom is involved in reactions with carbonyl compounds in the presence of Me$_3$SiOTf, ketones being more reactive than aldehydes (Eq. (46)) [48].

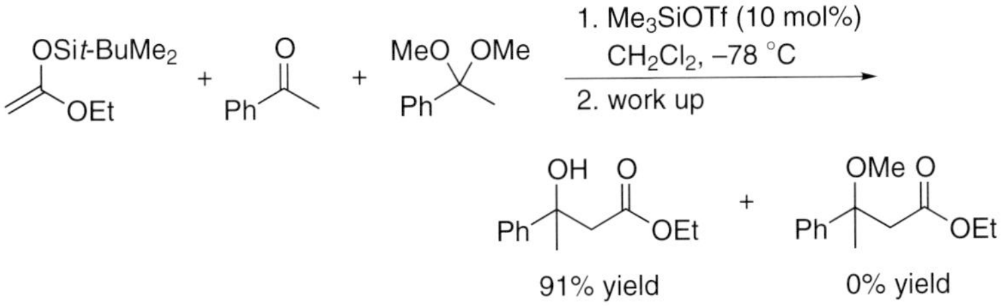

(45)

1. Me$_3$SiOTf (10 mol%)
   CH$_2$Cl$_2$, −78 °C
   —————————→
2. work up

Me$_3$SiO   OSiMe$_3$

OMe   + n-C$_5$H$_{11}$ +   H n-C$_7$H$_{15}$

OH O

n-C$_5$H$_{11}$ CO$_2$Me   + n-C$_7$H$_{15}$ CO$_2$Me

OH O

52% yield   not detected

(46)

2.3.4
Activation of Silicon Lewis Acids by Combination with Other Lewis Acids

Binding of silicon Lewis acid (R$_3$SiX) with another Lewis acid (LA) leads to the shift of the electron density from the silicon atom (confirmed by ^{29}Si NMR spectroscopic data). As a consequence, the resulting R$_3$SiX → LA complexes are much stronger donors of the silyl group than the starting SLA. Olah and colleagues demonstrated that reaction of Me$_3$SiBr with AlBr$_3$ produces the Me$_3$SiBr → AlBr$_3$ complex (^{29}Si NMR, $\delta = 62.7$) [49a]. Even SbF$_5$, one of the strongest Lewis acids, cannot abstract the fluoride anion from Me$_3$SiF and gives the Me$_3$SiF → SbF$_5$ complex (^{29}Si NMR, $\delta = 102$) rather than the silyl cation [49b].

The possibility of using R$_3$SiX → LA complexes as mediators in carbon–carbon bond-forming reactions was first demonstrated by Mukaiyama et al. in 1987 [51]. While quite inactive separately, Lewis acids Me$_3$SiCl and SnCl$_2$ taken together have properties of strong R$_3$SiX Lewis acids. Thus, aldehydes, α,β-unsaturated ketones, and acetals smoothly react with silyl enol ethers in the presence of this LA pair (Scheme 2.9). The Me$_3$SiCl–ZnCl$_2$ system can function analogously although it is less efficient than Me$_3$SiCl–SnCl$_2$ [50].

O   OH   O   Ph   O

Ph   Ph   Ph   Ph

        PhCHO ←   O
                                 Ph   →

            OSiMe$_3$ 15–20 mol% Me$_3$SiCl
                       8–12 mol% SnCl$_2$
         Ph   CH$_2$Cl$_2$, −78 °C

     ← HC(OMe)$_3$ EtCH(OMe)$_2$ →

O   OMe   O   OMe

Ph   OMe   Ph

Scheme 2.9

It has also been reported that R_3SiCl can be activated by addition of $InCl_3$ [51]. The reactivity of the R_3SiCl–$InCl_3$ mixture is highly dependent on the nature of the alkyl groups on the silicon atom. For example, the Me_3SiCl–$InCl_3$ system catalyzes the reactions of trimethylsilyl enol ethers both with aldehydes and acetals (Eq. (47)) yet only aldehydes react with *tert*-butyldimethylsilyl enol ethers in the presence of t-$BuMe_2SiCl$–$InCl_3$. This behavior enables selective nucleophilic addition at the carbonyl group in the presence of the acetal fragment (Eq. (48)).

(47)

68~93% yield

(48)

77% yield

Boron and aluminum compounds can also activate R_3SiX, leading to silicon species with very high catalytic activity. The high reactivity of these systems is probably associated with the complete transfer of the silyl group to the carbonyl oxygen atom to form the siloxycarbonium species $RCH=O^+SiR_3$.

The exothermic reaction of $B(OTf)_3$ with Me_3SiOTf gives the $Me_3SiB(OTf)_4$ adduct (^{29}Si NMR, $\delta = 62.0$). The ^{11}B NMR spectrum ($\delta = -3.17$, $\Delta v_{1/2} = 28$ Hz) corresponds to the $B(OTf)_4^-$ anion whereas the ^{13}C NMR spectrum shows the presence of only one trifluoromethyl group ($\delta = 118$, q, $^1J_{C,F} = 318$ Hz) [52a]. It is highly probable the trifluoromethyl group on this complex very rapidly migrates among all the triflate groups. Trace amounts of $Me_3SiB(OTf)_4$ are sufficient for reaction of aldehydes with silyl enol ethers. In the presence of an asymmetric center adjacent to the carbonyl group, the diastereoselectivity of the process can be changed by varying the volumes of the substituents on the silicon atom (Eq. (49)). Apparently, an increase in the size of the silyl group bound to the carbonyl oxygen leads to limitation of possible pathways of approach of the nucleophile, thereby improving the diastereoselectivity of the reaction (See also Eqs. (3) and (4) [8]) [52b]. A particularly useful property of $B(OTf)_3$ is its ability to form complexes with chlorosilanes R_3SiCl, giving silylating reagents which compare favorably with the $R_3SiOTf/B(OTf)_3$ system [52c]. The possibility of generating very strong silylating reagents based on sterically hindered chlorosilanes enables the use of these compounds instead of more expensive silyl triflates.

R_3Si	syn : anti
Me_3Si	8 : 1
i-Pr_3Si	97 : 1

(49)

A combination of Me_3SiOTf and sterically hindered organoaluminum compounds MAD or MABR is another example of the formation of very active R_3SiX [53a]. As follows from Eq. (50), these organoaluminum compounds coordinate the triflate anion more efficiently than $B(OTf)_3$. The $Me_3SiOTf/MABR$ system makes it possible to perform the reactions of silyl enol ethers even with poorly reactive carbonyl compounds, such as pivalaldehyde and methyl isopropyl ketone. The Me_3SiOTf/MAD or MABR/PhCHO combinations are also useful for initiating cationic polymerization of silyl enolates [53b].

cat.	yield, %
Me_3SiOTf	15
$Me_3SiOTf/B(OTf)_3$	43
Me_3SiOTf/MAD	62
$Me_3SiOTf/MABR$	76
MABR	7

(50)

Study of complex formation between benzaldehyde, with MAD, and Me_3SiOTf by ^{13}C NMR spectroscopy at -50 °C showed that addition of 2 equiv. Me_3SiOTf to the PhCHO $\rightarrow$ MAD adduct afforded a new electrophilic species of unknown nature. The ^{13}C NMR spectrum of the latter has a signal which is shifted downfield by approximately 3 ppm compared with the signal of the PhCHO $\rightarrow$ MAD complex. This species probably consists of benzaldehyde and two different Lewis acids, and it behaves as a true electrophile, which attacks the double bond of silyl enol ether [53b].

In addition to these examples, it should be noted that activation of R_3SiX might cause undesirable transformations. Thus, it is difficult to achieve high enantioselectivity when performing catalytic asymmetric aldol reactions of aldehydes with silyl enol ethers in the presence of chiral metal-centered Lewis acids. These difficulties are generally attributed to the effect of R_3SiX, which is generated in the early steps of the process and then promotes carbon–carbon cross-coupling yielding a racemic product [45b, 46, 54].

Noyori also demonstrated that aldehydes do not react with silyl enol ethers under the action of Me_3SiOTf [29, 41, 42]. In this connection it is reasonable to assume that the low enantioselectivity observed might result from formation of a complex between R_3SiX and a chiral Lewis acid. Under the action of this complex the trialkylsilyl fragment can transferred to the carbonyl group, producing a racemic product.

Representative Experimental Procedures

Typical Procedure for the Mukaiyama Aldol Reaction Catalyzed by $B(C_6F_5)_3$ [13].

An anhydrous solution of $B(C_6F_5)_3$ in toluene (81 μL, 0.02 mmol, 0.247 M) is added dropwise, at -78 °C, under argon, to a solution of aldehyde (1.0 mmol) and silyl enol ether (1.2 mmol) in dichloromethane (2 mL). The mixture is stirred for several hours at the same temperature and then 1 M HCl (10 mL) and THF (10 mL) are added. The reaction mixture is stirred for 0.5 h, poured into $NaHCO_3$ solution, extracted with diethyl ether, dried over $MgSO_4$ and concentrated, and the residue is purified by column chromatography on silica gel to give the corresponding aldol in high yield.

Preparation of 3,5-Bis(trifluoromethyl)phenylboron Dichloride (15) [24].

A solution of **13** (Lancaster Synthesis; ^{1}H NMR (C_6D_6, 300 MHz) δ (ppm) 7.81 (s, 1H), 8.01 (s, 2H); 3.86 g, 15 mmol) in benzene (30 mL) is heated under reflux with removal of water (CaH_2 in a Soxhlet thimble) for 2–5 h (oil bath: 100–105 °C) then cooled to room temperature and concentrated in vacuo to give trimeric anhydride **14** as a white solid (^{1}H NMR (C_6D_6, 300 MHz) δ 8.01 (s, 1H), 8.46 (s, 2H)). A 1 M solution of BCl_3 (30 mL, 30 mmol) in hexane and a 1 M solution of BBr_3 (30 mL, 30 mmol) in heptane are added separately to **14** at room temperature under argon. The two reaction mixtures are heated under reflux for 4 h (oil bath 100–105 °C) and 56 h (oil bath 105–110 °C), respectively, and the solvents are removed by distillation. Dichloroboron compound **15** is isolated as colorless oils by distillation under reduced pressure from the residues in ca. 40–50% yield: 38–40 °C (0.05–0.06 torr); ^{1}H NMR (C_6D_6, 300 MHz) δ 7.80 (s, 1H), 8.12 (s, 2H); ^{11}B NMR (C_6D_6, 96 MHz) δ 53.2; ^{13}C NMR (C_6D_6, 75.5 MHz) δ 123.1 (q, $J = 272.8$ Hz, 2C), 127.1 (s, 1C), 131.0 (q, $J = 33.5$ Hz, 2C), 134.8–135.2 (m, 1C), 135.5 (s, 2C); ^{19}F NMR (C_6D_6, 282 MHz) δ -64.3.

Preparation of Chiral Oxazaborolidine Catalyst 2f [24]. 15 (22.1 mg, 0.075 mmol) is added at room temperature, under argon, to a solution of *N*-(*p*-toluenesulfonyl)-(*S*)-tryptophan [23a] (32.3 mg, 0.09 mmol) in dichloromethane (0.75 mL). The mixture is stirred for 1 h then concentrated in vacuo to give **2f** as a white solid. This is dissolved in propionitrile and used for Mukaiyama aldol reactions. ^{1}H NMR (CD$_2$Cl$_2$, 300 MHz) δ 2.37 (s, 3H), 3.56 (dd, J = 2.6, 15.0 Hz, 1H), 3.83 (dd, J = 4.5, 15.0 Hz, 1H), 4.56–4.59 (m, 1H), 7.08–7.30 (m, 4H), 7.26 (d, J = 8.1 Hz, 2H), 7.54 (d, J = 8.1 Hz, 2H), 7.82 (d, J = 7.5 Hz, 1H), 7.95 (s, 1H), 8.04 (s, 2H), 8.22 (brs, 1H); ^{11}B NMR (CD$_2$Cl$_2$, 96 MHz) δ 33.8; ^{19}F NMR (C$_6$D$_6$, 282 MHz) δ −64.2.

Representative Procedure for the Mukaiyama Aldol Reaction Catalyzed by 2f (Method A, Table 2.1) [24]. Propionitrile (1 mL) is added at room temperature to **2f** (0.075 mmol, 6 mol%) prepared as described above. After being cooled to −78 °C, benzaldehyde (127 μL, 1.25 mmol) is added and a solution of 1-phenyl-1-(trimethylsiloxy)ethylene (308 μL, 1.5 mmol) in propionitrile (0.5 mL) is subsequently added dropwise over 2 min. The reaction mixture is stirred at −78 °C for 12 h and then quenched by addition of saturated aqueous NaHCO$_3$. The mixture is extracted with ether and the combined organic phases are dried over MgSO$_4$ and evaporated. The residue is dissolved in THF (2 mL) and 1 M aqueous HCl (2 mL), and the resulting solution is left to stand for 30 min. Saturated aqueous NaHCO$_3$ is added and the mixture is extracted with ether. The combined organic phases are dried over MgSO$_4$ and evaporated to furnish an oily residue. Silica gel chromatography (hexane–ethyl acetate, 4:1) affords 282 mg (>99% yield) of the known aldol product. The enantiomeric ratio and the absolute configuration are determined by HPLC analysis (Daicel OD-H column with hexane–*i*-PrOH, 20:1, flow rate 1.0 mL min^{-1}): t_R = 21.2 min ((*S*), minor enantiomer), 24.4 min ((*R*), major enantiomer).

Preparation of Me$_3$SiOTf [30b]. Allyltrimethylsilane (1.6 g, 14 mmol) is added dropwise, with stirring, to a solution of TfOH (1.5 g, 10 mmol) in dry dichloromethane (8 mL) and the reaction temperature is maintained between 15 and 20 °C for 1 h. The resulting mixture is transferred directly to a distillation apparatus and distillation under reduced pressure gives Me$_3$SiOTf as a colorless liquid (1.9 g, 85%), bp 52–53 °C at 31 Torr; ^{1}H NMR (CDCl$_3$) δ 0.50 ppm.

Preparation of Me$_3$SiNTf$_2$ [30b]. HNTf$_2$ (0.85 g) is reacted with a 3:1 molar excess of Me$_3$SiH in an FEP reactor. When the mixture is left to warm from −196 °C an initial rapid reaction occurs near 22 °C. After 4 h and occasional agitation, the homogeneous mixture is cooled slowly to −196 °C. A quantitative amount of hydrogen is recovered and the excess silane is pumped away at 10 °C, giving Me$_3$SiNTf$_2$ (0.98 g, 92%) as a colorless liquid of low volatility. ^{19}F NMR (CFCl$_3$) −77.47 (s) ppm; ^{1}H NMR 0.57 (s) ppm; major

m/e [Cl] 163 (Me$_3$SiOH$^+$), 147 (TfN$^+$), 77(?), 73 (Me$_3$Si$^+$) with weak ions at 282 (Tf$_2$NH$_2$$^+$) and 354 (M$^+$).

Preparation of Me$_3$SiB(OTf)$_4$ [30b]. TfOH (531 μL, 6 mmol) is added to BBr$_3$ (distilled from Al powder; 190 μL, 2 mmol) at 0 °C. After evolution of HBr has ceased the flask is evacuated for 1 h to give B(OTf)$_3$ as a viscous yellow liquid. In a separate flask, a solution of Me$_3$SiOTf is prepared by addition of TfOH (177 mL, 2 mmol) to a solution of allyltrimethylsilane (320 μL, 2 mmol) in dichloromethane (5 mL), and the resulting mixture is left to stand for 10 min. Addition of this solution to the B(OTf)$_3$ at 0 °C results in evolution of heat and formation of a pale yellow solution of Me$_3$SiB(OTf)$_4$ (0.4 M).

Typical Procedure for the Mukaiyama Aldol Reaction Catalyzed by Me$_3$SiNTf [45a]. Commercially available triflylimide (0.072 M solution in diethyl ether, 1.11 mL, 0.08 mmol) is added at −78 °C under argon to a solution of silyl enol ether (8.8 mmol) in diethyl ether (2 mL). After stirring the mixture for 15 min, aldehyde or ketone (1.0 M solution in diethyl ether, 8.0 mL, 8.0 mmol) is added dropwise over a period of 2 h at −78 °C. After stirring for 15 min at the same temperature, 1 M HCl (10 mL) and THF (10 mL) are added. The reaction mixture is stirred for 0.5 h, poured into NaHCO$_3$ solution, extracted with diethyl ether, dried over MgSO$_4$ and concentrated, and the residue is purified by column chromatography on silica gel to give the corresponding aldol in high yield.

References

1 (a) ISHIHARA, K. In *Lewis Acids in Organic Synthesis*; YAMAMOTO, H. Ed.; Wiley–VCH: Weinheim, 2000; Volume 1, pp. 89–190. (b) ISHIHARA, K. In *Lewis Acids Reagents*; YAMAMOTO, H. Ed.; Oxford University Press: Oxford, 1999; pp. 31–63.

2 WULF, W. D.; GILBERTSON, S. R. *J. Am. Chem. Soc.* **1985**, *107*, 503.

3 YAMAGO, S.; MACHII, D.; NAKAMURA, E. *J. Org. Chem.* **1991**, *56*, 2098.

4 NAKAMURA, E.; KUWAJIMA, I. *J. Am. Chem. Soc.* **1977**, *99*, 961.

5 SUGIMURA, H.; SHIGEKAWA, Y.; UEMATSU, M. *Synlett* **1991**, 153.

6 HEATHCOCK, C. H.; FLIPPIN, L. A. *J. Am. Chem. Soc.* **1983**, *105*, 1667.

7 EVANS, D. A.; GAGE, J. R. *Tetrahedron Lett.* **1990**, *31*, 5053.

8 DAVIS, A. P.; PLUNKETT, S. J.; MUIR, J. E. *Chem. Commun.* **1998**, 1797.

9 GUINDON, Y.; PRÉVOST, M.; MOCHIRIAN, P.; GUÉRIN, B. *Org. Lett.* **2002**, *4*, 1019.

10 KALBAKA, G. W.; TEJEDOR, D.; LI, N.-S.; MALLADI, R. R.; TROTMAN, S. *J. Org. Chem.* **1998**, *63*, 6438.

11 A review of B(C₆F₅)₃: PIERS, W. E.; CHIVERS, T. *Chem. Soc. Rev.* **1997**, *26*, 345.

12 For preparation of B(C₆F₅)₃, see: (a) MASSEY, A. G.; PARK, A. J. *J. Organomet. Chem.* **1964**, *2*, 245. (b) MASSEY, A. G.; PARK, A. J. *J. Organomet. Chem.* **1966**, *5*, 218.

13 (a) ISHIHARA, K.; HANAKI, N.; YAMAMOTO, H. *Synlett* **1993**, 577. (b) ISHIHARA, K.; FUNAHASHI, M.; HANAKI, N.; MIYATA, M.; YAMAMOTO, H. *Synlett* **1994**, 963. (c) ISHIHARA, K.; HANAKI, N.; FUNAHASHI, M.; MIYATA, M.; YAMAMOTO, H. *Bull. Chem. Soc. Jpn.* **1995**, *68*, 1721.

14 ISHIHARA, K.; KURIHARA, H.; YAMAMOTO, H. *Synlett* **1997**, 597.

15 (a) MORI, Y.; MANABE, K.; KOBAYASHI, S. *Angew. Chem. Int. Ed.* **2001**, *40*, 2816. (b) MORI, Y.; KOBAYASHI, J.; MANABE, K.; KOBAYASHI, S. *Tetrahedron* **2002**, *58*, 8263.

16 REETZ, M.; KUNISH, F.; HEITMANN, P. *Tetrahedron Lett.* **1986**, *27*, 4721.

17 (a) KIYOOKA, S.-I.; KANEKO, Y.; KOMURA, M.; MATSUO, H.; NAKANO, M. *J. Org. Chem.* **1991**, *56*, 2276. (b) KANEKO, Y.; MATSUO, T.; KIYOOKA, S. *Tetrahedron Lett.* **1994**, *35*, 4107. (c) KIYOOKA, S.; KANEKO, Y.; HARADA, Y.; MATSUO, T. *Tetrahedron Lett.* **1995**, *16*, 2821. (d) KIYOOKA, S.; KIRA, H.; HENA, M. A. *Tetrahedron Lett.* **1996**, *37*, 2597. (e) KIYOOKA, S.; HENA, M. A. *Tetrahedron: Asymmetry* **1996**, *7*, 2181. (f) KIYOOKA, S.; MAEDA, H. *Tetrahedron: Asymmetry* **1997**, *8*, 3371. (g) KIYOOKA, S.; MAEDA, H.; HENA, M. A.; UCHIDA, M.; KIM, C.-S.; HORIIKE, M. *Tetrahedron Lett.* **1998**, *39*, 8287.

18 (a) GOODMAN, J. M. *Tetrahedron Lett.* **1992**, *33*, 7219. (b) COREY, E. J.; ROHDE, J. J.; FISCHER, A.; AZIMIOARA, M. D. *Tetrahedron Lett.* **1997**, *38*, 33. (c) COREY, E. J.; ROHDE, J. J. *Tetrahedron Lett.* **1997**, *38*, 37. (d) COREY, E. J.; BARNES-SEEMAN, D.; LEE, T. W. *Tetrahedron Lett.* **1997**, *38*, 1699. (e) COREY, E. J.; BRANS-SEEMAN, D.; LEE, T. W. *Tetrahedron Lett.* **1997**, *38*, 4351. (f) COREY, E. J.; LEE, T. W. *Chem. Commun.* **2001**, 1321.

19 (a) FURUTA, K.; MARUYAMA, T.; YAMAMOTO, H. *J. Am. Chem. Soc.* **1991**, *113*, 1041. (b) FURUTA, K.; MARUYAMA, T.; YAMAMOTO, H. *Synlett* **1991**, 439. (c) ISHIHARA, K.; MARUYAMA, T.; MOURI, M.; GAO, Q.; FURUTA, K.; YAMAMOTO, H. *Bull. Chem. Soc. Jpn.* **1993**, *66*, 3483.

20 (a) PARMEE, E. R.; TEMPKIN, O.; MASAMUNE, S. *J. Am. Chem. Soc.* **1991**, *113*, 9365. (b) PARMEE, E. R.; HONG, Y.; TEMPKIN, O.; MASAMUNE, S. *Tetrahedron Lett.* **1992**, *33*, 1729.

21 (a) KIYOOKA, S.; KANEKO, Y.; KUME, K. *Tetrahedron Lett.* **1992**, *33*, 4927. (b) KIYOOKA, S.; YAMAGUCHI, T.; MAEDA, H.; KIRA, H.; HENA, M. A.; HORIIKE, M. *Tetrahedron Lett.* **1997**, *38*, 3553.

22 COREY, E. J.; CYWIN, C. L.; ROPER, T. D. *Tetrahedron Lett.* **1992**, *33*, 6907.

23 (a) COREY, E. J.; LOH, T.-P. *J. Am. Chem. Soc.* **1991**, *113*, 8966. (b) COREY, E. J.; LOH, T.-P.; ROPER, T. D.; AZIMIOARA, M. D.; NOE, M. C. *J. Am. Chem. Soc.* **1992**, *114*, 8290.

24 (a) ISHIHARA, K.; KONDO, S.; YAMAMOTO, H. *Synlett* **1999**, 1283. (b) ISHIHARA, K.; KONDO, S.; YAMAMOTO, H. *J. Org. Chem.* **2000**, *65*, 9125.

25 Reilly, M.; Oh, T. *Tetrahedron Lett.* **1995**, *36*, 221.

26 (a) Kinugasa, M.; Harada, T.; Fujita, K.; Oku, A. *Synlett* **1996**, 43. (b) Kinugasa, M.; Harada, T.; Egusa, T.; Fujita, K.; Oku, A. *Bull. Chem. Soc. Jpn.* **1996**, *69*, 3639. (c) Kinugasa, M.; Harada, T.; Oku, A. *J. Org. Chem.* **1996**, *61*, 6772. (d) Kinugasa, M.; Harada, T.; Oku, A. *J. Am. Chem. Soc.* **1997**, *119*, 9067. (e) Kinugasa, M.; Harada, T.; Oku, A. *Tetrahedron Lett.* **1998**, *39*, 4529. (f) Harada, T.; Egusa, T.; Kinugasa, M.; Oku, A. *Tetrahedron Lett.* **1998**, *39*, 5531. (g) Harada, T.; Egusa, T.; Oku, A. *Tetrahedron Lett.* **1998**, *39*, 5535. (h) Harada, T.; Nakamura, T.; Kinugasa, M.; Oku, A. *Tetrahedron Lett.* **1999**, *40*, 503.

27 Nevalainen, V.; Mansikka, T.; Kostiainen, R.; Simpura, I.; Kokkonen, J. *Tetrahedron: Asymmetry* **1999**, *10*, 1.

28 (a) Komura, K.; Nishitani, N.; Itsuno, S. *Polym. J.* **1999**, *31*, 1045. (b) Itsuno, S.; Komura, K. *Tetrahedron* **2002**, *58*, 8237.

29 Noyori, R.; Murata, S.; Suzuki, M. *Tetrahedron* **1981**, *37*, 3899.

30 (a) Dilman, A. D.; Loffe, S. *Chem. Rev.* **2003**, *103*, 733. (b) Oishi, M. In *Lewis Acids in Organic Synthesis*; Yamamoto, H. Ed.; Wiley–VCH: Weinheim, 2000; Volume 1, pp. 355–393. (c) Hosomi, A.; Miura, K. In *Lewis Acids Reagents*; Yamamoto, H. Ed.; Oxford University Press: Oxford, 1999; pp. 159–168.

31 Maerker, C.; Kapp, J.; Schleyer, P. v. R. In *Organosilicon Chemistry II*; Auner, N., Weis, J.; VCH: Weinheim, 1996; pp. 329–359.

32 Shin, S. K.; Beauchamp, J. L. *J. Am. Chem. Soc.* **1989**, *111*, 990.

33 (a) Schwarz, H. In *The Chemistry of Organic Silicon Compounds*; Patai, S., Rappoport, Z., Eds.; Wiley: Chichester, 1989; Part 1, pp 445–510. (b) Chojnowski, J.; Stanczyk, W. A. *Adv. Organomet. Chem.* **1990**, *30*, 243.

34 Lambert, J. B.; Zhao, Y.; Zhang, S. M. *J. Phys. Org. Chem.* **2001**, *14*, 370.

35 Lambert, J. B.; Zhao, Y. *Angew. Chem. Int. Ed. Engl.* **1997**, *36*, 400.

36 Lambert, J. B.; Lin, L. *J. Org. Chem.* **2001**, *66*, 8537.

37 Müller, T.; Zhao, Y.; Lambert, J. B. *Organometallics* **1998**, *17*, 278.

38 Hergott, H. H.; Simchen, G. *Liebigs Ann. Chem.* **1980**, 1718.

39 (a) Bassindale, A. R.; Stout, T. *J. Chem. Soc., Perkin Trans. 2* **1986**, 221. (b) Bassindale, A. R.; Lau, J. C.-Y.; Stout, T.; Tayor, P. G. *J. Chem. Soc. Perkin Trans. 2* **1986**, 227.

40 Mayr, H.; Gorath, G. *J. Am. Chem. Soc.* **1995**, *117*, 7862. (b) Kira, M.; Hino, T.; Sakurai, H. *Chem. Lett.* **1992**, 555. (c) Prakash, G. K. S.; Wang, Q.; Rasul, G.; Olah, G. A. *J. Organomet. Chem.* **1998**, *550*, 119. (d) Prakash, G. K. S.; Bae, C.; Rasul, G.; Olah, G. A. *J. Org. Chem.* **2002**, *67*, 1297.

41 Murata, S.; Suzuki, M.; Noyori, R. *Tetrahedron* **1988**, *44*, 4259.

42 (a) Murata, S.; Suzuki, M.; Noyori, R. *J. Am. Chem. Soc.* **1980**, *102*, 3248. (b) Murata, S.; Suzuki, M.; Noyori, R. *Tetrahedron Lett.* **1980**, *21*, 2527.

43 Mukai, C.; Hashizume, S.; Nagami, K.; Hanaoka, M. *Chem. Pharm. Bull.* **1990**, *38*, 1509.

44 KAWAI, M.; ONAKA, M.; IZUMI, Y. *Bull. Chem. Soc. Jpn* **1988**, *61*, 1237.

45 (a) ISHIHARA, K.; HIRAIWA, Y.; YAMAMOTO, H. *Synlett* **2001**, 1851. (b) ISHIHARA, K.; HIRAIWA, Y.; YAMAMOTO, H. *Chem. Commun.* **2002**, 1564.

46 HOLLIS, T. K.; BOSNICH, B. *J. Am. Chem. Soc.* **1995**, *117*, 4570.

47 (a) OOI, T.; TAYAMA, E.; TAKAHASHI, M.; MARUOKA, K. *Tetrahedron Lett.* **1997**, *38*, 7403. (b) CHEN, J.; SAKAMOTO, K.; ORITA, A.; OTERA, J. *J. Org. Chem.* **1998**, *63*, 9739. (c) OTERA, J.; CHEN, J. *Synlett* **1996**, 321.

48 MOLANDER, G. A.; CAMERON, K. O. *J. Org. Chem.* **1991**, *56*, 2617.

49 (a) OLAH, G. A.; FIELD, L. D. *Organometallics* **1982**, *1*, 1485. (b) OLAH, G. A.; HEILIGER, L.; LI, X.-Y.; PRAKASH, G. K. S. *J. Am. Chem. Soc.* **1990**, *112*, 5991.

50 IWASAWA, N.; MUKAIYAMA, T. *Chem. Lett.* **1987**, 463.

51 MUKAIYAMA, T.; OHNO, T.; HAN, J. S.; KOBAYASHI, S. *Chem. Lett.* **1991**, 949.

52 (a) DAVIS, A. P.; JASPARS, M. *Angew. Chem. Int. Ed. Engl.* **1992**, *31*, 470. (b) DAVIS, A. P.; PLINKETT, S. J. *J. Chem. Soc., Chem. Commun.* **1995**, 2173. (c) DAVIS, A. P.; MUIR, J. E.; PLUNKETT, S. J. *Tetrahedron Lett.* **1996**, *37*, 9401.

53 (a) OISHI, M.; ARATAKE, S.; YAMAMOTO, H. *J. Am. Chem. Soc.* **1998**, *120*, 8271. (b) OISHI, M.; YAMAMOTO, H. *Macromolecules* **2001**, *34*, 3512.

54 (a) CARREIRA, E. M.; SINGER, R. A. *Tetrahedron Lett.* **1994**, *35*, 4323. (b) DENMARK, S. E.; CHEN, C.-T. *Tetrahedron Lett.* **1994**, *35*, 4327. (c) For detailed mechanistic discussion of catalytic asymmetric Mukaiyama aldol reaction, see: CARREIRA, E. M. In *Comprehensive Asymmetric Catalysis*, JACOBSEN, E. N., PFALTZ, A.; YAMAMOTO, H., Eds.; Springer: Heidelberg, 1999; Vol. 3, pp 997–1065.

3
Copper Lewis Acids

Jeffrey S. Johnson and David A. Nicewicz

3.1
Introduction

Copper complexes serve as structurally diverse Lewis acids that promote additions of enolates and latent enolates to carbonyl compounds. The exact mode of activation depends on the complex: many copper(II) complexes are known to effectively activate the electrophilic component in aldol additions whereas copper(I) complexes are implicated in aldol reactions that feature nucleophile activation (Scheme 3.1). Irrespective of the mechanistic details, when the metal complex carries stereochemical information in its ligand framework, chirality transfer to the nascent carbinol stereogenic center can be nearly complete.

This review will survey nucleophilic addition of enolates and latent enolates to carbonyl compounds catalyzed by copper Lewis acids. Particular attention will be paid to stereoselective variants and the development of stereochemical models to account for observed enantiomeric enrichment. Applications to natural product synthesis will be highlighted. A distinction is drawn between carbonyl activation in a Mukaiyama aldol sense and nucleophile activation via a metalloenolate; because each of these reaction-types do involve Lewis acid–Lewis base interactions, however, both reaction families will be included in this chapter. Coverage will focus on catalytic examples.

3.2
Early Examples

The ability of Cu(II) ion to promote the addition of acetone to aromatic aldehydes in crossed-aldol condensation reactions was demonstrated by Iwata and Emoto in 1974 [1]. Subsequent extension to a regioselective crossed aldol reaction with 2-butanone was later described by Irie and Watanabe [2]. Both of these early examples employ more than one equivalent of Cu(II) source relative to the aldehyde.

As a forerunner to his pioneering Au(I) work, Ito reported in 1985 that

Modern Aldol Reactions. Vol. 2: Metal Catalysis. Edited by Rainer Mahrwald
Copyright © 2004 WILEY-VCH Verlag GmbH & Co. KGaA, Weinheim
ISBN: 3-527-30714-1

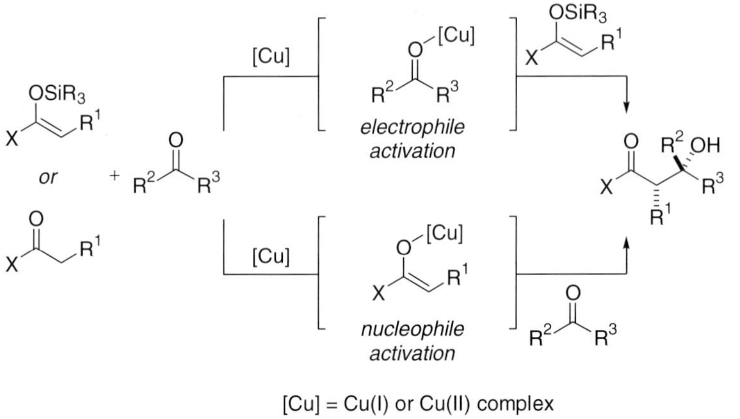

Scheme 3.1
Modes of activation for Cu-catalyzed aldol reactions.

catalytic quantities of a Cu(I) catalyst could be employed to promote addition of ethyl isocyanoacetate (**1**) to α,β-unsaturated aldehydes (Eq. (1)) [3]. The reactions are selective for formation of the *trans*-4,5-disubstituted oxazoline adducts (**3**). A footnote of that paper indicates that enantioselective variants of this reaction are possible employing ($-$)-ephedrine as a scalemic additive.

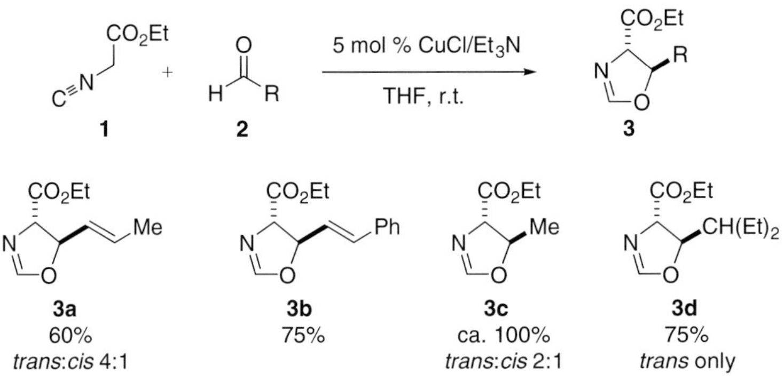

3.3
Mukaiyama Aldol Reactions with Cu(II) Complexes

3.3.1
Enolsilane Additions to (Benzyloxy)acetaldehyde

3.3.1.1 Scope and Application
In 1996 Evans and coworkers reported highly enantioselective additions of latent enolates to (benzyloxy)acetaldehyde (**7**) catalyzed by enantiomerically pure pyridyl bis(oxazoline) Cu(II) complexes (**4**, hereafter (pybox)CuL$_n$) [4,

5]. The reaction is a Mukaiyama aldol addition in which the aldehyde is activated toward nucleophilic addition by the electropositive Cu(II) center. The adduct is a β-silyloxy ester derivative that is readily desilylated under acidic conditions for the purpose of analyzing the enantiomeric enrichment of the product (Eq. (2)). Simultaneous investigations revealed that a bidentate C_2-symmetric bis(oxazoline) ligand is also an effective chiral control element, albeit with slightly reduced levels of enantiocontrol (Eq. (3)). The pendant phenyl substituent is optimal for the pybox ligand, and the *tert*-butyl group is most effective among those surveyed for the bis(oxazoline) scaffold.

$$(2)$$

$$(3)$$

Catalyst preparation depends on the identity of the counter-anion, which has a marked effect both on rate and selectivity. Bis(oxazoline)Cu(OTf)$_2$ (**5**, hereafter (box)Cu(OTf)$_2$) and (pybox)Cu(OTf)$_2$ complexes are prepared simply by mixing equimolar quantities of the ligand and Cu(OTf)$_2$ in CH$_2$Cl$_2$. The corresponding (ligand)Cu(SbF$_6$)$_2$ complexes are synthesized via anion metathesis of the (ligand)CuCl$_2$ complexes with two equivalents of AgSbF$_6$. Filtration of the resulting AgCl salt gives a clear blue or green solution of the active catalyst complex.

(Ph-pybox)Cu(SbF$_6$)$_2$-catalyzed additions to (benzyloxy)acetaldehyde are highly enantioselective for several acetate-type nucleophiles derived from thio- and oxo-esters (Figure 3.1). Less flexibility is possible with the electrophile. *p*-Methoxybenzyloxyacetaldehyde is an excellent substrate for the addition, but butoxyacetaldehyde is somewhat less selective. Enantiocontrol is significantly less for aldehydes nominally incapable of chelation.

The Chan diene (**13**) and dioxolanone-derived nucleophile (**14**) both serve as effective acetoacetate nucleophile equivalents in asymmetric catalyzed additions to benzyloxyacetaldehyde (Scheme 3.2). The former example was optimized to employ only 2 mol% chiral catalyst to deliver multigram

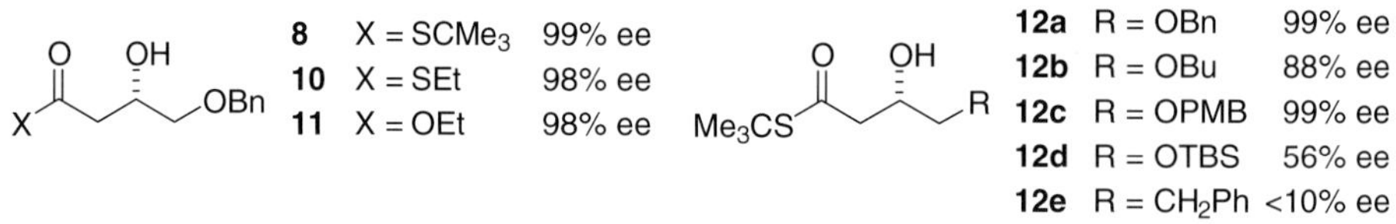
8 X = SCMe3 99% ee
10 X = SEt 98% ee
11 X = OEt 98% ee
12a R = OBn 99% ee
12b R = OBu 88% ee
12c R = OPMB 99% ee
12d R = OTBS 56% ee
12e R = CH2Ph <10% ee

Fig. 3.1
Enantioenriched aldol adducts derived from (Ph-pybox)Cu$(SbF_6)_2$-catalyzed reactions (Eq. (2)).

quantities of essentially optically pure material (**15**). On these scales the reaction must be initiated at a low temperature, because the reaction is highly exothermic. The δ-hydroxy-β-keto ester product has been diastereoselectively reduced to afford either the *syn* (**16**) or *anti* (**17**) diol product in good yield.

The Chan diene addition product provides a useful entry into polyacetate building blocks. To this end, the enantiomeric (*R*,*R*-Ph-pybox)Cu$(SbF_6)_2$

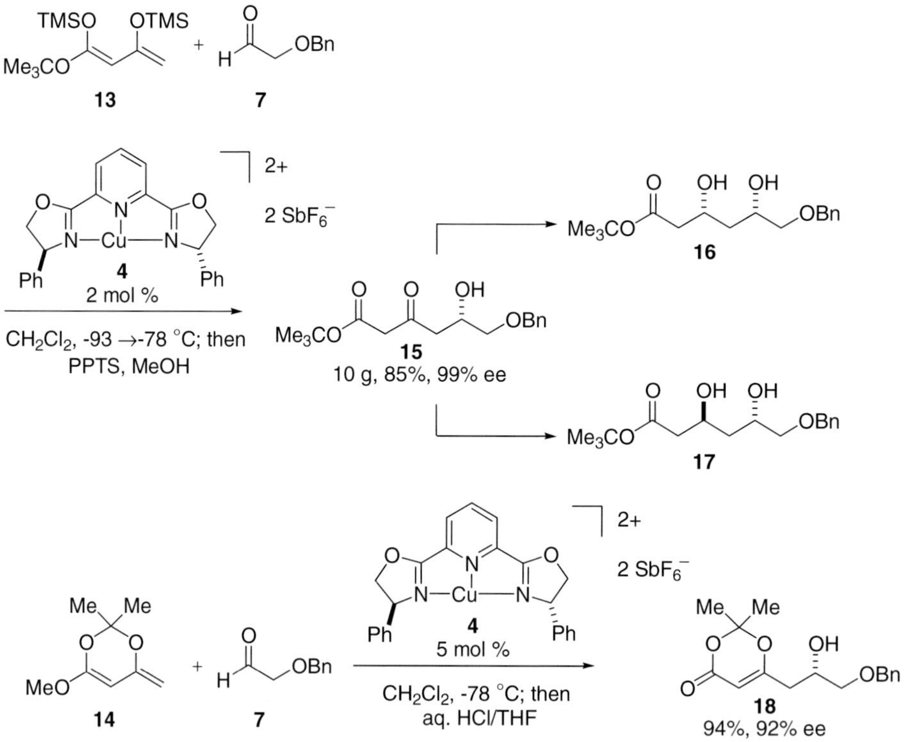
13
7
4
2 mol %
CH2Cl2, -93 →-78 °C; then
PPTS, MeOH
15
10 g, 85%, 99% ee
16
17
14
7
4
5 mol %
CH2Cl2, -78 °C; then
aq. HCl/THF
18
94%, 92% ee

Scheme 3.2
Enantioselective aldol reactions of acetoacetate nucleophile equivalents catalyzed by (Ph-pybox)Cu$(SbF_6)_2$ complexes.

complex has been used to deliver the needed aldol enantiomer (*ent*-**15**) for ultimate transformation to the cytostatic natural product phorboxazole B [6], and the bryostatin family of antitumor agents (Scheme 3.3) [7]. It is noteworthy that in the former reaction a common asymmetric aldol product provides a common starting material for two different pyran rings.

A stereoselective catalyzed vinylogous aldol addition was developed for application to the asymmetric synthesis of callepeltoside A (Scheme 3.4) [8]. The reaction makes use of an air-stable hydrated catalyst, $[((R,R)$-Ph-pybox)Cu(OH$_2$)$_2$](SbF$_6$)$_2$ (**22**), to effect the formation of the δ-hydroxy-α,β-unsaturated ester **21** with complete *E* selectivity and excellent enantiocontrol.

Substituted (propionate-type) silylketene acetals also add to (benzyloxy)-acetaldehyde with high diastereo- and enantiocontrol under the influence of (pybox)Cu(SbF$_6$)$_2$ catalysis (Figure 3.2). A range of cyclic and acyclic nucleophiles participate in diastereo- and enantioselective aldol reactions to give the *syn* aldol diastereomer in all cases but one (**23a**–**23e**). The *syn* selectivity predominates irrespective of the geometry of the starting silylketene acetal. The only exception to this trend is 2-trimethylsilyloxyfuran, which affords the *anti* diastereomer **23f** in good chemical and optical yield.

Kunieda and coworkers reported a modified catalyst system in 1999 that probes the effect of the backbone spacer connecting the two oxazoline rings and steric congestion about the metal center [9]. Anthracene-based bis(oxazoline)Cu(II) complexes were prepared and tested in the addition of *t*-butyl thioester silylketene acetal to (benzyloxy)acetaldehyde (Eq. (4)). The methylene-bridged complex **24·Cu(OTf)**$_2$ strongly favors formation of the *R* enantiomer, whereas extending the linking chain by one CH$_2$ group (**25·Cu(OTf)**$_2$) results in a selectivity turnover to favor the *S* enantiomer. The authors propose that a change in aldehyde binding geometry could result from the structural perturbation. In neither reaction is selectivity superior to that of the Evans system.

ligand	8S:8R
24a	1:24
24b	1:3.3
25a	9:1
25b	3.2:1

(4)

24a, R = H
24b, R = Me

25a, R = H
25b, R = Me

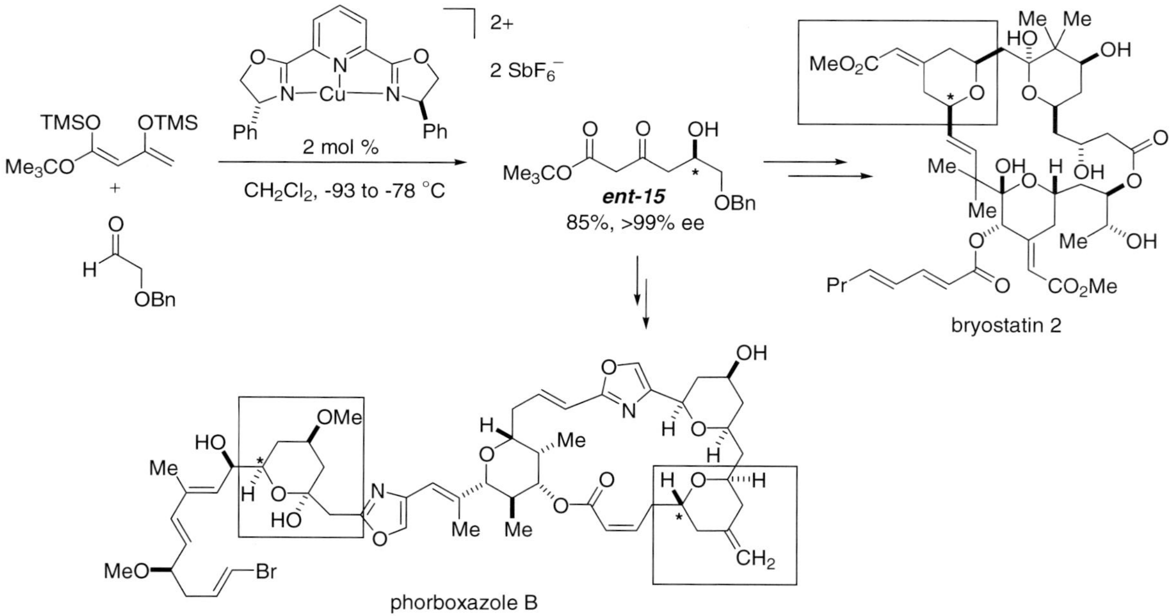

Scheme 3.3

Application of enantioselective (Ph-pybox)
Cu(SbF$_6$)$_2$-catalyzed aldol reactions to
pyran-containing natural products.

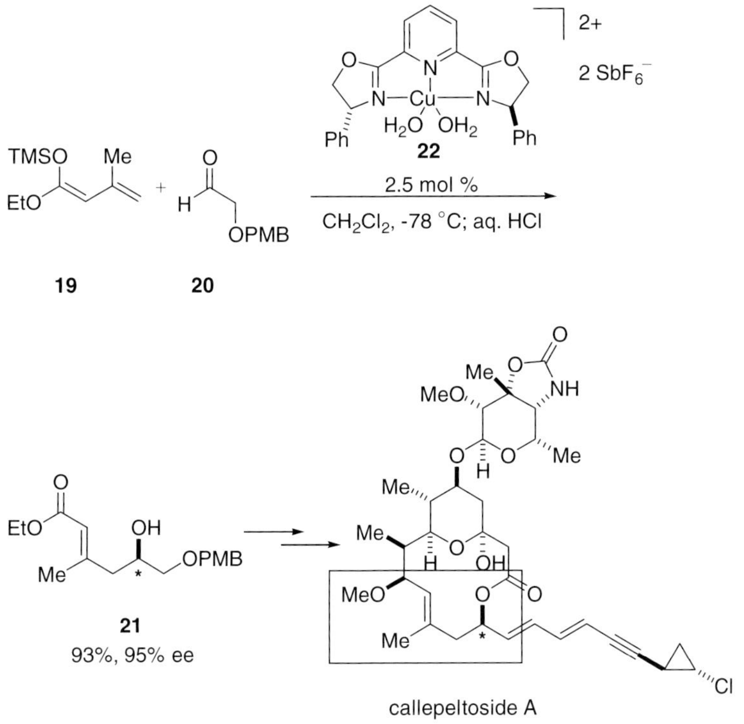

Scheme 3.4
Application of enantioselective (Ph-pybox)
Cu(SbF$_6$)$_2$-catalyzed vinylogous aldol
reactions to callepeltoside A.

3.3.1.2 Mechanism and Stereochemistry

The proposed mechanism for the (pybox)Cu(SbF$_6$)$_2$ catalyzed addition reaction involves activation of the chelating electrophile by the metal center (**26**), nucleophilic addition (**26** → **27**), silylation of the metal aldolate (**27** → **28**), and release of the neutral product (**28** → **29** + **4**) (Scheme 3.5).

The silicon-transfer step of this mechanism has a significant intermolecular component, as evidenced by double-labeling experiments (Scheme 3.6). The identity of the species responsible for intermolecular silyl transfer is not known, although the metal aldolate and Me$_3$SiSbF$_6$ are potential candidates. What is apparent from the enantiomeric enrichment of the aldols in the crossover experiment is that no stereochemical "leakage" occurs in this process: any potential achiral aldol catalyst [10] is not competitive with the chiral cationic Cu(II) complex.

The asymmetric aldol reaction catalyzed by (Ph-pybox)Cu(SbF$_6$)$_2$ has a significant positive non-linear effect when complexes are prepared from enantioimpure pybox ligands. Experimental evidence points to a catalytically inactive heterochiral dimer as the source of this non-linear effect (reservoir

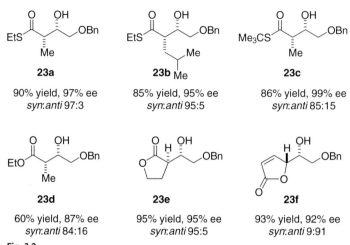

23a

90% yield, 97% ee
syn:anti 97:3

23b

85% yield, 95% ee
syn:anti 95:5

23c

86% yield, 99% ee
syn:anti 85:15

23d

60% yield, 87% ee
syn:anti 84:16

23e

95% yield, 95% ee
syn:anti 95:5

23f

93% yield, 92% ee
syn:anti 9:91

Fig. 3.2
Aldol adducts derived from enantio- and
diastereoselective (Ph-pybox)Cu(SbF$_6$)$_2$-
catalyzed aldol reactions by use of
substituted silylketene acetals.

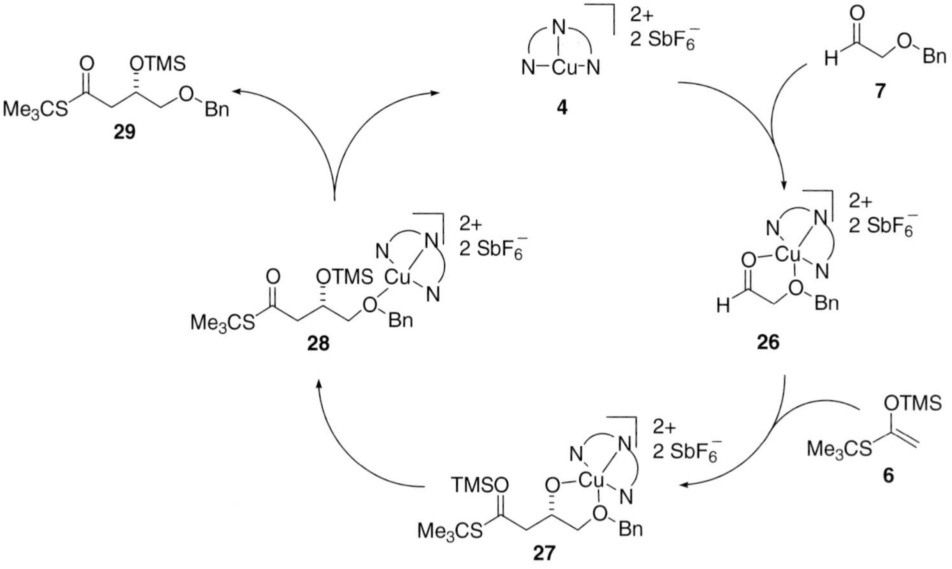

Scheme 3.5
Proposed mechanism for (Ph-pybox)
Cu(SbF$_6$)$_2$-catalyzed enantioselective aldol
reactions.

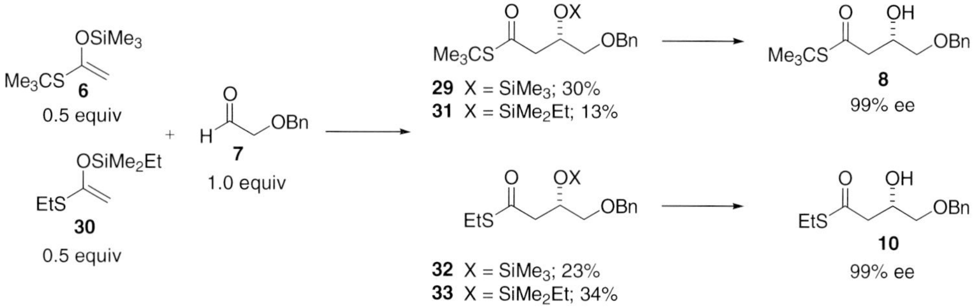

Scheme 3.6
Crossover experiments to verify
intermolecular silicon transfer in
enantioselective (Ph-pybox)Cu(SbF$_6$)$_2$-
catalyzed aldol reactions.

effect) [11]. Corroboration was obtained via crystallization of the hetero-chiral dimer **34**, demonstrating that its formation is indeed feasible (Figure 3.3). Semiempirical calculations support the notion that the homochiral dimer is less stable than the heterochiral dimer, accounting for the positive non-linear effect.

Substantial insight into the mechanism of asymmetric induction has been obtained via crystallization of monomeric [(pybox)CuL$_n$](SbF$_6$)$_2$ complexes. An X-ray structure of [(*i*-Pr-pybox)Cu(OH$_2$)$_2$](SbF$_6$)$_2$ (**35**) reveals square py-ramidal geometry with one water molecule occupying the coordination site in the ligand plane and the second water molecule occupying the axial position (Figure 3.4). Neither counter-ion is within the coordination sphere of the metal. It is revealing that the Cu–O bond length in the ligand plane is considerably shorter than the C–O$_{axial}$ bond length (1.985(7) compared with 2.179(7) Å). For maximum electrophile activation, aldehyde coordina-tion should occur in the ligand plane. The presence of the axial binding site provides a second "contact point" for the chelating carbonyl compound and introduces an additional element of substrate organization.

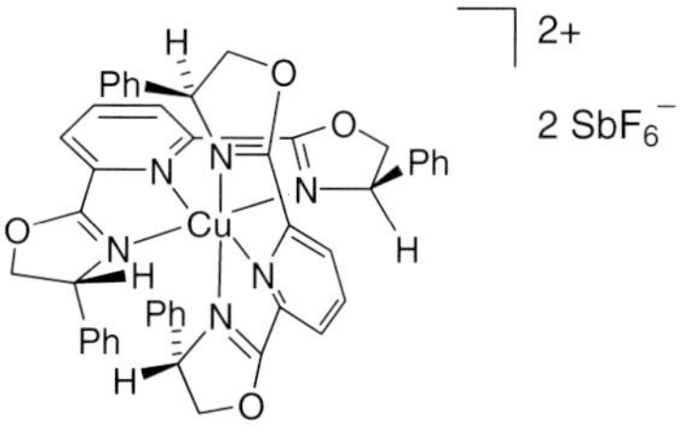

Fig. 3.3
Structure of heterochiral dimer [((*R,R*)-Ph-pybox)Cu((*S,S*)-Ph-pybox)](SbF$_6$)$_2$ (**34**).

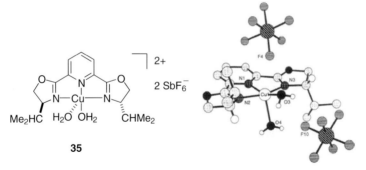

35

Fig. 3.4
Structure of [(*i*-Pr-pybox)Cu(OH$_2$)$_2$](SbF$_6$)$_2$ (**35**).

Ultimate corroboration of this mode of activation was obtained via crystallization of the catalyst–substrate complex [(Ph-pybox)Cu(BnOCH$_2$CHO)]-(SbF$_6$)$_2$ (**36**). The aldehyde coordinates to form a five-membered ring chelate, with the ether oxygen occupying the axial position (Figure 3.5). The aromatic ring of the benzyloxy group is ca. 3.5 Å removed from the aromatic pyridyl ring in an offset face–face arrangement, a π–π interaction that might explain the superior selectivities observed for *p*-PMBOCH$_2$CHO and BnOCH$_2$CHO compared with *n*-BuOCH$_2$CHO. Thus coordinated, the aldehyde *re* face is shielded by the proximal phenyl ring of the pybox ligand; addition to the *si* face is predicted and experimentally observed.

This model predicts that (*S*)- and (*R*)-α-benzyloxypropionaldehyde will behave as matched and mismatched substrates in the addition. In accord with this proposed transition state assembly, the (*S*) isomer (R$_1$ = Me, R$_2$ = H) undergoes a highly efficient and diastereoselective addition (2 h, dr = 98.5:1.5) whereas the (*R*) isomer (R$_1$ = H, R$_2$ = Me) is a sluggish reaction partner and poorly diastereoselective (12 h, dr = 50:50).

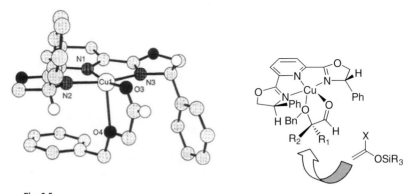

Fig. 3.5
X-ray structure of [(Ph-pybox)Cu
(BnOCH$_2$CHO)](SbF$_6$)$_2$ (**36**) and stereochemical model for enantioselective
additions.

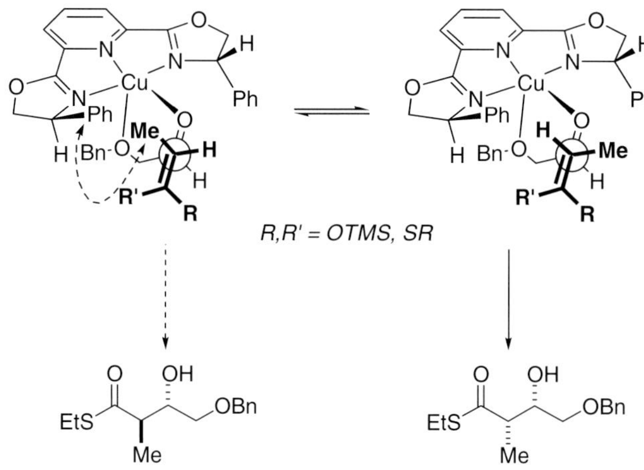

Fig. 3.6
Stereochemical models for *syn*-selective
aldol reactions catalyzed by (Ph-pybox)
Cu(SbF$_6$)$_2$.

The *syn* diastereoselectivity is accounted for by an open antiperiplanar
transition structure that minimizes *gauche*, dipole, and other through-space
effects (Figure 3.6).

Representative Experimental Procedures

Preparation of (S,S)-Ph-pybox)Cu(SbF$_6$)$_2$ (4). In a nitrogen atmosphere box
(S,S)-bis(phenyloxazolinyl)pyridine (18.5 mg, 0.05 mmol) and CuCl$_2$ (6.7
mg, 0.05 mmol) were placed in an oven-dried round-bottomed flask containing a magnetic stirring bar. In a nitrogen atmosphere box AgSbF$_6$
(34.4 mg, 0.10 mmol) was placed in an oven-dried round-bottomed flask
containing a magnetic stirring bar. The flasks were fitted with serum caps
and removed from the nitrogen atmosphere box. The flask containing the
ligand–CuCl$_2$ mixture was charged with CH$_2$Cl$_2$ (1.0 mL). The resulting
suspension was stirred rapidly for 1 h to give a fluorescent green suspension. AgSbF$_6$ (in 0.5 mL CH$_2$Cl$_2$) was added via a cannula with vigorous
stirring, followed by a 0.5 mL rinse. The resulting mixture was stirred rapidly for 3 h in the absence of light and filtered through an oven-dried glass
pipet tightly packed with cotton to remove the white AgCl precipitate, yielding active catalyst as a clear blue solution.

**Catalyzed Addition of Silylketene Acetals to Benzyloxyacetaldehyde Using (S,S)-
Ph-pybox)Cu(SbF$_6$)$_2$.** Benzyloxyacetaldehyde (70.0 µL, 0.50 mmol), followed
by a silylketene acetal (0.60 mmol), were added to a −78 °C solution of 4 in
CH$_2$Cl$_2$. The resulting solution was stirred at either −78 or −50 °C until the
aldehyde was completely consumed (15 min to 48 h) as determined by TLC
(30% EtOAc–hexanes). The reaction mixture was then filtered through a 1.5

cm × 8 cm plug of silica gel with Et$_2$O (50 mL). Concentration of the ether solution gave a clear oil, which was dissolved in THF (10 mL) and 1 M HCl (2 mL). After standing at room temperature for 15 min, this solution was poured into a separatory funnel and diluted with Et$_2$O (10 mL) and H$_2$O (10 mL). After mixing, the aqueous layer was discarded, and the ether layer was washed with saturated aqueous NaHCO$_3$ (10 mL) and brine (10 mL). The resulting ether layer was dried over anhydrous MgSO$_4$, filtered, and concentrated to provide the hydroxy esters.

3.3.2
Enolsilane Additions to α-Keto Esters

3.3.2.1 Scope and Application

Dialkylketones are typically poor electrophiles in traditional aldol bond constructions, but the presence of a strong electron-withdrawing group in α-keto esters engenders reactivity that more closely resembles that of aldehydes. Evans and coworkers described the first catalytic, enantioselective enolsilane addition to pyruvate esters [12, 13]. The most effective catalyst with regard to yield and enantiocontrol is the (*t*-Bu-box)Cu(OTf)$_2$ complex **5** (Eq. (5)) and its corresponding hydrated derivative **5**·(H$_2$O)$_2$ (Eq. (6)). The latter is an air-stable solid with identical reactivity when used in the presence of a desiccant. In contrast to the pybox system the cationic complex [(*t*-Bu-box)Cu](SbF$_6$)$_2$ results in reduced enantioselectivity [14].

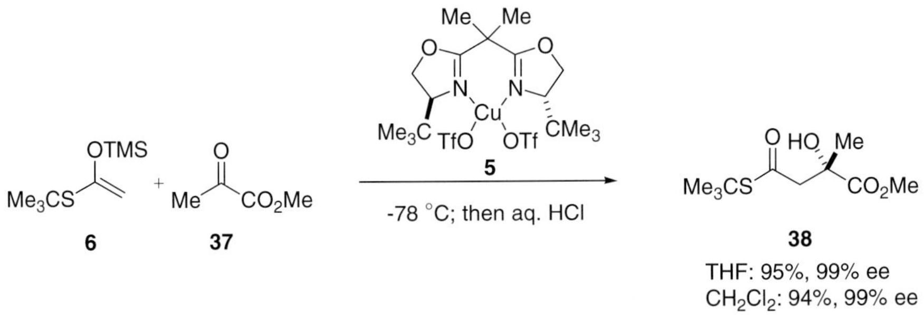

THF: 95%, 99% ee
CH$_2$Cl$_2$: 94%, 99% ee

(5)

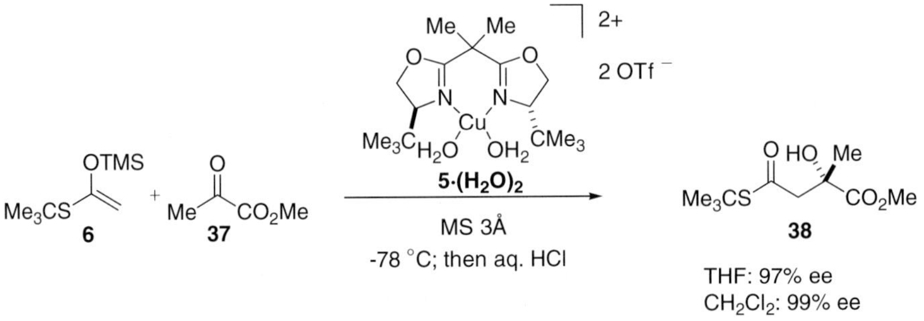

THF: 97% ee
CH$_2$Cl$_2$: 99% ee

(6)

The addition reactions can be effectively performed in a range of solvents, including THF, Et$_2$O, CH$_2$Cl$_2$, PhMe, hexane, and PhCF$_3$. The enantiomeric excess is >94% for addition of the *tert*-butyl thioacetate silylketene acetal to methyl pyruvate in all of these solvents. Catalyst loadings down to 1 mol% are feasible. The temperature–enantioselectivity profile has been studied and shown to be relatively flat (99% ee at −78 °C; 92% ee at +20 °C).

Interestingly, the catalytic reaction in THF, a relatively good donor solvent, is significantly faster than the identical reaction in CH$_2$Cl$_2$. Control experiments with stoichiometric quantities of (*t*-Bu-box)Cu(OTf)$_2$ demonstrate that the actual addition step is faster in CH$_2$Cl$_2$, a fact consistent with the predicted deactivation of the Lewis acidic center in THF via solvent coordination. Accordingly, THF must play a role in promoting catalyst turnover. One postulated role of THF in the catalytic cycle is to act as a silicon shuttle, forming a more reactive silylating species (e.g. [THF-SiMe$_3$]OTf). The "silicon shuttle" hypothesis predicts significant intermolecular crossover, which is experimentally borne out by double-labeling experiments in analogy to those described above for benzyloxyacetaldehyde additions.

Silylation of the putative metal aldolate by an exogenous Si(+) source results in significant rate accelerations. For example, a catalyzed pyruvate addition that requires 14 h in the absence of an additive is complete in 0.5 h in the presence of 1.0 equiv. TMSOTf (Eq. (7)). The presence of stoichiometric quantities of this Lewis acid does not erode the selectivity of the reaction. The Cu(II) complex again reacts to complete exclusion of the achiral complex.

$$(7)$$

The scope of the reaction with regard to the carboalkoxy and acyl moieties (electrophile) includes a range of substituents (Figure 3.7). α-Branched substrates (e.g. *i*-PrC(O)CO$_2$Me) result in low π-facial selectivity (**39e**) but comprise the only subset of poorly selective α-keto esters. Enolsilanes derived from acetone and acetophenone are effective and selective nucleophiles in additions to methyl pyruvate (**39g–h**). Propionate silylketene acetals are also usually effective (**39i**). As in the [(pybox)Cu](SbF$_6$)$_2$-catalyzed additions to benzyloxyacetaldehyde, good *syn* diastereoselectivity is observed. The only

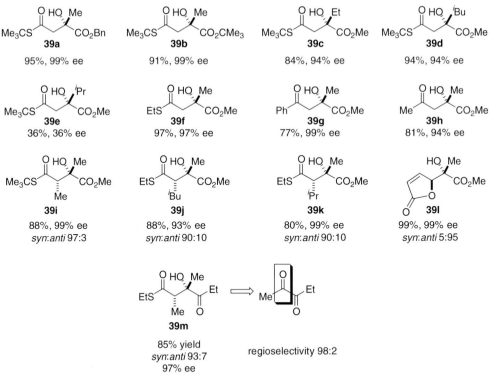

Fig. 3.7
Aldol products derived from enantio-
selective additions catalyzed by (t-Bu-box)
Cu(OTf)$_2$ (**5**).

exception to this trend is again 2-trimethylsilyloxyfuran, for which *anti* diastereoselectivity is high (**39l**). 2,3-Pentanedione also participates in selective
aldol reactions with silylketene acetals. In addition to diastereo- and enantioselectivity issues faced in other examples this electrophile contains a
subtle regiochemical issue between two nominally similar carbonyl groups.
In practice, the (t-Bu-box)Cu(OTf)$_2$ complex performs the subtle discrimi-
nation between the two groups and effects a highly regio- and stereoselective aldol reaction with the acetyl group to give **39m**.

Verdine has described the application of this aldol methodology to the
enantio- and diastereocontrolled synthesis of α-hydroxy-α-methyl-β-amino
acids (**40**) in a sequence that uses the carbothioalkoxy group as an amine
surrogate via a Curtius rearrangement (Scheme 3.7) [15]. Thus, the desired
protected β-amino acid can be obtained in four steps with the needed stereochemical relationships established in the (t-Bu-box)Cu(OTf)$_2$-catalyzed aldol
addition.

The asymmetric pyruvate addition can be effected with a complex derived

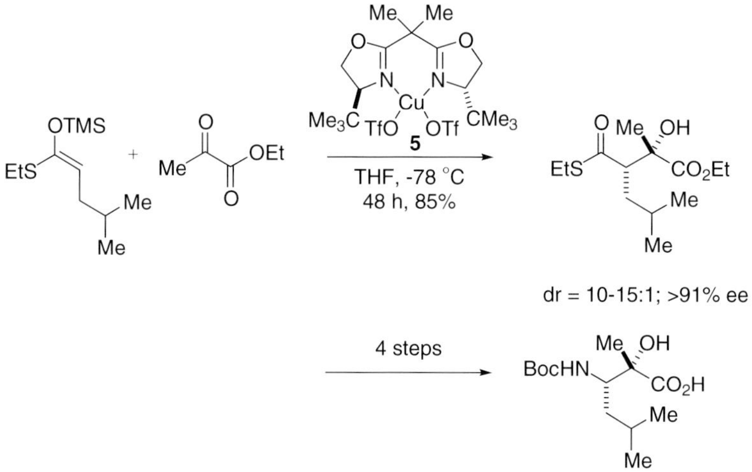

Scheme 3.7
Synthesis of α-hydroxy-α-alkyl-β-amino acids
from enantioselective pyruvate aldol
reactions catalyzed by (*t*-Bu-box)Cu(OTf)$_2$
(**5**).

from Cu(OTf)$_2$ and a polystyrene-bound bis(oxazoline) ligand (**41**) with selectivity approaching that of the solution reaction (Eq. (8)) [16]. As with many solid-supported complexes, catalytic activity was significantly less than the soluble variant. Nonetheless, Salvadori and co-workers demonstrated that the ligand could be reused in multiple reaction cycles with no loss of activity provided that additional Cu(OTf)$_2$ was added to the reaction mixture. In the absence of additional Cu(OTf)$_2$, recycling is still possible, with

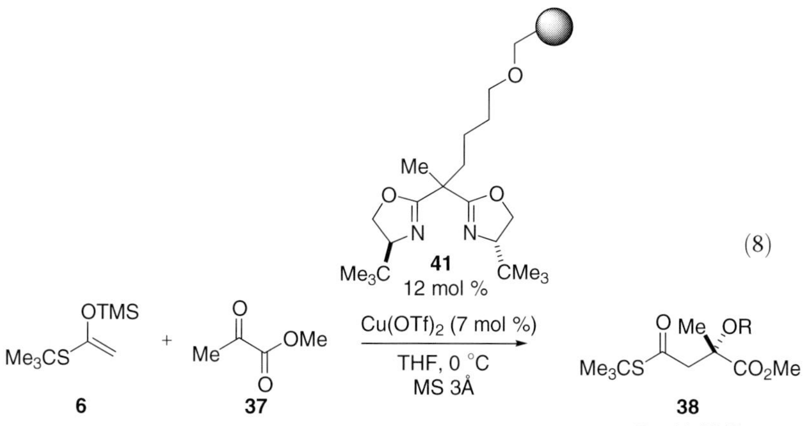

(8)

7 cycles; reaction time = 1-4 h, ee = 88-93%

the consequence of extended reaction times in subsequent cycles. It is interesting to note that the relative amounts of the silylated and unsilylated aldol products vary from run to run, but the enantioselectivity is relatively constant. A post-catalytic cycle desilylation seems most reasonable.

Jørgenson and co-workers have extended the α-keto ester additions to keto malonate substrates (Eq. (9)) [17]. In these asymmetric additions, the tertiary carbinol is not a stereogenic center; in essence the chiral complex induces asymmetry on the nucleophile. For a range of enolsilane nucleophiles, enantiocontrol in the addition step is moderate to excellent. The optimal promoter for these additions is the (Ph-box)Cu(OTf)$_2$ complex (**42**). In all instances but one the (*E*)-enolsilane was employed; the (*Z*)-enolsilane derived from propiophenone gave excellent results (**43e**).

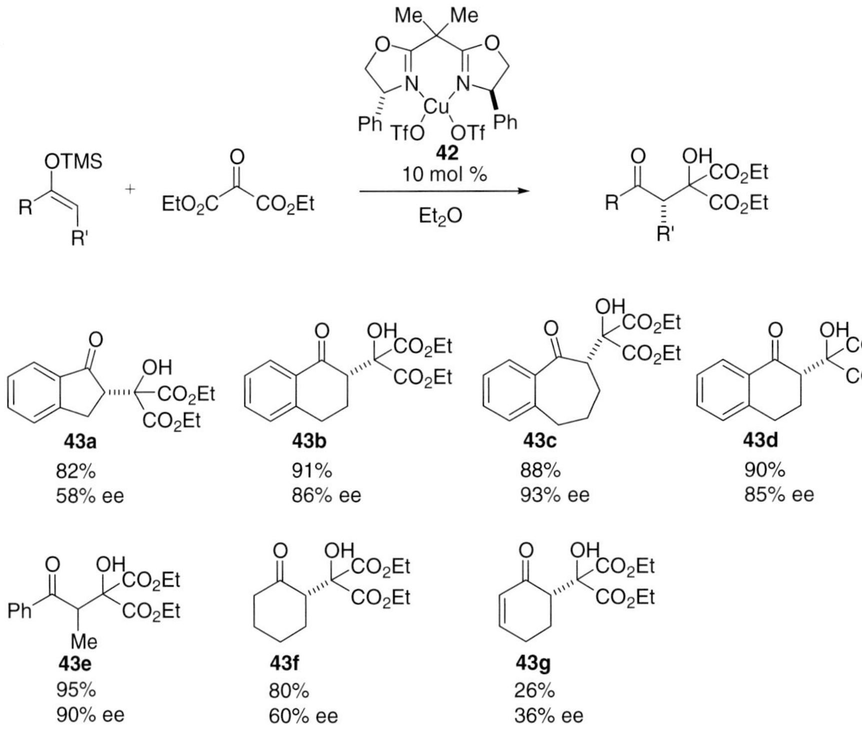

Me Me
O O
N N
Cu
Ph TfO OTf Ph
42
10 mol %
Et2O
OTMS
R
R'
+
EtO2C CO2Et
O
O OH
CO2Et
CO2Et
R
R'
OH
CO2Et
CO2Et
43a
82%
58% ee
43b
91%
86% ee
43c
88%
93% ee
43d
90%
85% ee
Ph
Me
43e
95%
90% ee
43f
80%
60% ee
43g
26%
36% ee

(9)

Dalko and Cossy have employed the Danishefsky diene in additions to ethyl pyruvate catalyzed by an uncharacterized complex prepared by mixing enantiopure stilbene diamine **44** and cyclobutanone **45** (1:1), followed by complexation with Cu(OTf)$_2$ (Eq. (10)) [18]. Cyclobutanone was optimal with regard to yield and enantioselectivity for the ketones and aldehydes surveyed. Reactant stoichiometry and premixing time were found to have a significant effect on enantioselectivity and reaction efficiency. The reaction affords a mixture of both the silylated and desilylated acyclic aldol product

(48), in addition to the cyclized dihydropyrone (47). Whether the dihydropyrone is formed by a concerted or stepwise mechanism is yet to be determined. In practice, the acyclic aldols are easily cyclized to the dihydropyrone in the presence of trifluoroacetic acid for the purpose of determining the enantiomeric excess. This catalyst system is noteworthy for the simplicity with which the active catalyst is assembled (in situ).

3.3.2.2 Mechanism and Stereochemistry

The mechanism of Cu(II)-catalyzed additions to α-keto esters is thought to proceed via a Mukaiyama aldol pathway, with the difunctional electrophile undergoing bidentate activation by the Cu(II) Lewis acid (49). This coordination event lowers the LUMO of the ketone to a point that facilitates addition of the silylketene acetal (49 → 50). Silylation of the Cu(II) aldolate via an intra- or intermolecular silicon transfer gives the neutral metal-coordinated adduct (52) that decomplexes to regenerate the catalytically active Lewis acid and release the product, 53 (Scheme 3.8) [13].

A distorted square-planar metal center is implicated in all reactions involving (t-Bu-box)CuL$_n$ [19]. This is suggested both by X-ray crystallographic studies of the hydrated complex [(t-Bu-box)Cu(OH$_2$)$_2$](SbF$_6$)$_2$ and by PM3 calculations designed to probe the structure of activated intermediates. The X-ray structure reveals that the coordinated water molecules are tilted out of the ligand plane by approximately 30° (Figure 3.8). This is a steric effect, as water molecules in the corresponding [(i-Pr-box)Cu(OH$_2$)](SbF$_6$)$_2$ complex are nearly coplanar with the ligand (approximately 7° out of plane).

By inspection, replacing the water molecules with the oxygen atoms of the pyruvate ester should result in a complex in which the enantiotopic faces of the carbonyl are significantly differentiated. This has been con-

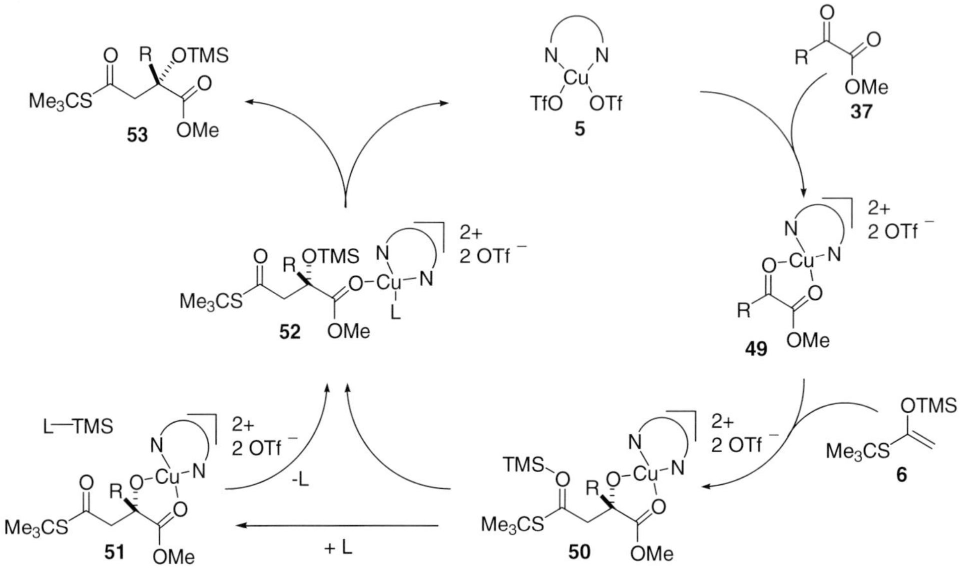

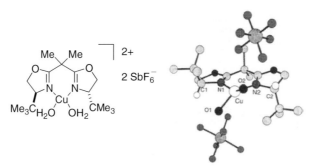

Scheme 3.8
Proposed mechanism for (t-Bu-box)Cu(OTf)₂-catalyzed enantioselective aldol reactions.

firmed by PM3 calculations. The pyruvate ester additions are all consistent with the stereochemical model shown in Figure 3.9. The bulky *t*-butyl group effectively shields the *re* face of the ketone, directing nucleophilic addition to the *si* face. This complexation mode is now well established with this family of catalysts.

The diastereoselectivity in additions of substituted enolsilanes to α-keto esters can be rationalized by an open, antiperiplanar transition structure that minimizes steric interactions between the enolsilane substituent and

Fig. 3.8
X-ray crystal structure of [(t-Bu-box)Cu(H₂O)₂](SbF₆)₂.

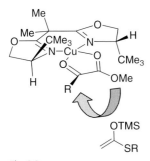

Fig. 3.9
Model for enantioselective addition to α-keto esters catalyzed by (t-Bu-box)Cu(OTf)$_2$.

the pendant ligand substituent (Figure 3.10). The disposition of the –OTMS and –SR groups are less important in this model, a point that is supported by the relative insensitivity of reaction diastereoselectivity as a function of enolsilane geometry.

General Experimental Procedure

In an inert atmosphere box, (S,S)-bis($tert$-butyloxazoline) (15 mg, 0.050 mmol) and Cu(OTf)$_2$ (18 mg, 0.050 mmol) were placed in an oven-dried

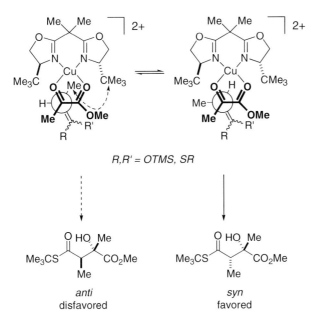

Fig. 3.10
Stereochemical models for *syn*-selective aldol reactions catalyzed by (t-Bu-box)Cu(OTf)$_2$.

10-mL round-bottomed flask containing a magnetic stirring bar. The flask was fitted with a serum cap, removed from the inert atmosphere box, and charged with solvent (1.5–3.0 mL). The resulting suspension was stirred rapidly for 4 h with CH_2Cl_2 to give a slightly cloudy bright green solution or 1 h with THF to give a clear dark green solution. The catalyst was cooled to −78 °C, and the pyruvate (0.50 mmol) was added, followed by the silylketene acetal (0.60 mmol). The resulting solution was stirred at −78 °C until the pyruvate was completely consumed (0.5 to 24 h) as determined by TLC (2.5% Et_2O–CH_2Cl_2). The reaction mixture was then filtered through a 2 cm × 4 cm plug of silica gel with Et_2O (60 mL). Concentration of the Et_2O solution gave the crude silyl ether which was dissolved in THF (5 mL) and treated with 1 M HCl (1 mL). After being stirred at room temperature for 1–5 h this solution was poured into a separatory funnel and diluted with Et_2O (20 mL) and H_2O (10 mL). After mixing the aqueous layer was discarded and the ether layer was washed with saturated aqueous $NaHCO_3$ (10 mL) and brine (10 mL). The resulting ether layer was dried over anhydrous Na_2SO_4, filtered, and concentrated to provide the hydroxy esters. Purification was achieved by flash chromatography.

3.3.3
Enolsilane Additions to Unfunctionalized Aldehydes

In 1998, Kobayashi made the counterintuitive observation that the Lewis acid-catalyzed addition of enolsilanes to aldehydes could be conducted in wet organic solvents (e.g. 10% H_2O in THF) [20]. The initial study documented that a wide range of metal salts are effective in promoting Mukaiyama aldol reactions in an aqueous environment. It is particularly relevant to this chapter that $Cu(ClO_4)_2$ acts as a catalyst (Eq. (11)), but is not particularly efficient (one turnover in 12 h at ambient temperature). The carbonyl addition pathway is clearly faster than Lewis or Brønsted acid-catalyzed decomposition of the enolsilane.

This discovery led to the development of an enantioselective variant. Implicit requirements for aqueous enantioselective Mukaiyama aldol reactions include a strong association between the chiral ligand and the metal center that is not disrupted by water, and/or a ligand–metal complex that is considerably more active than the corresponding hydrated complex, $M^m(OH_2)_nX_m$. Given the documented activity of $Cu(ClO_4)_2$ in water, attention was directed to bis(oxazoline) ligands, known to have strong affinity for Cu(II). The optimal catalyst with regard to both chemical and optical yield was the (*i*-Pr-box)Cu(OTf)$_2$ complex; H$_2$O–EtOH (1:9) was identified as the best solvent. Under the optimized reaction conditions a range of substituted enolsilanes underwent asymmetric catalyzed addition to aromatic, heteroaromatic, alkenyl, and aliphatic aldehydes with moderate to good enantioselectivity (Eq. (12)) [21]. The absolute stereochemistry of these aldol adducts is unfortunately not known, so speculation about the mechanism of asymmetric induction is premature at this time.

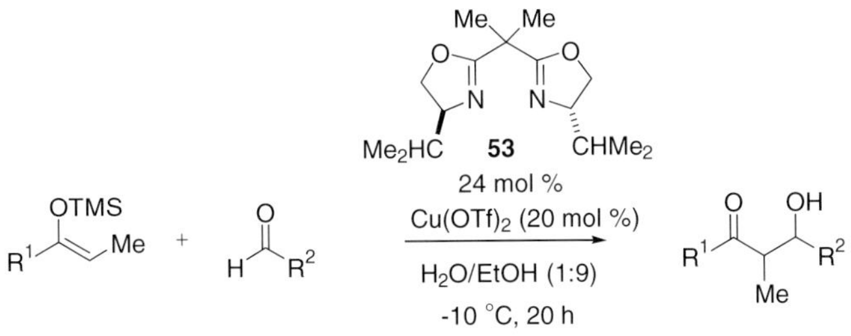

$$ \text{(12)} $$

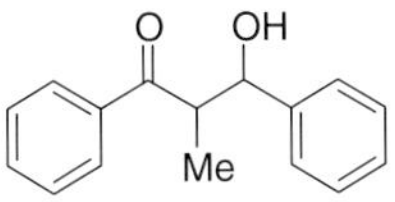

54a
74%, syn/anti = 3.2/1
67% ee (syn)

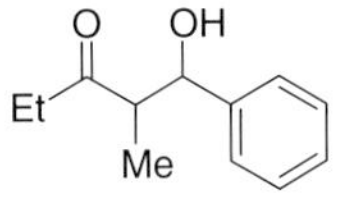

54b
81%, syn/anti = 3.5/1
81% ee (syn)

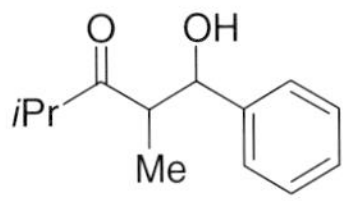

54c
95%, syn/anti = 4.0/1
77% ee (syn)

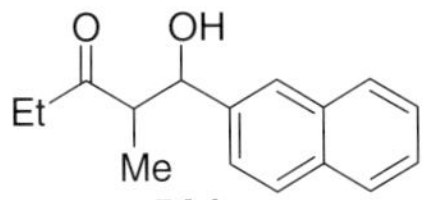

54d
91%, syn/anti = 4.0/1
79% ee (syn)

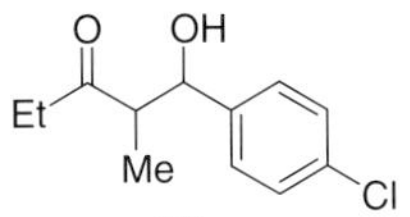

54e
88%, syn/anti = 2.6/1
76% ee (syn)

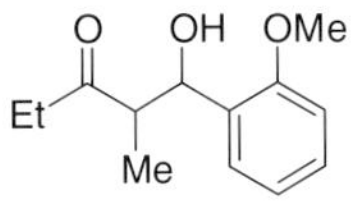

54f
87%, syn/anti = 2.9/1
75% ee (syn)

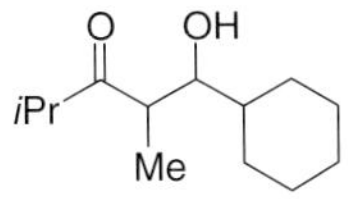

54g
77%, syn/anti = 4.6/1
42% ee (syn)

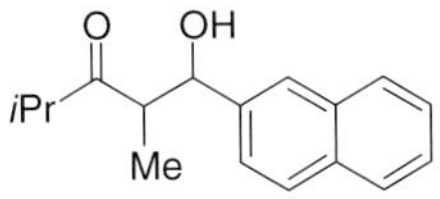

54h
97%, syn/anti = 4.0/1
81% ee (syn)

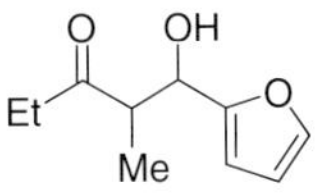

54i
86%, syn/anti = 4.0/1
76% ee (syn)

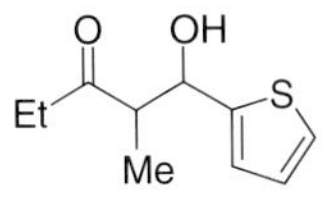

54j
78%, syn/anti = 5.7/1
75% ee (syn)

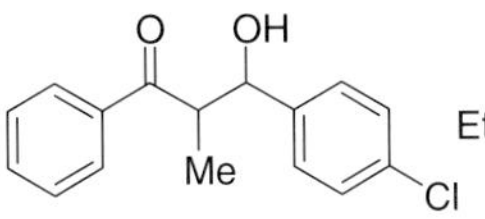

54k
56%, syn/anti = 1.6/1
67% ee (syn)

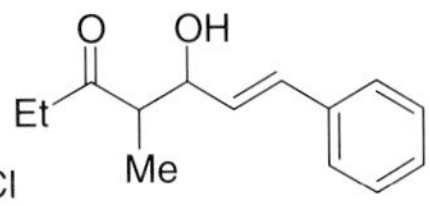

54l
94%, syn/anti = 2.3/1
57% ee (syn)

Subsequent experiments demonstrated that pure H$_2$O, rather than mixtures of H$_2$O and organic solvent could be used as the solvent, either by the use of an additive (Triton-X100, Eq. (13)) or a lipophilic Cu(II) salt in conjunction with a fatty acid additive (Eq. (14)) [22, 23].

$$(13)$$

$$(14)$$

3.4
Additions Involving In-Situ Enolate Formation

A continuing goal of organic chemists is the development of "direct" reactions in which the compounds undergoing reaction are activated in situ. The Mukaiyama aldol reaction, despite its broad utility, is not an example of a direct reaction, because preformation of an enolsilane in a separate step is a necessary requirement. Direct enolization and subsequent aldol reaction have been achieved in a handful of asymmetric Cu(II)-catalyzed reactions.

3.4.1
Pyruvate Ester Dimerization

Additions to pyruvate esters without pre-activation of the nucleophilic reactant have been explored by Jørgenson and co-workers. Ethyl pyruvate is enantioselectively dimerized in the presence of a chiral Cu(II) Lewis acid and catalytic quantities of a trialkylamine base to afford diethyl 2-hydroxy-2-

methyl-4-oxoglutarate, **55** (Eq. (15)) [24]. Formation of the aldol was achieved with good enantiocontrol by use of (*t*-Bu-box)Cu(OTf)$_2$ as catalyst in conjunction with a dialkylaniline base. Subtle interplay between the identity of the solvent, the counter-anion, and base were observed. The initial aldol adduct cyclizes in the presence of base and TBS-Cl to afford a highly substituted γ-lactone **56** in moderate yield and with high enantioselectivity (Eq. (16)). The scope of this reaction beyond use of ethyl pyruvate was not described.

$$(15)$$

>80% conversion, 93% ee

$$(16)$$

48%, 96% ee

3.4.2
Addition of Nitromethane to *α*-Keto Esters

The enantioselective addition of nitroalkanes to carbonyl compounds (Henry reaction) has been documented by Jørgenson and co-workers [25]. With nitromethane as solvent, a combination of the (*t*-Bu-box)Cu(OTf)$_2$ complex **5** (20 mol%) and Et$_3$N (20 mol%) effect room-temperature aldol reactions with a range of *α*-keto esters (Eq. (17)). The reaction is especially enantioselective with alkyl and aromatic groups on the ketone moiety. β,γ-Unsaturated-*α*-keto esters react with nitromethane to afford aldol products in good yield, but enantiocontrol is significantly lower than for aromatic or aliphatic substrates. Product partitioning is completely selective, however, for the 1,2-mode of addition compared with 1,4-conjugate addition. This selectivity is not observed when the reaction is conducted in the absence of (*t*-Bu-box)Cu(OTf)$_2$, leading the authors to propose that the nitronate anion might be coordinated to the metal center during the catalytic reaction. Triethylamine is uniquely suited as catalytic base in this reaction: *N*-

methylmorpholine, dimethylaniline, tribenzylamine, pyridine, ethyl diiso-propylamine, and potassium carbonate are all inferior with regard to both yield and enantiocontrol. Enantioselectivity is optimal for the t-Bu-box li-gand and triflate counter-ion (as compared with hexafluoroantimonate).

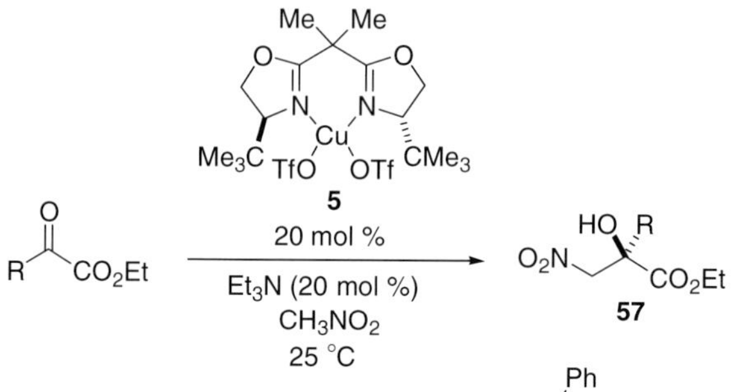

$$ (17) $$

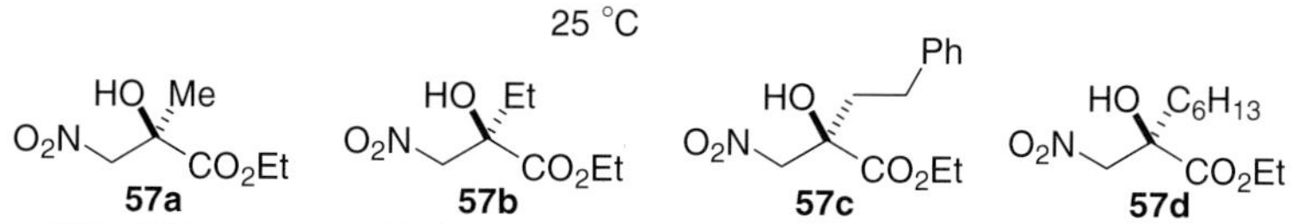

57a
95%, 92% ee

57b
46%, 90% ee

57c
47%, 77% ee

57d
91%, 93% ee

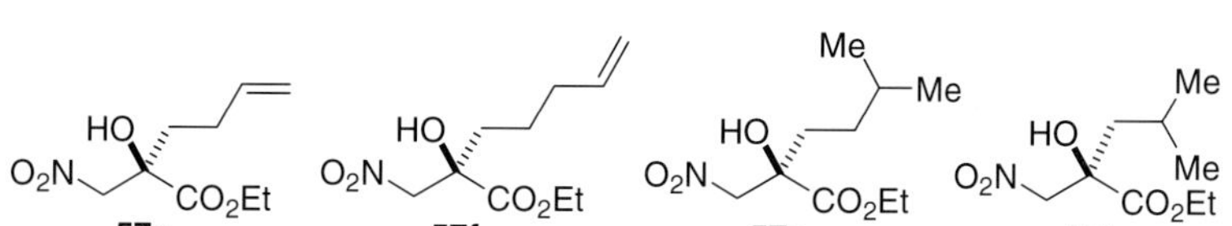

57e
97%, 94% ee

57f
92%, 94% ee

57g
90%, 94% ee

57h
99%, 92% ee

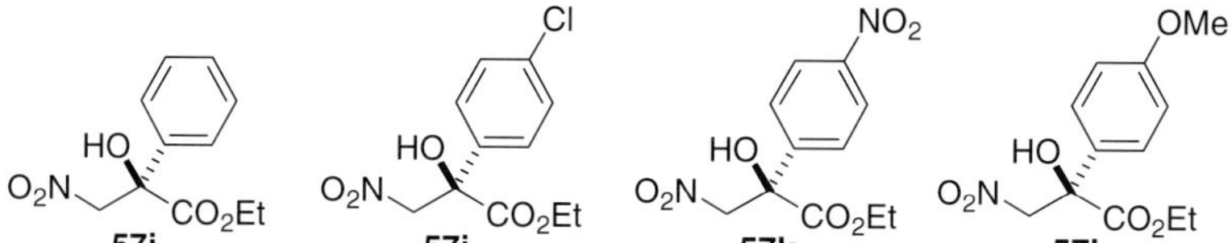

57i
81%, 86% ee

57j
91%, 88% ee

57k
99%, 93% ee

57l
68%, 57% ee

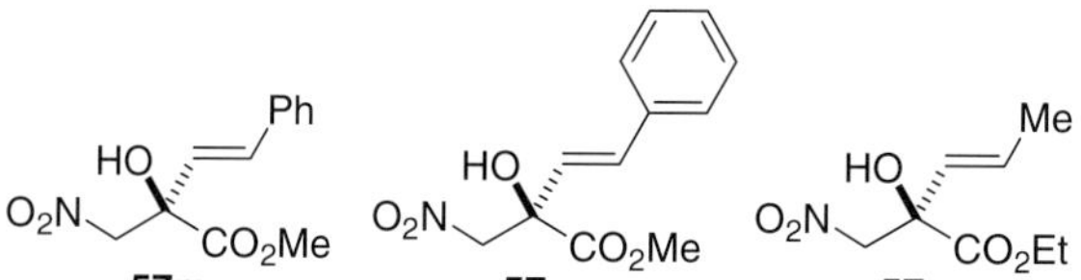

57m
95%, 35% ee

57n
95%, 30% ee

57o
>96%, 60% ee

Reactant stoichiometry is critical in the catalyzed Henry reaction. With a 20% loading of (t-Bu-box)Cu(OTf)$_2$ catalyst optimum enantiocontrol (92% ee for the reaction of nitromethane with ethyl pyruvate) is realized by use of 20 mol% Et$_3$N. In contrast, using 15 mol% Et$_3$N under otherwise iden-tical conditions leads to a product enantiomeric excess of only 56%. Use of 25 mol% Et$_3$N leads to a product with 73% ee. The diminished enantio-selectivity when employing excess base relative to Lewis acid can be accounted for by a racemic pathway – Et$_3$N alone catalyzes non-selective nitro-aldol addition. The reason for reduced selectivity in the presence of excess Lewis acid (relative to base) is less clear.

Another distinctive feature of the (*t*-Bu-box)Cu(OTf)$_2$-catalyzed Henry reaction is reversed π-facial selectivity relative to that normally obtained with this catalyst (vide supra). Indeed, the usual distorted square-planar arrangement (cf. Figures 3.8 and 3.9) cannot correctly account for the stereochemistry observed (the absolute stereochemistry of **57j** was determined by X-ray crystallography). Jørgenson and co-workers instead propose that the ketone carbonyl and the nitronate anion coordinate to the Cu(II) center in the ligand plane, with the carboethoxy group bound in the axial position to complete the square pyramidal complex (Figure 3.11). The Zimmerman–Traxler chair-like transition structure is proposed on the basis of the observation of complete 1,2-selectivity in additions to β,γ-unsaturated-α-keto

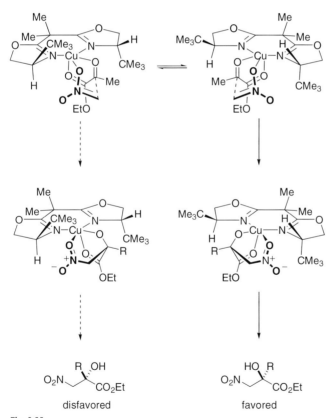

Fig. 3.11
Proposed stereochemical model for
enantioselective Henry reactions catalyzed
by (*t*-Bu-box)Cu(OTf)$_2$.

esters. One possible explanation of this selectivity is that the electropositive metal center directs addition to the carbonyl, rather than the C=C π bond. The strong preference for one of the two diastereomeric transition structures illustrated is less clear, because both require interaction of one reaction component (either α-keto ester or nitronate) with the bulky *tert*-butyl group of the ligand.

3.4.3
Malonic Acid Half Thioester Additions to Aldehydes

Shair and coworkers have developed a mild aldol addition based on the decarboxylative Claisen condensation that is the key step in polyketide biosynthesis [26]. Malonic acid half thioesters (MAHT) are employed as enolate equivalents in addition reactions with aldehydes catalyzed by Cu(2-ethylhexanoate)$_2$ and 5-methoxybenzimidazole. This metal salt and additive were optimized after initial screening experiments that identified Cu(OAc)$_2$ and imidazole as promising leads for the desired transformation.

The decarboxylative aldol addition is performed at ambient temperature under ambient atmosphere in wet solvent to afford moderate to excellent yields of the β-hydroxy thioesters (Figure 3.12). The substrate scope is good with regard to the electrophile. A common problem with direct aldol reactions is self-condensation of enolizable aldehydes; the reaction is not observed in this instance, demonstrating the mildness of the reaction. Good levels of diastereocontrol are observed when α-methyl MAHT are used. These diastereomer ratios are kinetic values, not thermodynamic, as judged by resubmission experiments.

In the absence of aldehyde, no decarboxylation is observed, suggesting that a Cu(II)-thioacetate enolate is not on the reaction pathway. The authors propose that enolization of the MAHT by the amine base might be required for the reaction to proceed. No incorporation of a second aldehyde into an isolated aldol adduct is observed under the catalyzed reaction conditions – retro-aldol reactions do not occur.

General Experimental Procedure

5-Methoxybenzimidazole (0.11 mmol, 0.22 equiv.) and Cu(2-ethylhexanoate)$_2$ (0.1 mmol, 0.2 equiv.) were added to a stirred solution of malonic acid half benzylthioester (0.5 mmol, 1.0 equiv.) in THF (5 mL; solvent stored in a vial without protection from the air before use) at 23 °C. After the reaction became homogeneous (ca. 1 min) aldehyde (0.5 mmol, 1.0 equiv.) was added. The solution was stirred at 23 °C for the prescribed time and quenched with 0.5 M HCl solution. The resulting solution was diluted with EtOAc and washed successively with 0.5 M HCl, saturated aq. NaHCO$_3$, and brine. The organic layer was dried over Na$_2$SO$_4$, filtered through cotton,

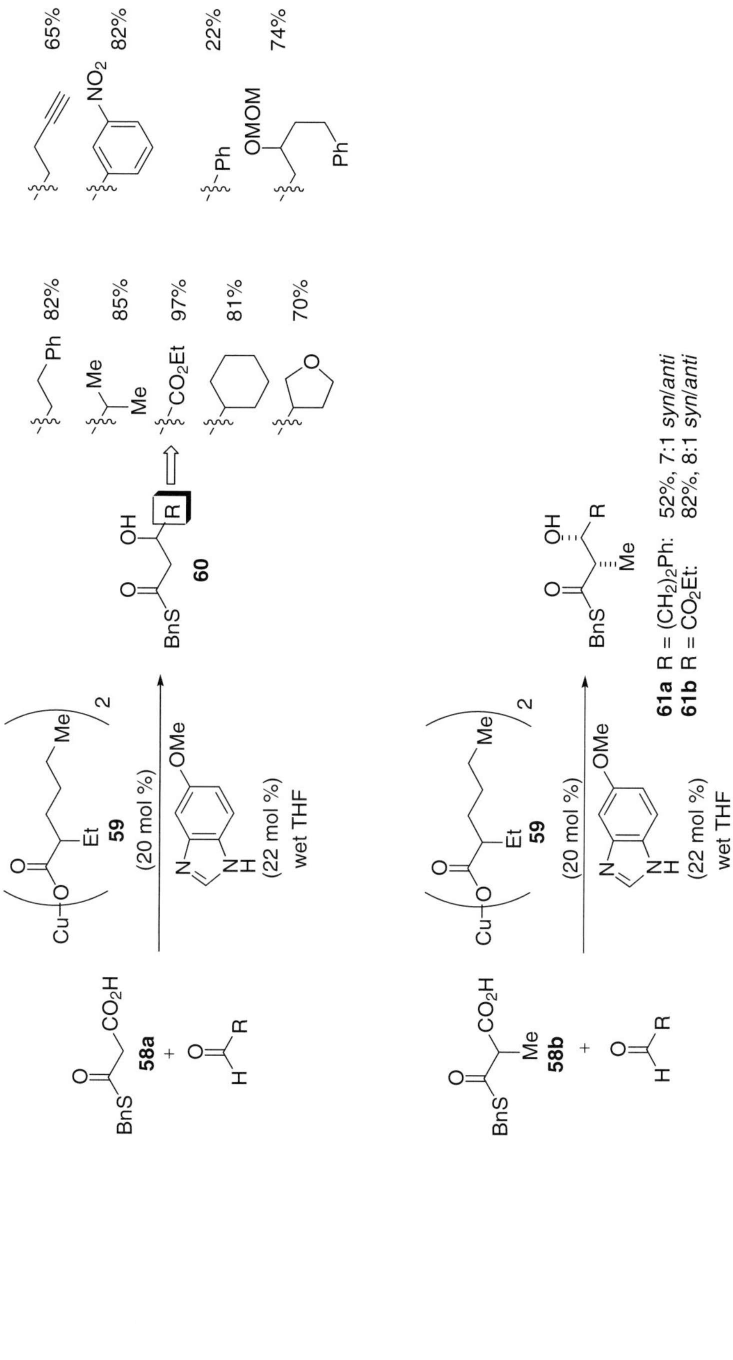

Fig. 3.12
Direct aldol additions of MAHT catalyzed by Cu(II) complexes.

and concentrated under reduced pressure. The product was purified by continuous gradient flash column chromatography.

3.4.4
Dienolate Additions to Aldehydes

3.4.4.1 Scope and Application

Activation of silyl enolates toward aldol additions can be achieved by desilylation. Carreira and coworkers developed this approach in highly enantioselective additions of silyl dienolates to aromatic, heteroaromatic, and α,β-unsaturated aldehydes in the presence of an (S-Tol-BINAP)CuF$_2$ catalyst [27]. The copper fluoride catalyst is generated in situ by treatment of S-Tol-BINAP with Cu(OTf)$_2$, followed by addition of a crystalline, anhydrous fluoride source, (Bu$_4$N)Ph$_3$SiF$_2$ (TBAT). When as little as 2 mol% is used, (S-Tol-BINAP)CuF$_2$ (**63**) catalyzes enantioselective addition of a silyl dioxolanone-derived dienolate to a range of aldehydes at -78 °C (Eq. (18)). Selectivity ranged from good to excellent (83–95% ee) for all substrates except α-methylcinnamaldehyde, for which the level of enantioinduction was somewhat lower (65% ee). Aliphatic aldehydes are also selective electrophiles, but alkylnals suffered from low yields ($< 40\%$).

(18)

a: 92%, 94% ee   b: 86%, 93% ee   c: 98%, 95% ee   d: 91%, 94% ee

e: 93%, 94% ee   f: 83%, 85% ee   g: 82%, 90% ee

h: 48%, 91% ee   i: 81%, 83% ee   j: 74%, 65% ee

(*S*-Tol-BINAP)CuF$_2$-catalyzed addition of silyl dienolates to aldehydes has found synthetic utility in the construction of the polyol subunit of amphotericin B (Scheme 3.9) [28]. Both key fragments of the polyol chain (C$_1$–C$_{13}$) were derived from the same aldol reaction of furfural and the trimethylsilyl dienolate, differing only in the antipode of (Tol-BINAP)CuF$_2$ catalyst used. It is noteworthy that the furfural aldol adduct can be obtained in >99% ee after a single recrystallization of the aldol product.

Chiral copper enolate methodology has also been employed for the total synthesis of leucascandrolide A [29]. Crotonaldehyde reacts with the aforementioned trimethylsilyl dienolate in the presence of 2 mol% (*R*-Tol-BINAP)CuF$_2$ to afford the allyl alcohol adduct in 91% ee and 42% yield (Scheme 3.10). Yields were hampered in this instance because of crotonaldehyde polymerization, as noted by the authors. Further elaboration of this aldol product furnished a highly convergent synthesis of leucascandrolide A.

3.4.4.2 Mechanistic Considerations [30]

The hard fluoride anion is an effective desilylating agent. When mismatched with a soft Cu(II) cation the fluoride anion is designed to serve as an in-situ means of enolate formation. This is achieved by desilylation of the enolsilane, then by metalation. Supporting evidence for copper enolate formation was obtained by independent synthesis – when silyldienolate **62** is subjected to MeLi (10 mol%) followed by (*S*-BINAP)Cu(OTf)$_2$ and benzaldehyde the expected aldol adduct is obtained in good yield and enantioselectivity. Similar results are obtained by employing (Bu$_4$N)Ph$_3$SiF$_2$ as the desilylating reagent.

Mechanistic studies by Carreira and coworkers have concluded that the catalytically active species is the Cu(I) dienolate depicted in Scheme 3.11. Cu(II) complexes are known to undergo a one-electron reduction in the presence of enolsilanes. Thus desilylation of the silyl enolate via the hard fluoride anion of the copper complex followed by copper metalation gives the putative Cu(I) dienolate **66**. The same (*S*-Tol BINAP)Cu(dienolate) has been observed independently via desilylation of the trimethylsilyl dienolate by (Bu$_4$N)Ph$_3$SiF$_2$, followed by treatment with (*S*-Tol-BINAP)Cu(ClO$_4$). Formation of the (*S*-Tol BINAP)Cu(dienolate) species and its disappearance after addition of benzaldehyde was closely monitored by IR spectroscopy (ReactIR). Regeneration of the catalytically active species is achieved by desilylation of another molecule of silyldienolate by the resulting Cu(I) alkoxide complex (**68** → **69** + **66**). Reactivity is also observed in the presence of catalytic quantities of (*S*-Tol-BINAP)Cu(O^{t}Bu), with identical yields and selectivity as with the corresponding Cu(I) or Cu(II) fluoride catalysts, providing corroborating evidence for the proposed mechanism.

General Procedure. A mixture of Cu(OTf)$_2$ (3.6 mg, 0.010 mmol, 2 mol%) and (*S*)-Tol-BINAP (7.5 mg, 0.011 mmol, 2.2 mol%) in 2 mL THF was stirred at 23 °C in an inert gas atmosphere for 10 min to yield a clear yellow

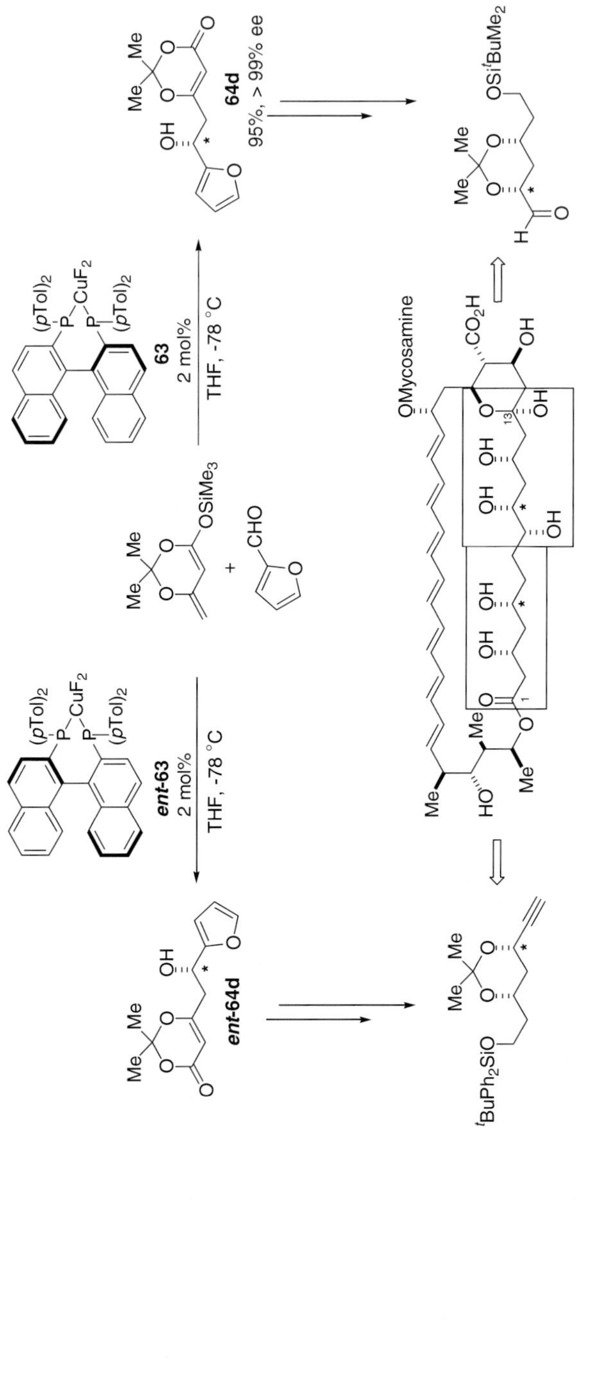

Scheme 3.9
Application of (*S*-Tol-BINAP)CuF$_2$-catalyzed additions of silyl dienolates to the polyol subunits of amphotericin B.

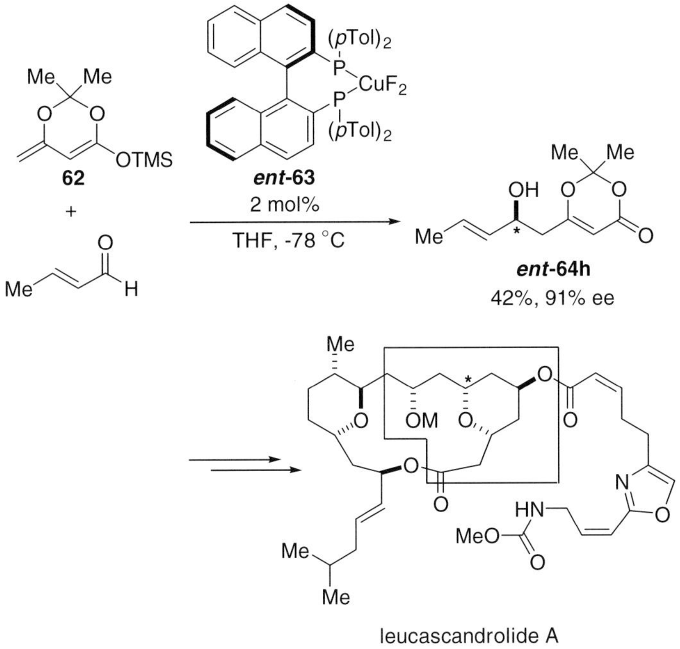

Scheme 3.10
Application of an (*S*-Tol-BINAP)CuF$_2$-
catalyzed addition of a silyl dienolate to the
leucascandrolide A.

solution. A solution of Ph$_3$SiF$_2$(Bu$_4$N) (10.8 mg, 0.02 mmol, 4 mol%) in 0.5 mL THF was added via a cannula and stirring was continued for 10 min. The mixture was cooled to −78 °C and the dioxenone-derived dienolate (0.16 mL, 0.75 mmol) was added dropwise, followed by a solution of the aldehyde (0.50 mmol) in 0.5 mL THF. The progress of the reaction was monitored by TLC (reaction times 0.5–8 h). On completion trifluoroacetic acid (0.2 mL) was added at −78 °C and the solution was left to warm to 23 °C. Stirring was continued an another hour. The reaction mixture was diluted with ether (5 mL) and a saturated aqueous solution of NaHCO$_3$ was added dropwise until evolution of gas ceased. The organic layer was washed with brine (3 mL), dried over Na$_2$SO$_4$, and concentrated in vacuo. Purification of the crude material by chromatography on silica gel with 3:1 ether–hexanes afforded the aldol adduct. The enantiomeric excess of the alcohol products was determined by HPLC analysis using a racemic sample as reference.

3.4.5
Enantioselective Cu(II) Enolate-Catalyzed Vinylogous Aldol Reactions

The traditional challenge associated with vinylogous aldol reactions is competition between reactivity at the α or γ positions of the vinyl enolate. For-

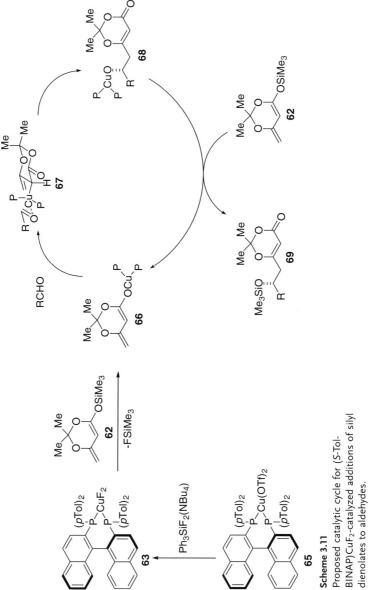

Scheme 3.11

Proposed catalytic cycle for (*S*-Tol-BINAP)CuF$_2$-catalyzed additions of silyl dienolates to aldehydes.

mation of the α-aldolate product has been suppressed by employing bulky Lewis acids such as aluminum tris(2,6-diphenyl)phenoxide (ATPH). Campagne and Bluet also discouraged α-aldolate formation and rendered the vinylogous aldol enantioselective by use of Carreira's catalyst system (Eq. (19)) [31]. In the presence of 10 mol% (S-Tol-BINAP)CuF$_2$ (63), the authors observed only the γ-aldol products 70 in moderate enantioselectivity at ambient temperature. The authors propose that the catalytic cycle is analogous to that of the Carreira system, although no mechanistic studies of this system have yet been reported.

3.5
Conclusions

Copper complexes enable mechanistically diverse and synthetically useful approaches to the synthesis of β-hydroxy ketones. The Cu(II)-catalyzed asymmetric Mukaiyama aldol reaction developed by Evans provides facile stereocontrolled access to a range of aldol adducts derived from chelating electrophiles. Subsequent extension of this system in the context of "green" chemistry, described by Kobayashi, enabled access to aldols from unfunctionalized aldehydes. Recent efforts have focused on the use of copper complexes to effect direct aldol unions by way of in-situ enolization. Under this general mechanistic umbrella, reports of addition of nitromethane to pyruvate esters, malonic half thioester additions to aldehydes, and desilylative dienolate additions have been described. It is both remarkable and exciting that these mechanistically dissimilar reactions are all catalyzed by the same metal. Given that nearly all of the examples from this chapter were reported after 1995 it is reasonable to expect continued interest and development in these fundamental bond constructions.

References

1 M. IWATA, S. EMOTO, *Chem. Lett.* **1974**, 959–960.

2 K. IRIE, K.-i. WATANABE, *Chem. Lett.* **1978**, 539–540.

3 Y. ITO, T. MATSUURA, T. SAEGUSA, *Tetrahedron Lett.* **1985**, *26*, 5781–5784.

4 D. A. EVANS, J. A. MURRY, M. C. KOZLOWSKI, *J. Am. Chem. Soc.* **1996**, *118*, 5814–5815.

5 D. A. EVANS, M. C. KOZLOWSKI, J. A. MURRY, C. S. BURGEY, K. R. CAMPOS, B. T. CONNELL, R. J. STAPLES, *J. Am. Chem. Soc.* **1999**, *121*, 669–685.

6 D. A. EVANS, D. M. FITCH, T. E. SMITH, V. J. CEE, *J. Am. Chem. Soc.* **2000**, *122*, 10033–10046.

7 D. A. EVANS, P. H. CARTER, E. M. CARREIRA, A. B. CHARETTE, J. A. PRUNET, M. LAUTENS, *J. Am. Chem. Soc.* **1999**, *121*, 7540–7552.

8 D. A. EVANS, E. HU, J. D. BURCH, G. JAESCHKE, *J. Am. Chem. Soc.* **2002**, *124*, 5654–5655.

9 H. MATSUNAGA, Y. YAMADA, T. IDE, T. ISHIZUKA, T. KUNIEDA, *Tetrahedron: Asymmetry* **1999**, *10*, 3095–3098.

10 T. K. HOLLIS, B. BOSNICH, *J. Am. Chem. Soc.* **1995**, *117*, 4570–4581.

11 C. GIRARD, H. B. KAGAN, *Angew. Chem. Int. Ed.* **1998**, *37*, 2923–2959.

12 D. A. EVANS, M. C. KOZLOWSKI, C. S. BURGEY, D. W. C. MACMILLAN, *J. Am. Chem. Soc.* **1997**, *119*, 7893–7894.

13 D. A. EVANS, C. S. BURGEY, M. C. KOZLOWSKI, S. W. TREGAY, *J. Am. Chem. Soc.* **1999**, *121*, 686–699.

14 For recent modifications to the bis(oxazoline) ligand and application to the pyruvate addition, see: H. L. VAN LINGEN, J. K. W. VAN DE MORTEL, K. F. W. HEKKING, F. L. VAN DELFT, T. SONKE, F. P. J. T. RUTJES, *Eur. J. Org. Chem.* **2003**, 317–324.

15 R. ROERS, G. L. VERDINE, *Tetrahedron Lett.* **2001**, *42*, 3563–3565.

16 S. ORLANDI, A. MANDOLI, D. PINI, P. SALVADORI, *Angew. Chem. Int. Ed.* **2001**, *40*, 2519–2521.

17 F. REICHEL, X. M. FANG, S. L. YAO, M. RICCI, K. A. JORGENSEN, *Chem. Commun.* **1999**, 1505–1506.

18 P. I. DALKO, L. MOISAN, J. COSSY, *Angew. Chem. Int. Ed.* **2002**, *41*, 625–628.

19 J. S. JOHNSON, D. A. EVANS, *Acc. Chem. Res.* **2000**, *33*, 325–335.

20 S. KOBAYASHI, S. NAGAYAMA, T. BUSUJIMA, *J. Am. Chem. Soc.* **1998**, *120*, 8287–8288.

21 S. KOBAYASHI, S. NAGAYAMA, T. BUSUJIMA, *Tetrahedron* **1999**, *55*, 8739–8746.

22 K. MANABE, S. KOBAYASHI, *Chem. Eur. J.* **2002**, *8*, 4095–4101.

23 S. KOBAYASHI, K. MANABE, *Acc. Chem. Res.* **2002**, *35*, 209–217.

24 K. JUHL, N. GATHERGOOD, K. A. JORGENSEN, *Chem. Commun.* **2000**, 2211–2212.

25 C. CHRISTENSEN, K. JUHL, R. G. HAZELL, K. A. JORGENSEN, *J. Org. Chem.* **2002**, *67*, 4875–4881.

26 G. LALIC, A. D. ALOISE, M. S. SHAIR, *J. Am. Chem. Soc.* **2003**, *125*, 2852–2853.

27 J. KRUEGER, E. M. CARREIRA, *J. Am. Chem. Soc.* **1998**, *120*, 837–838.

28 J. KRUGER, E. M. CARREIRA, *Tetrahedron Lett.* **1998**, *39*, 7013–7016.

29 A. FETTES, E. M. CARREIRA, *Angew. Chem. Int. Ed.* **2002**, *41*, 4098–4101.

30 B. L. PAGENKOPF, J. KRUGER, A. STOJANOVIC, E. M. CARREIRA, *Angew. Chem. Int. Ed.* **1998**, *37*, 3124–3126.

31 G. BLUET, J. M. CAMPAGNE, *J. Org. Chem.* **2001**, *66*, 4293–4298.

4
Tin-promoted Aldol Reactions and their Application to Total Syntheses of Natural Products

Isamu Shiina

4.1
Introduction

Stereoselective aldol reactions are frequently used for synthesis of complicated natural and unnatural oxygenated products, because β-hydroxy carbonyl groups are now easily prepared by several effective methods. Among the three stable valences of tin, the stannic and stannous species are generally used for effective formation of the desired aldol adducts from two starting materials. These tin-promoted reactions are divided into two types according to the principles: (i) directed Mukaiyama aldol reaction of silyl enolates with carbonyl compounds promoted by Sn(IV) or Sn(II) Lewis acids, and (ii) the crossed aldol addition of *C*- or *O*-enolates with Sn(IV) or Sn(II) to other carbonyl components. This review first covers Sn(IV)-mediated aldol reactions of enol silyl ethers (ESE) or ketene silyl acetals (KSA) with carbonyl compounds or acetals, which have been developed as powerful tools for stereoselective synthesis of β-hydroxy or β-alkoxy carbonyl groups. Chiral diamine–Sn(II) complex-promoted aldol and related addition reactions for preparation of a variety of optically active polyoxy compounds will be the second subject discussed. Finally, recent applications of the reactions to highly enantioselective syntheses of optically active natural products will be described.

4.2
Tin-promoted Intermolecular Aldol Reactions

4.2.1
Achiral Aldol Reactions

In 1973, Mukaiyama and Narasaka developed an acid-catalyzed aldol reaction of silyl enolates with electrophiles and revealed that Lewis acids such as $TiCl_4$, $SnCl_4$, $AlCl_3$, $BF_3 \cdot OEt_2$, and $ZnCl_2$ promoted the reaction quite

Modern Aldol Reactions. Vol. 2: Metal Catalysis. Edited by Rainer Mahrwald
Copyright © 2004 WILEY-VCH Verlag GmbH & Co. KGaA, Weinheim
ISBN: 3-527-30714-1

effectively, affording a variety of β-hydroxy ketones from ESE and carbonyl compounds (Eq. (1)) [1]. The synthetic capacity of KSA in the new aldol reaction was also reported in 1975, and the corresponding β-hydroxy- and β-siloxycarboxylic esters were obtained in good combined yields by use of $TiCl_4$ (Eq. (2)).

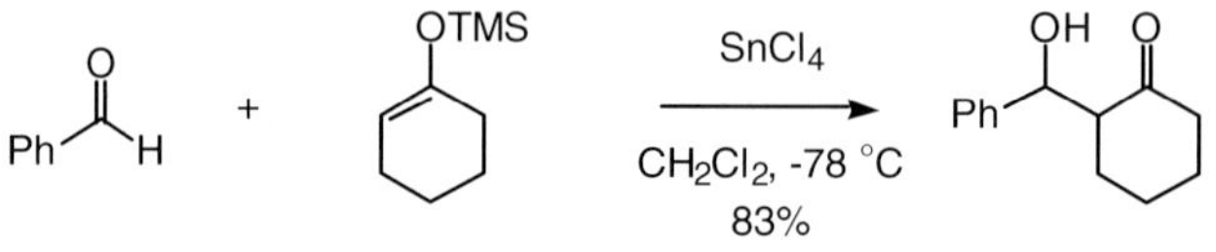

(1)

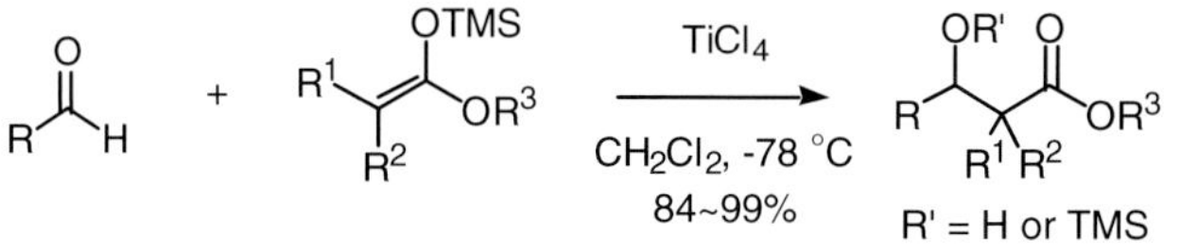

(2)

Although it is mentioned in their reports that $TiCl_4$ seems to be superior to other Lewis acids with regard to yield, $SnCl_4$ was also a popular reagent because of its mild activity and good chelation ability. For example, Wissner applied the $SnCl_4$-mediated aldol reaction of tris(trimethylsiloxy)ethene with several aldehydes to the synthesis of α,β-dihydroxycarboxylic acids (Eq. (3)) [2] and Ricci and Taddai prepared a bicyclic γ-lactone in good yield by aldol addition of 2,5-disiloxyfuran to two molar amounts of acetone using $SnCl_4$ (Eq. (4)) [3].

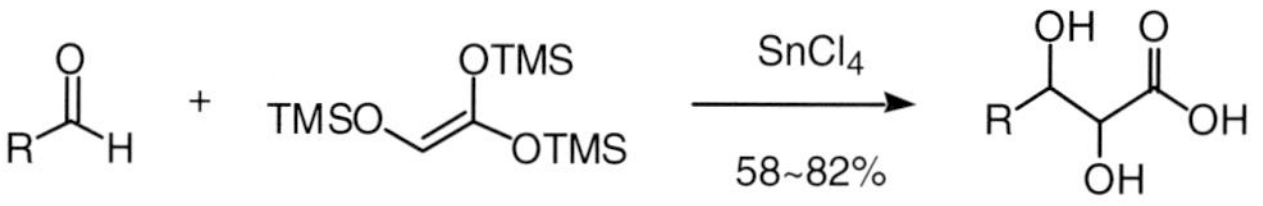

(3)

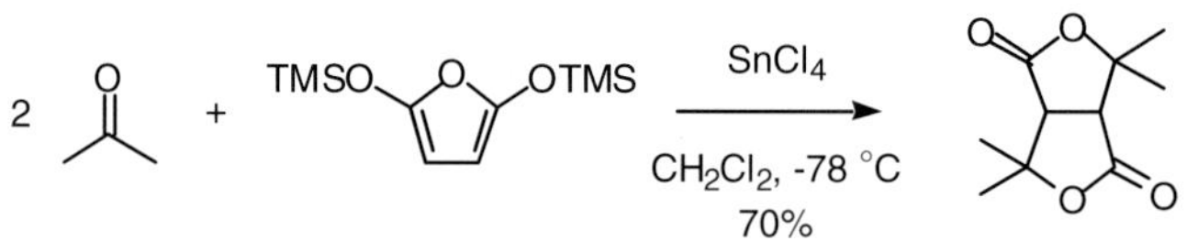

(4)

In 1983, Kuwajima and Nakamura reported a novel method for generation of α-stannylketones from ESE and $SnCl_4$ and studied the properties of the new metallic species in the reaction with carbonyl compounds giving aldol adducts (Eq. (5)) [4]. This facile method for preparing trichlorostannyl enolates was successfully employed in the regioselective synthesis of aldols, as shown in Eq. (6) [5].

Interestingly, *syn* selectivity was observed in this alternative method, in contrast with the *anti* selectivity obtained in the direct $SnCl_4$-promoted aldol reaction of ESE with the electrophiles (Eqs. (5) and (7)) [1]. Therefore, different mechanisms were proposed for these reactions, and it was assumed that the silyl nucleophiles could directly attack the carbonyl compounds activated by the Lewis acid at low temperature.

Structural features and reactivity of the Sn(IV) C- or O-enolates have been investigated [6, 7]. Yamamoto and Stille independently studied the aldol reaction of stannyl enolates derived from ketones with aldehydes, and showed that stereoselectivity depended on the substituents on the tin and the reaction temperature (Eqs. (8) and (9)) [8, 9].

Mukaiyama and Iwasawa developed a facile method for the generation of Sn(II) enolates in situ from the corresponding carbonyl compounds with Sn(OTf)$_2$ and a tertiary amine (Eqs. (10) and (11)) [10], and excellent *syn* selectivity of the aldol reaction was observed in the course of their studies of Sn(II) enolate chemistry (described in a later section).

$$(10)$$

$$(11)$$

4.2.2
The Reaction of Silyl Enolates with Aldehydes or Ketones

Diastereoselective addition of ESE and KSA to aldehydes using SnCl$_4$ were systematically studied by Heathcock, Reetz, and Gennari, who produced a variety of synthetic intermediates. As shown in Eqs. (12)–(15), Heathcock and Reetz independently examined the stereoselectivity of the Mukaiyama aldol reaction of ESE with many kinds of aldehyde, promoted by SnCl$_4$ [11, 12]. Their results can be summarized:

1. good 2,3-*anti* or 2,3-*syn* asymmetric induction was observed in the reaction between achiral simple or α-heteroatom-substituted aliphatic aldehydes and ESE derived from ethyl ketones (Eqs. (12) and (13));

$$(12)$$

$$(13)$$

2. high 3,4-*syn* asymmetric induction was observed in the reaction between α-heteroatom-substituted aliphatic aldehydes and ESE derived from methyl ketones (Eq. (14)); and

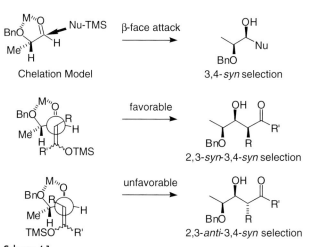

3. good 2,3-*syn* and high 3,4-*syn* asymmetric induction was observed in the reaction between α-heteroatom-substituted aliphatic aldehydes and ESE derived from ethyl ketones (Eq. (15)).

The observed excellent 3,4-*syn* selectivity for the α-branched aldehyde was explained by the formation of a Lewis acid–aldehyde complex (so-called chelation model, Scheme 4.1). Good stereoselectivity was not achieved, however, when KSA was employed, even in the reaction with α-benzyloxypropionaldehyde, except when tetrasubstituted KSA were used (Eqs. (16)–(18)).

Scheme 4.1
Chelation model for producing 3,4-*syn* aldols.

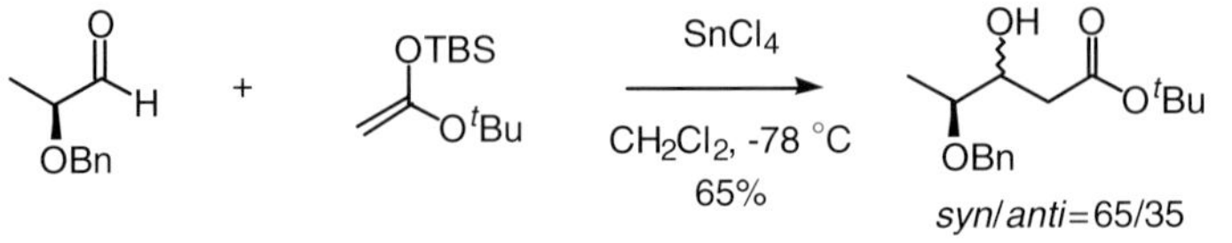

SnCl4
CH2Cl2, -78 °C
65%
syn/anti=65/35

(16)

SnCl4
CH2Cl2, -78 °C
>85%
syn/anti=>97/3

(17)

SnCl4
CH2Cl2, -78 °C
93%
2,3-syn/anti=5/95
3,4-syn/anti=>99/1

(18)

Gennari further studied the Mukaiyama aldol addition of KSA to aldehydes and found that *S*-^{t}Bu propanethioate or ethanethioate is a quite suitable precursor of the required KSA for stereoselective reactions [13]. Although the reaction of KSA derived from *S*-^{t}Bu propanethioate with simple achiral aliphatic aldehydes gave poor 2,3-diastereoselectivity (Eq. (19)), it reacted with α-alkoxy aldehydes highly stereoselectively to afford 2,3-*syn*-3,4-*syn* isomers as shown in Eq. (20).

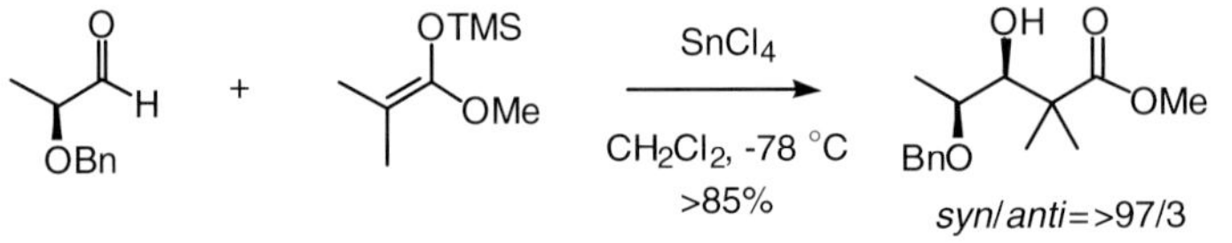

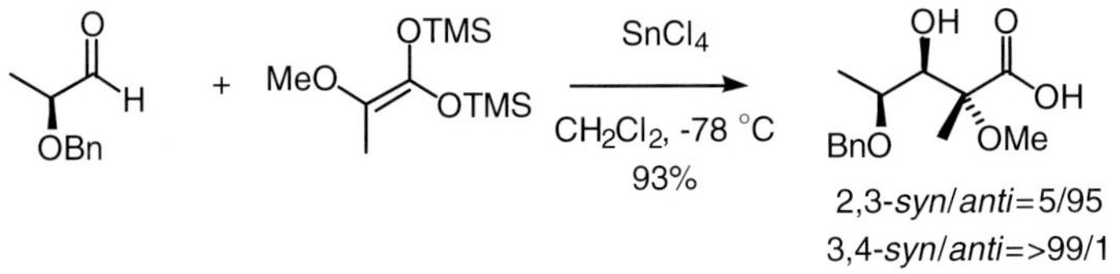

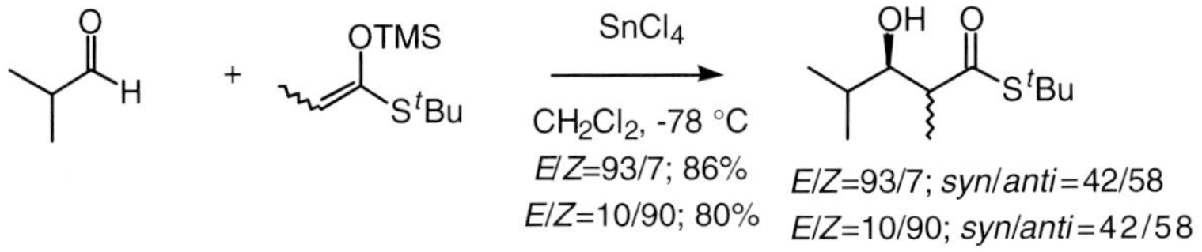

SnCl4
CH2Cl2, -78 °C
E/Z=93/7; 86%
E/Z=10/90; 80%
E/Z=93/7; syn/anti=42/58
E/Z=10/90; syn/anti=42/58

(19)

SnCl4
CH2Cl2, -78 °C
E/Z=>95/5; 89%
E/Z=5/>95; 90%
E/Z=>95/5; 2,3-syn/anti=97/3
3,4-syn exclusively
E/Z=5/>95; 2,3-syn/anti=76/24
3,4-syn exclusively

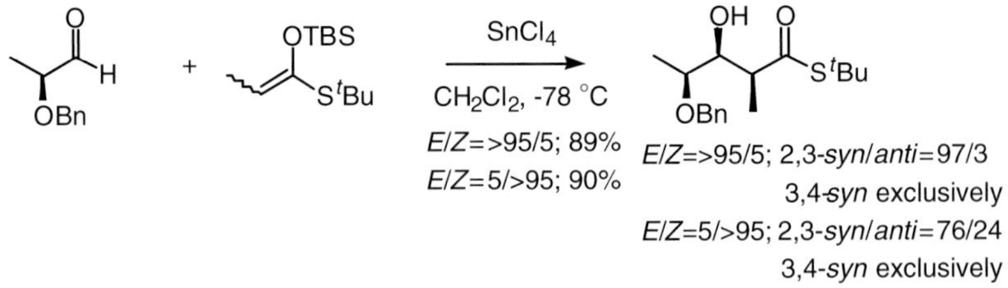

(20)

Furthermore, KSA derived from *S*-^{t}Bu ethanethioate also gave a 3,4-*syn* adduct preferentially in good yield (Eq. (21)). The highly 3,4-*syn* asymmetric induction by chelation control using $SnCl_4$ is quite effective for the construction of natural complex molecules (as is described at the end of this section).

$$\text{(21)}$$

It is worthy of note that α-stannylthioesters, instantly generated from KSA by treatment with $SnCl_4$, also react with aldehydes to afford the corresponding aldol adducts, but the stereoselectivity of this reaction is sometimes very different from that in the reaction involving KSA and an $SnCl_4$–aldehyde complex (compare Eqs. (19) and (22), and Eqs. (20) and (23)). The order of addition of KSA and aldehydes to a solution of Lewis acid catalysts such as $SnCl_4$ should therefore be selected carefully in accordance with the stereoselectivity desired.

$$\text{(22)}$$

$$\text{(23)}$$

Further examples of the construction of multi-functional β-hydroxy carbonyl compounds using $SnCl_4$ are given in Eqs. (24)–(28). Although sulfur-substituted ESE derived from methylthioacetone reacted with α-alkoxy aldehyde to give a 3,4-*syn* adduct preferentially (Eq. (24)) [14], the reaction of KSA prepared from methyl α-methylthiopropionate afforded an almost equimolar mixture of the corresponding 3,4-*syn* and *anti* diol groups (Eq. (25)) [15].

A 2,3-*anti*-3,4-*syn* α-amino carboxylic ester was stereoselectively prepared by reaction of amino-substituted KSA with α-benzyloxypropionaldehyde (Eq. (26)), and the adduct was converted to the corresponding γ-lactone which is known as a synthetic intermediate of L-daunosamine and L-vancosamine [16].

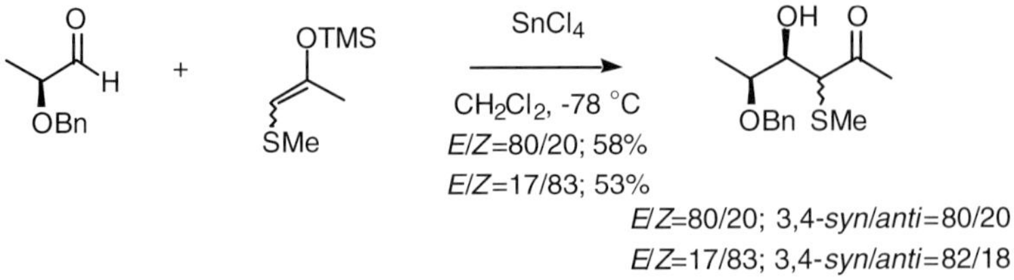

OTMS
SMe
SnCl4
CH2Cl2, -78 °C
E/Z=80/20; 58%
E/Z=17/83; 53%
OH O
OBn SMe
E/Z=80/20; 3,4-syn/anti=80/20
E/Z=17/83; 3,4-syn/anti=82/18

(24)

MeS
OTMS
OMe
E/Z=25/75
SnCl4
CH2Cl2, -78 °C
i) NaIO4
ii) Δ
65%, syn/anti=52/48

(25)

Bn2N
OTMS
OtBu
SnCl4
CH2Cl2
-78 to -40 °C
50%
2,3-syn/anti=16/84
3,4-syn/anti=>98/2
a known intermediate in the
synthesis of aminosugars
L-Daunosamine
L-Vancosamine

(26)

It is notable that the reaction of KSA derived from ethyl acetate with an α-amino aldehyde in the presence of $SnCl_4$ gave a 3,4-*syn* amino alcohol under chelation-control conditions (Eq. (27)) [17] whereas the opposite diastereoselectivity was observed when an α-phenylthio aldehyde was used as the electrophile (Eq. (28)) [18] and a 3,4-*syn* aldol adduct was obtained when $TiCl_4$ was employed for the latter reaction instead of $SnCl_4$. The 3,4-*anti* selectivity in the reaction promoted by $SnCl_4$ was therefore explained, as an exception, by the non-chelation model.

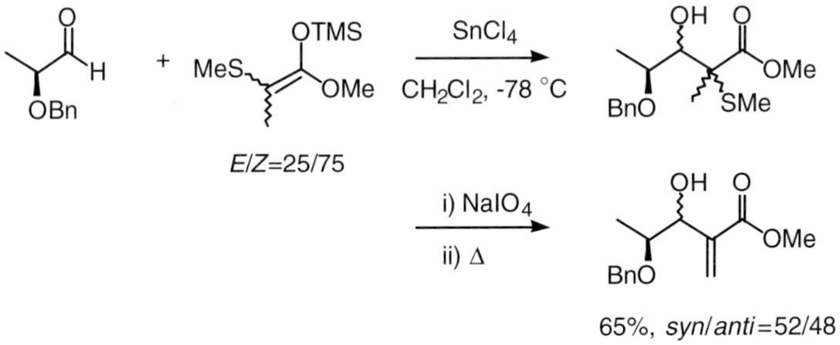

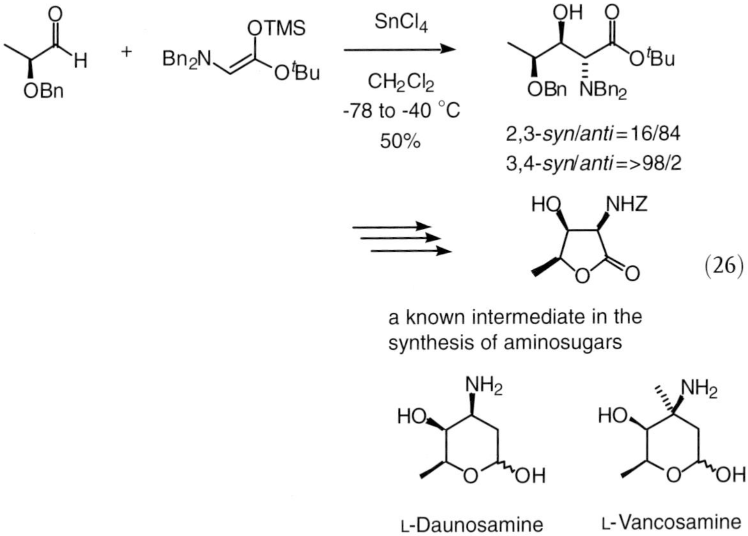

$$(27)$$

$$(28)$$

The SnCl$_4$-promoted diastereoselective aldol reaction has been applied to the synthesis of some parts of complex molecules such as oligopeptides, oligosugars, and polyoxyamides (Eqs. (29)–(36)). Joullié obtained a 2,3-*syn*-3,4-*syn* thioester by reaction of KSA derived from *S*-^{t}Bu propanethioate with an α-alkoxy aldehyde, as shown in Eq. (29) [19]. The prepared intermediate was successfully converted to the macrocyclic peptides didemnin A, B, and C in 1990.

$$(29)$$

Danishefsky employed a trisubstituted butadiene for reaction with an aldehyde connected to a ribonucleoside, with promotion by SnCl$_4$ (Eq. (30)); subsequent desilylation of the adduct afforded the corresponding dihydropyran which was transformed to tunicamycins [20].

Cox and Gallagher employed a cyclic ESE as a nucleophile for reaction with ribosyl aldehyde, as depicted in Eq. (31); the aldol isomers formed could be used as precursors of tetracyclic hemiketals [21]. Akiyama and

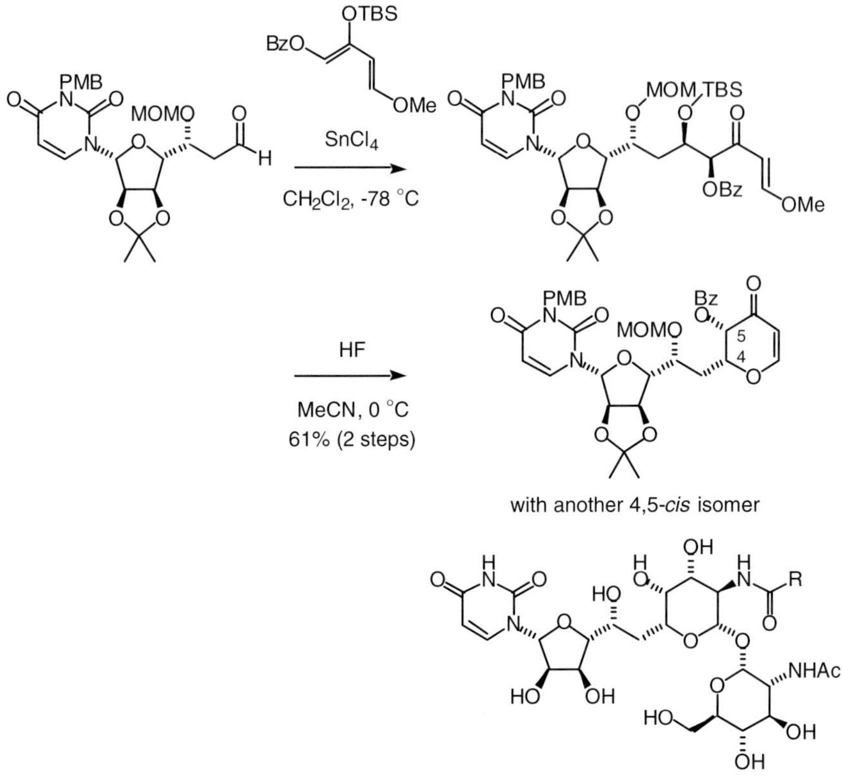

Tunicamycins, R = unsaturated alkyl chains

(30)

Ozaki developed a new chiral auxiliary group in the diastereoselective synthesis of optically active aldols with a tertiary hydroxyl group (Eq. (32)) [22]. For example, (R)-dimethyl citramalate was synthesized from an adduct produced by the SnCl$_4$-accelerated aldol reaction of KSA derived from ethyl acetate with the chiral pyruvate.

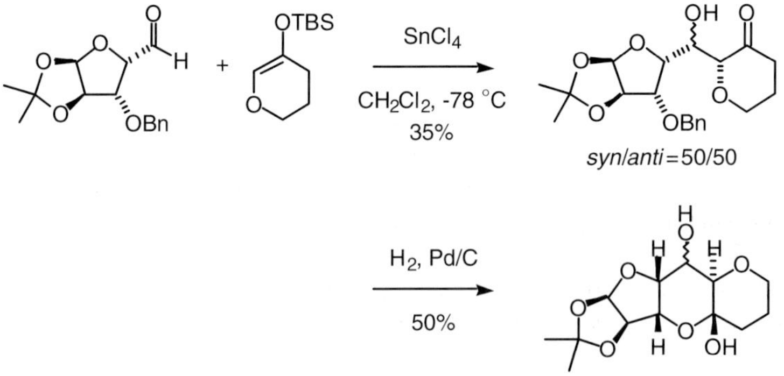

syn/anti = 50/50

(31)

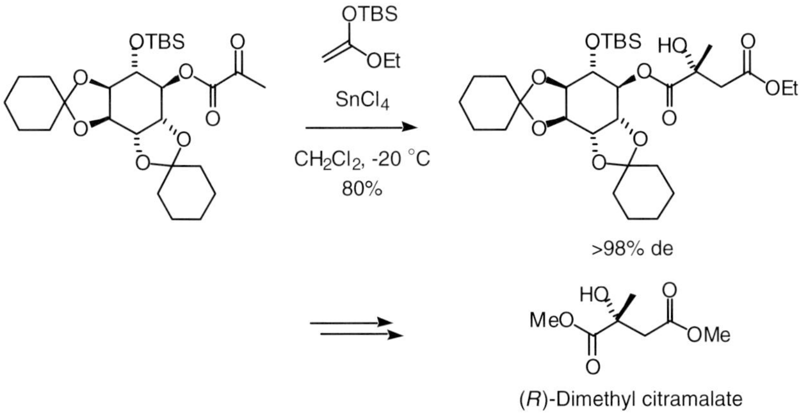

(R)-Dimethyl citramalate

(32)

Mukai and Hanaoka reported the formal synthesis of AI-77B in which they successfully used the stereoselective direct aldol reaction of KSA derived from S-^{t}Bu ethanethioate with an α-alkoxy aldehyde (Eq. (33)) [23].

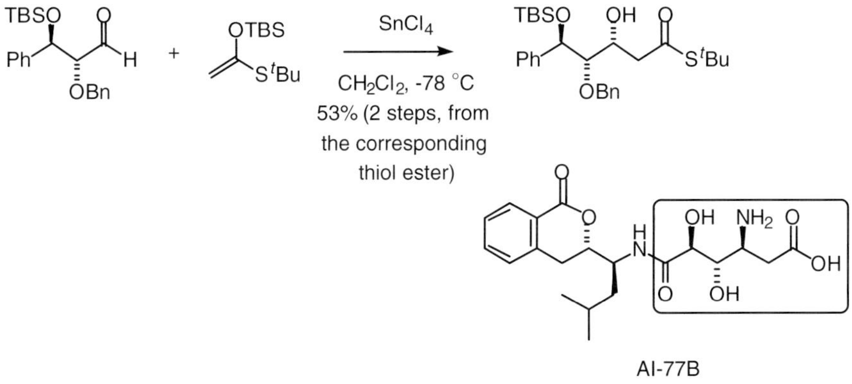

AI-77B

(33)

They also used an α-trichlorostannyl thioester generated from KSA with $SnCl_4$ in a reaction with an α-alkoxy aldehyde to produce a 2,3-*anti*-3,4-*syn* aldol group which was employed as an intermediate in the total synthesis of bengamide E, as shown in Eq. (34) [24]. Recently, Boeckman re-applied this methodology for the practical preparation of bengamide B, E, and Z (Eq. (35)) [25].

Mukai and Hanaoka also used $SnCl_4$ as catalyst in the reaction of the KSA derived from S-^{t}Bu ethanethioate with an α,β-dibenzyloxy aldehyde possessing a Co complex part, to produce a new synthetic intermediate of

2,3-*syn*/*anti*=8/92
3,4-*syn* exclusively

Bengamide E   (34)

R = Bn; 2,3-*syn*/*anti*=8/92
R = 2-naphthylmethyl; 2,3-*syn*/*anti*=11/89

Bengamide B, E, Z   (35)

bengamide E under direct Mukaiyama aldol-reaction conditions (Eq. (36)) [26]. In contrast, non-metalated α,β-dibenzyloxy aldehyde stereorandomly reacted with the same KSA to give a mixture of isomers, and a similar result was also observed in Liu's recent research (Eq. (37)) [27].

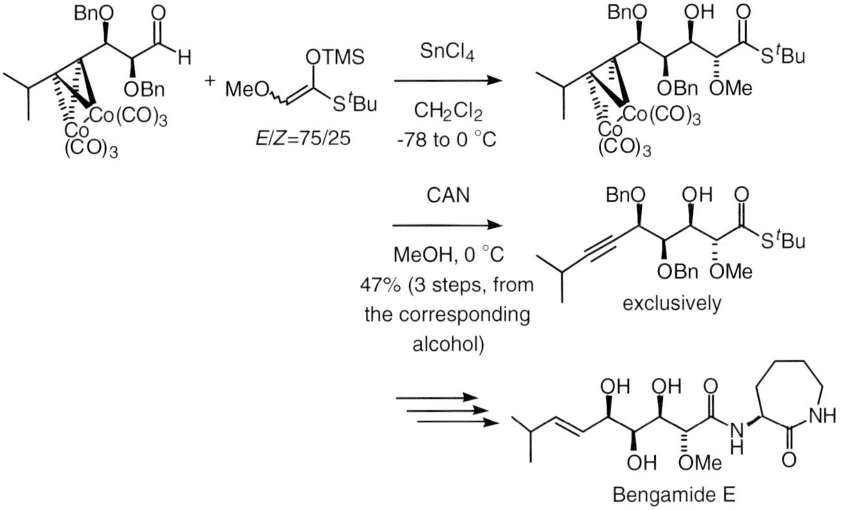

Bengamide E

(36)

no selectivity

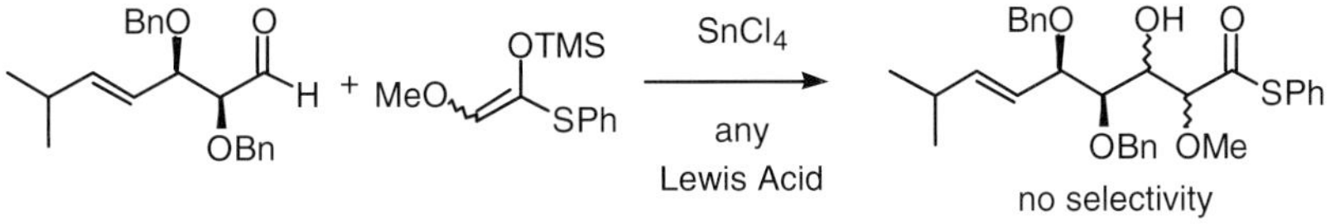

(37)

4.2.3
The Reaction of Silyl Enolates with Acetals

Mukaiyama reported that acetals are also activated by Lewis-acid catalysts and are effectively coupled with nucleophilic ESE and KSA to give the desired β-alkoxy carbonyl compounds in high yields (Eq. (38)) [28]. Kuwajima showed that trichlorostannyl C- or O-enolates could be used as nucleophiles in reactions with acetals and aldehydes (Eq. (39)) [5].

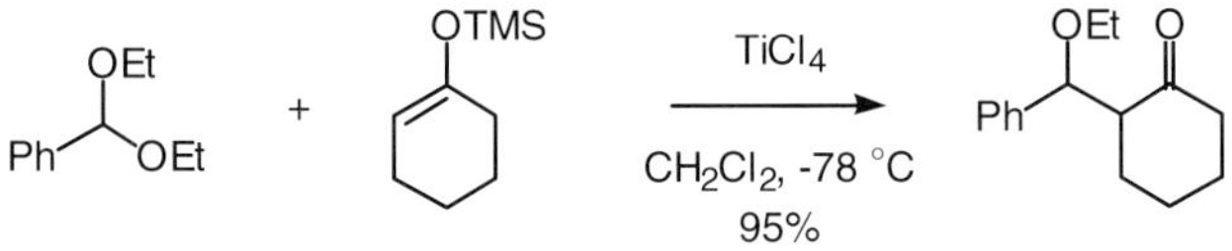

(38)

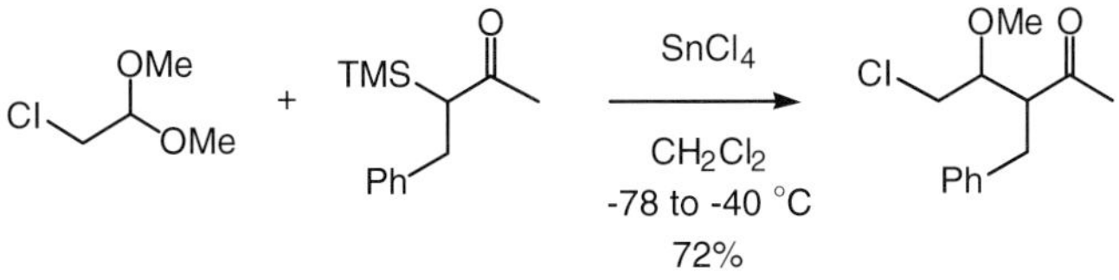

(39)

Some examples of reactions using ESE and KSA with acetals to produce the corresponding *β*-alkoxy carbonyl compounds are given in Eqs. (40)–(43) [29–32]. This method is useful for the synthesis of protected aldols directly from the silyl enolates and acetals.

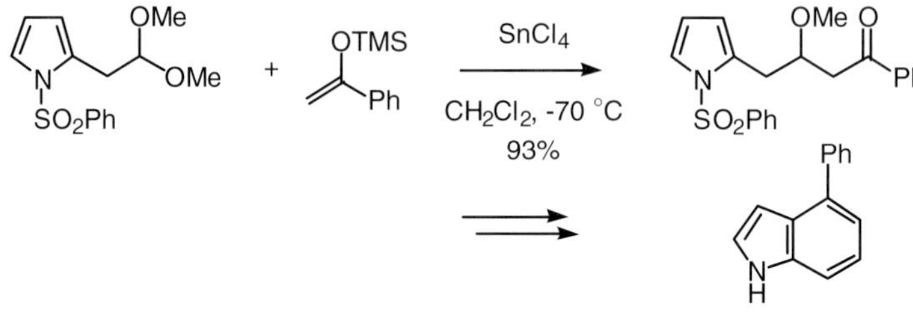

(40)

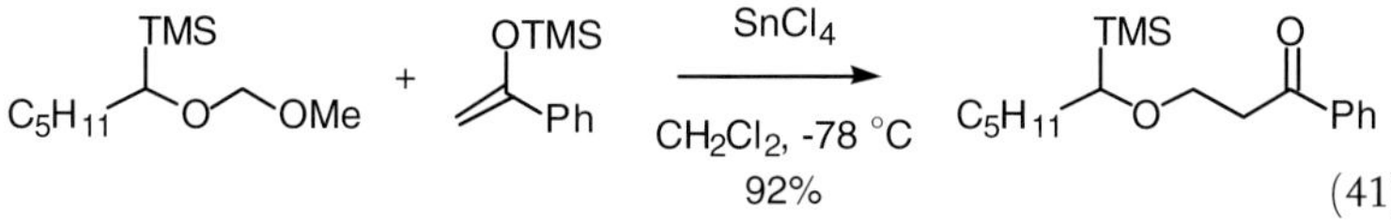

(41)

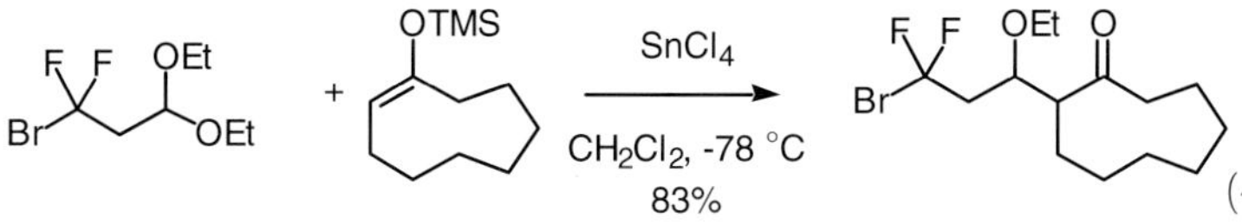

(42)

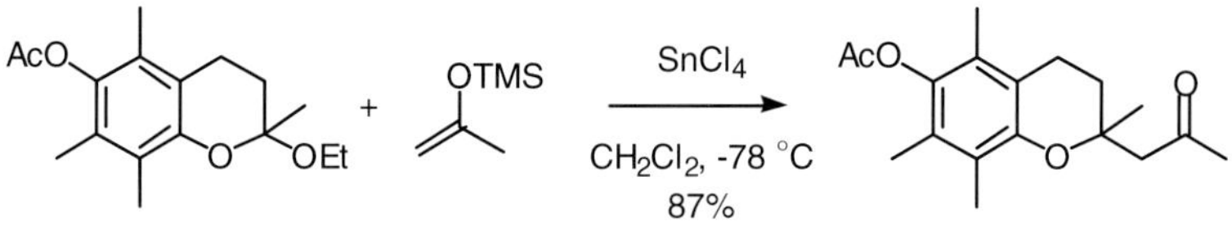

(43)

Although little systematic research has been performed on the stereoselectivity of this reaction, some approaches to the use of chiral acetal parts for asymmetric synthesis have been reported. Kishi successfully accomplished the preparation of an optically active synthetic intermediate of aklavinone by reaction of α-stannylketone with a chiral aromatic aldehyde acetal as depicted in Eq. (44) [33]. A similar achiral acetal was also used for synthesis of racemic 4-demethoxydaunomycinone by Rutledge in 1986 (Eq. (45)) [34].

Rutledge's and Yamamoto's approaches are noteworthy for the stereoselective synthesis of optically active *β*-alkoxy carbonyl compounds in $SnCl_4$-promoted aldol reactions (Eqs. (46) and (47)) [35, 36].

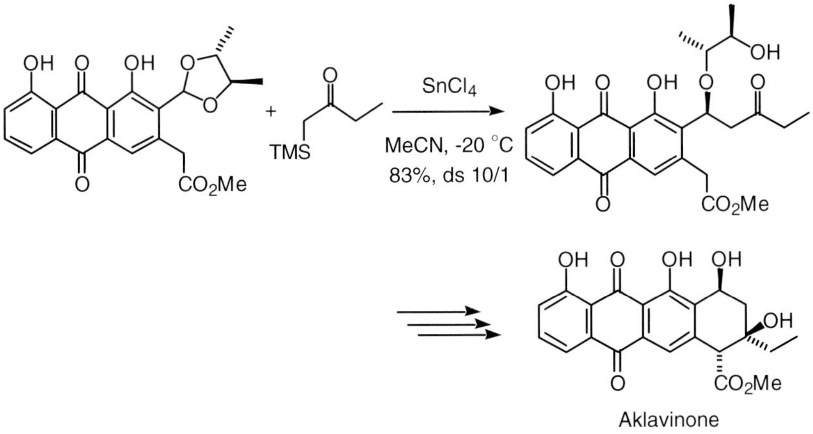

Aklavinone

(44)

rac-4-demethoxydaunomycinone

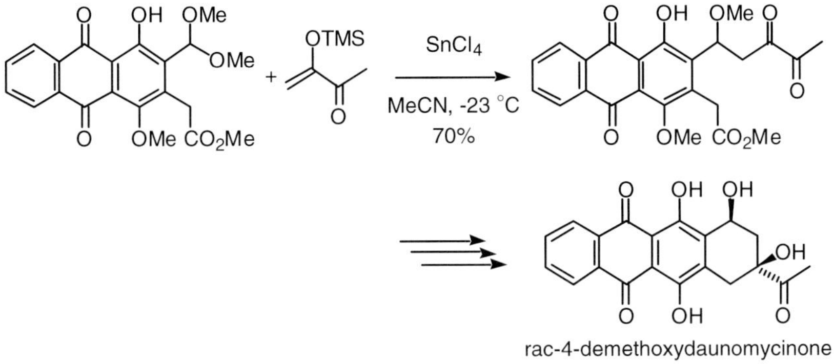

(45)

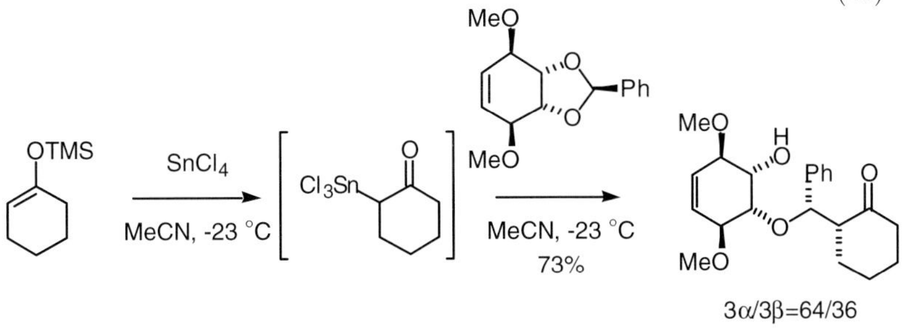

$3\alpha/3\beta=64/36$

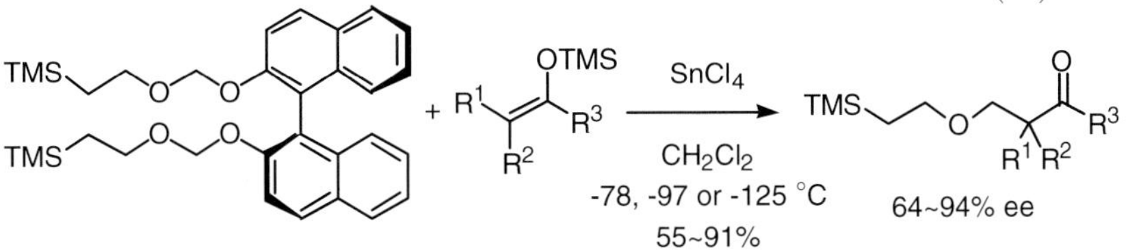

(46)

$64\sim94\%$ ee

$55\sim91\%$

(47)

Otera discovered that Sn(IV) Lewis acids such as $Bu_2Sn(OTf)_2$, $Bu_2Sn(ClO_4)_2$, Bu_3SnClO_4, and $(C_6F_5)_2SnBr_2$ are suitable activators for the aldol reaction of ESE and KSA with aldehydes [37]. Interestingly, it was proved by elaborate research that reactions of ESE with ketones and of KSA with acetals do not proceed under the influence of these catalysts. It was also shown that α,β-unsaturated aldehydes are much more reactive than aromatic and saturated aliphatic aldehydes. They expanded this concept to "parallel recognition", in which some reaction patterns proceed exclusively, affording the desired adducts only, on treatment of several different nucleophiles and electrophiles in one pot. When an ambient electrophile is used in this new strategy ESE and KSA, respectively, react with one of the electrophilic points in the substrate to produce a single adduct in high yield with perfect selectivity (Eqs. (48) and (49)).

$$(48)$$

$$(49)$$

4.2.4
Reaction of Dienol Silyl Ethers

Siloxyheteroaromatic compounds function as nucleophilic dienol silyl ethers with carbonyl compounds or acetals under the influence of a Lewis acid catalyst. For example, Takei showed that 2-siloxyfurans react with aldehydes and acetals to give the corresponding γ-substituted γ-lactones in high yields (Eqs. (50) and (51)) [38].

$$(50)$$

$$(51)$$

2-Siloxypyrrole has also been used as a suitable nucleophile in the $SnCl_4$-promoted aldol reaction with aldehydes shown in Eq. (52). Several polyoxy α-amino acids and carbocyclic amines were synthesized by Rassu and Casiraghi using a stereoselective aldol reaction of 2-siloxypyrrole with protected glyceraldehyde in the presence of $SnCl_4$ [39].

$$(52)$$

Baldwin successfully used the reaction of functionalized 2-siloxypyrrole with 2-methylpropionaldehyde for synthesis of an intermediate of lactacystin, a natural γ-lactam (Eq. (53)) [40].

$$(53)$$

Lactacystin

4.3
Tin-promoted Intramolecular Aldol Reactions

4.3.1
The Intramolecular Aldol Reaction of Silyl Enolates

Not only intermolecular additions of ESE and KSA to acetals, but intramolecular reactions of acetals with a silyl enolate moiety are also quite effective, especially for synthesis of strained cyclic compounds. Kocienski fully studied the intramolecular reaction of ESE to give medium-sized compounds (Eq. (54)), and this method was even found to be applicable to the synthesis of an eight-membered carbocycle (Eq. (55)) [41]. Paquette also succeeded in preparing a bicyclic eight-membered ring compound as shown in Eq. (56) [42].

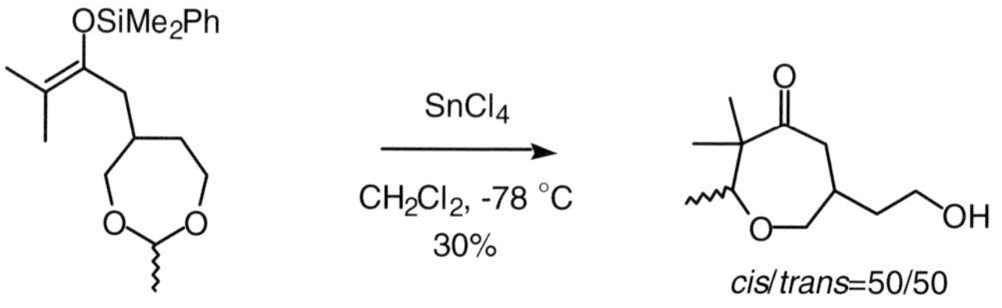

(54)

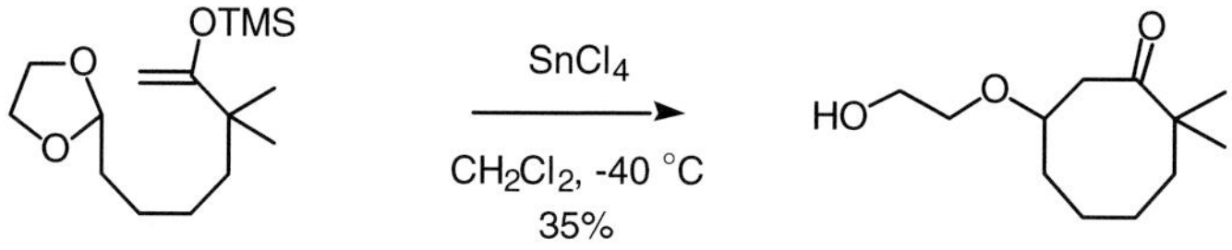

cis/trans=50/50

(55)

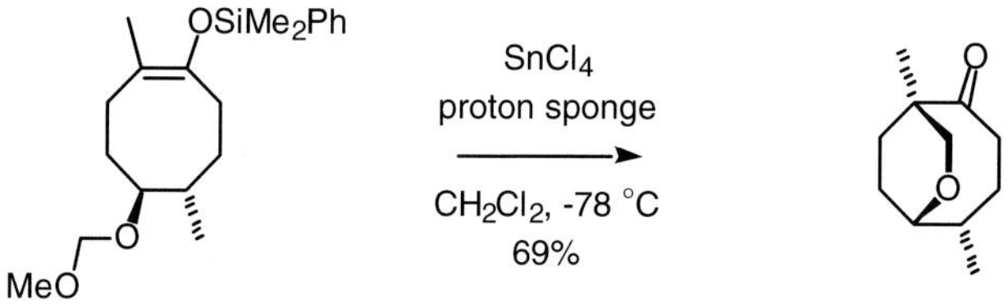

(56)

Tatsuta recently reported a total synthesis of pyralomicin 1c in which SnCl$_4$-catalyzed cyclization was effectively employed for formation of a 6-membered ring core as shown in Eq. (57) [43].

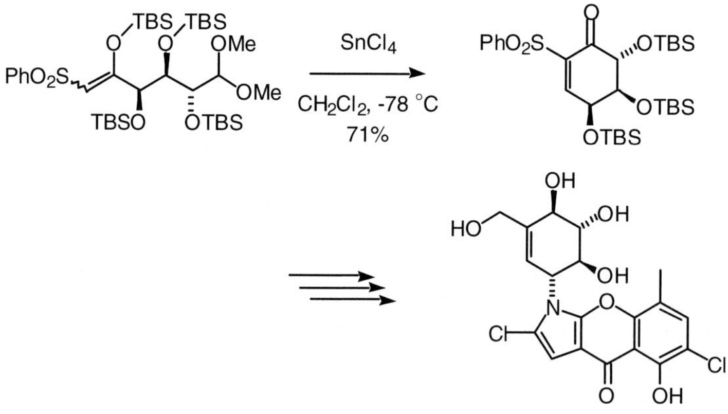

(57)

Pyralomicin 1c

[1,3] rearrangement of ESE with anomeric carbon was developed by Ley in 1998, and several 6- and 5-membered C-glycosides have been prepared by $SnCl_4$ acceleration (Eqs. (58) and (59)) [44].

$$C_6H_{13} \cdots \text{(structure)} \xrightarrow[\text{79~86\%}]{\begin{array}{c} SnCl_4 \\ CH_2Cl_2, -30\,^\circ C \end{array}} C_6H_{13} \cdots \text{(structure)} \quad (58)$$

R = Ph, C≡CPh, ^tBu

α/β=75/25

$$\text{(structure, OTMS)} \xrightarrow[\substack{CH_2Cl_2 \\ 56\%}]{SnCl_4} HO \text{(structure)} C_7H_{15} \quad (59)$$

4.3.2
Reaction of Dienol Silyl Ethers or γ-Silyl-α,β-enones

In 1986, Kuwajima established a method for generation of γ-stannyl-α,β-enones as nucleophilic species. It was found that these reactions with acetals proceeded smoothly to afford the corresponding coupling products in good yields (Eq. (60)) [45].

$$\text{TMS} \text{(structure)} \xrightarrow[CH_2Cl_2, 0\,^\circ C]{SnCl_4} \left[Cl_3Sn \text{(structure)} \right] \quad (60)$$

γ-stannylenone

$$\text{(structure, OMe/OMe)} \xrightarrow[\substack{CH_2Cl_2, 0\,^\circ C \\ 71\%}]{\text{γ-stannylenone}} \text{(structure, OMe)}$$

Remarkable results from $SnCl_4$-induced aldol cyclization using dienol silyl ethers have been observed in synthetic studies on taxane diterpenoids [46]. In the initial approach for construction of the basic skeleton of a taxane ring using a γ-stannyl-α,β-enone generated from a γ-silyl-α,β-enone by Si–Sn metal exchange, the yield of the desired tricyclic compound was unsatisfactory (Eq. (61)). Cyclization of acetals with a dienol silyl ether moiety promoted by $SnCl_4$ occurred rapidly, however, to afford the aromatic taxanes in high yields (Eqs. (62) and (63)). Kuwajima recently accomplished the total synthesis of paclitaxel (taxol), using this intermediate as the main component with the eight-membered ring core.

$$(61)$$

$$(62)$$

Paclitaxel (Taxol®)

$$(63)$$

4.4
Chiral Diamine–Sn(II) Complex-promoted Aldol Reactions

Enantioselective aldol addition is one of the most powerful tools for construction of new carbon–carbon bonds with control of the absolute configurations of new chiral centers, and the utility of this reaction has been demonstrated by several applications to the synthesis of natural products such as carbohydrates, macrolide and polyether antibiotics, etc. In the asymmetric aldol reactions reported chiral auxiliary groups are usually attached to the reacting ketone-equivalent molecules. Until the early 1980s there had been no example of an aldol-type reaction in which two achiral carbonyl compounds were used to form a chiral molecule with the aid of a chiral ligand.

Chiral auxiliaries derived from (*S*)-proline seemed to be particularly attractive, because they have conformationally rigid pyrrolidine rings. Chiral diamines derived from (*S*)-proline, especially, are successfully employed for creation of an efficient chiral environment because almost all the main and transition metals having vacant d orbitals are capable of accepting a bidentate ligand. An intermediate derived from the chiral ligand and an organometallic reagent would have a conformationally restricted *cis*-fused five-membered ring chelate and would afford optically active organic compounds by reaction with appropriate substrates.

4.4.1
Asymmetric Aldol and Related Reactions of Sn(II) Enolates

Enantioselective aldol reaction via Sn(II) enolates coordinated with chiral diamines was explored by Mukaiyama and Iwasawa in 1982 [10c, 47]. In the presence of chiral diamine **1a**, various optically active aldol adducts were produced by reactions between aromatic ketones and aldehydes (Eqs. (64) and (65)). This is the first example of the formation of crossed aldol products in high optical purity, using chiral diamines as chelating agents, starting from two achiral carbonyl compounds.

$$(64)$$

$$(65)$$

This procedure is successfully applied to the reactions of carboxylic acid derivatives such as thioamides and thione esters (Eqs. (66) and (67)) [48]. 3-Acetylthiazolidine-2-thiones are quite suitable substrates for the Sn(II) enolate-mediated asymmetric aldol reaction, and various optically active *β*-hydroxy 3-acylthiazolidine-2-thiones are obtained by use of chiral diamine **1a** (Eq. (68)) [49].

$$\text{(66)}$$

$$\text{syn/anti}=92/8$$
$$85\% \text{ ee (syn)}$$

$$\text{(67)}$$

$$\text{syn/anti}=78/22$$
$$90\% \text{ ee (syn)}$$

$$\text{(68)}$$

$$65 \text{~>} 90\% \text{ ee}$$

A complex formed from chiral diamine **2a** with the Sn(II) enolate of 3-acetylthiazolidine-2-thione reacts with some *α*-keto esters to afford aldol adducts with tertiary hydroxyl groups and high ee, as shown in Eq. (69) [50].

$$\text{(69)}$$

$$85 \text{~>} 95\% \text{ ee}$$

$$65 \text{~} 80\%$$

When 3-(2-benzyloxyacetyl)thiazolidine-2-thione is treated under these reaction conditions the corresponding *anti*-diol groups are produced with good diastereoselectivity and high enantioselectivity by addition of chiral diamine **1a** (Eq. (70)) [51].

$$syn/anti = 19/81\!\sim\!7/93$$
$$87\!\sim\!94\%\ ee\ (anti)$$

$$(70)$$

Because Sn(II) enolates of thioesters are generated by the reaction of Sn(II) thiolates with ketenes, the optically active β-hydroxy thioesters are also readily synthesized by way of an aldol reaction with aldehydes in the presence of Sn(OTf)$_2$ and chiral diamine **1a** (Eq. (71)) [52].

$$(71)$$

Mukaiyama and Iwasawa also developed an enantioselective Michael Addition reaction using Sn(OTf)$_2$ with chiral diamines [48, 53]. For example, Sn(II) enolate of methyl ethanedithioate reacts with benzalacetone in the presence of chiral diamine **2a** and trimethylsilyl trifluoromethanesulfonate (TMSOTf) to give the corresponding Michael adduct in 82% yield with good enantioselectivity (Eq. (72)).

$$(72)$$

The Michael adducts were obtained from trimethylsilyl enethioate and α,β-unsaturated ketones in high yields with moderate to good enantioselectivity by use of a catalytic amount of chiral diamine–Sn(OTf)$_2$ complex (Eq. (73)) [48, 54]. As shown in Scheme 4.2, Si–Sn metal exchange occurs rapidly to generate chiral Sn(II) enolate and TMSOTf in situ, because the silicon–sulfur bond is rather weak and tin has high affinity for sulfur. Activation of the α,β-unsaturated ketone by TMSOTf would lead to the Michael reaction, affording the silyl enolate of the Michael adduct and regeneration of the chiral diamine–Sn(II) complex. To preclude the competitive

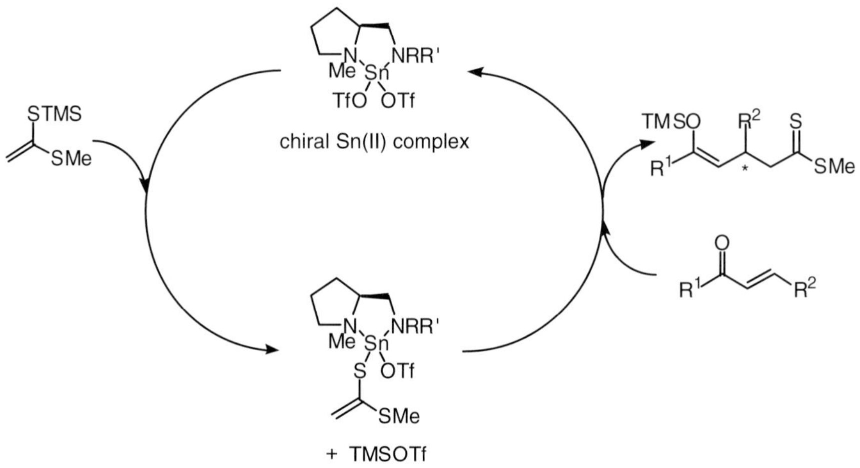

Scheme 4.2
Catalytic asymmetric Michael reaction using chiral Sn(II) enolate.

direct reaction of KSA with α,β-unsaturated ketone under the influence of TMSOTf, the concentration of the KSA is kept low by slow addition of a solution of the KSA to the reaction mixture.

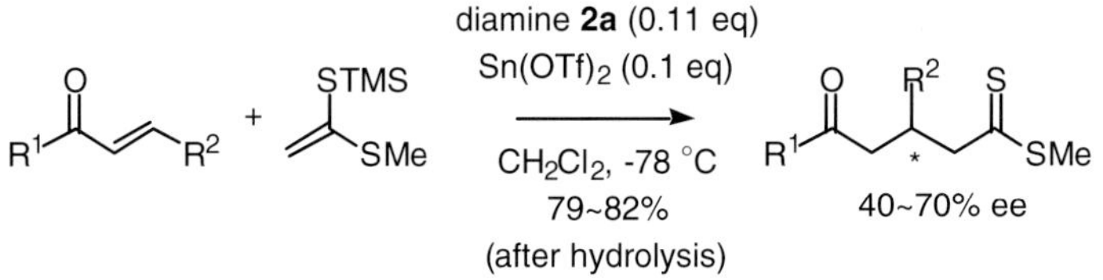

These chiral enolate preparations and reactions with several electrophiles giving the optically active aldols have been applied to the synthesis of natural compounds such as β-lactam antibiotics. References in reviews by Mukaiyama et al. [10c, 48, 55] describe advanced studies on Sn(II) enolate chemistry.

4.4.2
Chiral Diamine–Sn(II) Complex-promoted Aldol Reactions

Chiral Sn(II) Lewis acids prepared in situ by coordination of chiral pyrrolidine derivatives to Sn(OTf)$_2$ were developed by Mukaiyama and Kobayashi in 1989 to promote the asymmetric aldol reaction of ESE or KSA with carbonyl compounds. Some chiral Lewis acids had already been reported and fruitful results were observed in the field of the Diels–Alder and related reactions, in particular, in the late 1980s. The chiral Lewis acids employed for

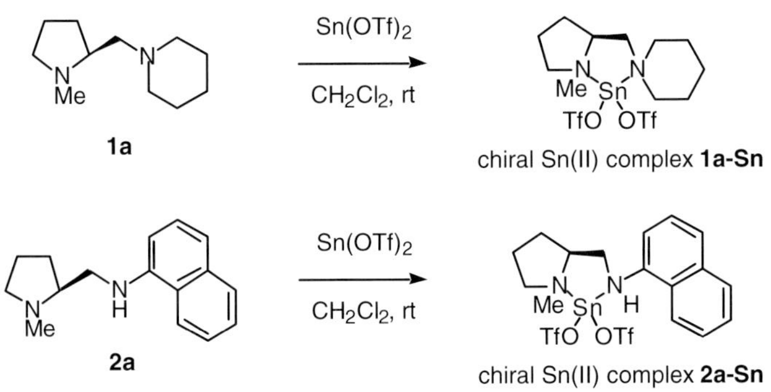

Scheme 4.3
Chiral Sn(II) Lewis acids generated from Sn(OTf)$_2$ with diamines.

these reactions were rather strong and hard acidic metals such as aluminum and titanium. Chiral Sn(II) Lewis acids prepared in situ by chelation of a chiral diamine to Sn(OTf)$_2$, might, on the other hand, be effective because Sn(II) is a soft metal and the complex has one vacant d orbital to be coordinated with oxygen in the carbonyl group of an aldehyde without losing the favorable asymmetric environment (Scheme 4.3).

On this basis a variety of efficient asymmetric aldol reactions between achiral silyl enolates and achiral carbonyl compounds have been developed. Some chiral diamines used in the asymmetric aldol reaction of KSA with carbonyl compounds promoted by Sn(OTf)$_2$ are listed in Scheme 4.4.

4.4.3
Asymmetric Aldol Reaction of Silyl Enolates

Asymmetric aldol reaction of a KSA derived from *S*-Et ethanethioate with aldehydes achieves high ee by employing a chiral promoter, the combined use of Sn(OTf)$_2$ coordinated with the chiral diamine (**1a** or **2a**), and tributyltin fluoride (Eq. (74)) [56].

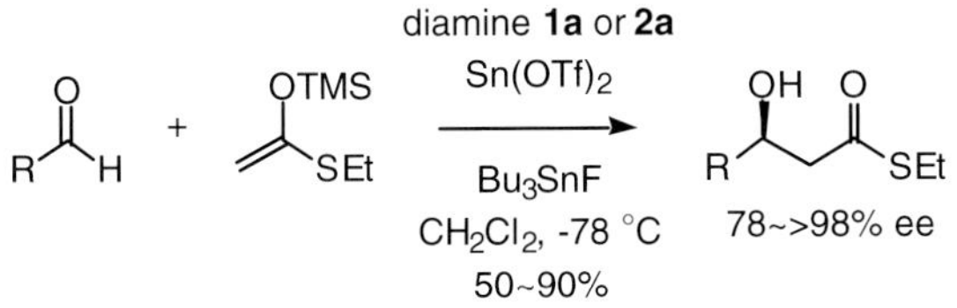

$$\text{(74)}$$

The complex consisting of Sn(OTf)$_2$ with **2a** is quite effective for reaction of KSA generated from *S*-Et propanethioate with aldehydes to afford the corresponding *syn* aldol adducts with excellent diastereoselectivity and enantioselectivity (Eq. (75)) [56b, 57]. A highly enantioselective aldol reac-

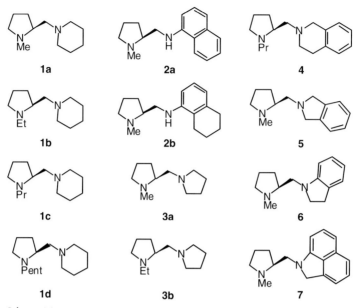

Scheme 4.4
Useful chiral diamines for asymmetric aldol reaction.

tion of the KSA of benzyl acetate with achiral aldehydes can be conducted using the chiral promoter formed from $Sn(OTf)_2$ with **1b** (Eq. (76)) [58].

$$
\underset{R}{\overset{O}{\overset{\|}{C}}}{}_H \;+\; \underset{SEt}{\overset{OTMS}{\diagup}} \quad\xrightarrow[\substack{Bu_3SnF \\ [Bu_2Sn(OAc)_2] \\ CH_2Cl_2,\ -78\ ^\circ C \\ 48\sim91\% \\ [70\sim96\%]}]{\substack{diamine\ \mathbf{2a} \\ Sn(OTf)_2}}\quad \underset{R}{\overset{OH\quad O}{\diagdown}}{}_{SEt}
$$

diamine **2a**
Sn(OTf)₂

Bu₃SnF
[Bu₂Sn(OAc)₂]
CH₂Cl₂, -78 °C
48~91%
[70~96%]

syn/anti = >99/1
>98% ee (syn)

(75)

$$
\underset{R}{\overset{O}{\overset{\|}{C}}}{}_H \;+\; \underset{OBn}{\overset{OTBS}{\diagup}} \quad\xrightarrow[\substack{Bu_3SnF \\ mesitylene \\ CH_2Cl_2,\ -95\ ^\circ C \\ 51\sim79\%}]{\substack{diamine\ \mathbf{1b} \\ Sn(OTf)_2}}\quad \underset{R}{\overset{OH\quad O}{\diagdown}}{}_{OBn}
$$

diamine **1b**
Sn(OTf)₂

Bu₃SnF
mesitylene
CH₂Cl₂, -95 °C
51~79%

89~>98% ee

(76)

In the presence of a promoter including the chiral diamine **1d**, the KSA of a thioester reacts with α-ketoesters to afford the corresponding aldol-type adducts, 2-substituted malates, in good yields with excellent ee (Eq. (77)) [59].

$$
\underset{\substack{R=Me,\ ^iPr,\ Ph}}{\overset{O}{\underset{R}{\bigparallel}}\!\!\!\diagdown\!\!CO_2Me} \;+\; \overset{OTMS}{\underset{SEt}{\bigvee}} \quad \xrightarrow[\substack{Bu_3SnF\\CH_2Cl_2,\ -78\ ^\circ C\\74{\sim}81\%}]{\substack{diamine\ \mathbf{1d}\\Sn(OTf)_2}} \quad \underset{92{\sim}>98\%\ ee}{\overset{MeO_2C\ \ OH\ O}{\underset{R}{\bigparallel}}\!\!\!\diagdown\!\!SEt} \tag{77}
$$

In the course of developments in asymmetric synthesis using $Sn(OTf)_2$ it has also been revealed that a tetrahydrothiophene ligand, an analog of the chiral diamine, also affords an asymmetric environment around the $Sn(II)$ metal suitable for promoting the enantioselective aldol reaction giving the desired adducts with high selectivity (Eq. (78)) [60].

$$
\underset{R}{\overset{O}{\bigparallel}}\!\!\!\diagdown\!\!H \;+\; \overset{OTMS}{\underset{SEt}{\bigvee}} \quad \xrightarrow[\substack{Bu_2Sn(OAc)_2\\CH_2Cl_2,\ -78\ ^\circ C\\48{\sim}98\%}]{\substack{Sn(OTf)_2}} \quad \underset{\substack{syn/anti=86/14{\sim}100/0\\63{\sim}93\%\ ee\ (syn)}}{\overset{OH\ \ O}{\underset{R}{\bigparallel}}\!\!\!\diagdown\!\!SEt} \tag{78}
$$

4.4.4
Catalytic Asymmetric Aldol Reaction

Catalytic asymmetric synthesis is an extremely desirable method for producing optically active compounds from achiral substrates. If asymmetric amplification could be realized by employing a catalytic amount of a chiral source, substantial amounts of optically active compounds could be synthesized in a convenient and rational way. The promoter, consisting of a chiral diamine and $Sn(OTf)_2$, has mild acidity which accelerates the asymmetric aldol reaction of KSA with aldehydes; these reactions always required a stoichiometric amount of the chiral diamines, however. From experimental examination of the mechanism of the stoichiometric asymmetric aldol reaction, Mukaiyama and Kobayashi considered the possibility of catalytic use of the chiral diamine–$Sn(OTf)_2$ complex. Their hypothesis was:

- the reaction first produces $Sn(II)$ alkoxides and TMSOTf (Scheme 4.5);
- if the substrates (KSA and aldehyde) are added quickly to a solution of the chiral diamine–$Sn(OTf)_2$ complex, the initially formed TMSOTf promotes the achiral asymmetric process to produce the racemic aldol from the remaining substrates; and
- if the substrates are added slowly to a solution of the chiral diamine–$Sn(OTf)_2$ complex, in accordance with the reaction rate of KSA with aldehyde, transmetalation from $Sn(II)$ on the formed $Sn(II)$ alkoxide to Si would occur by sequential reaction with TMSOTf in situ.

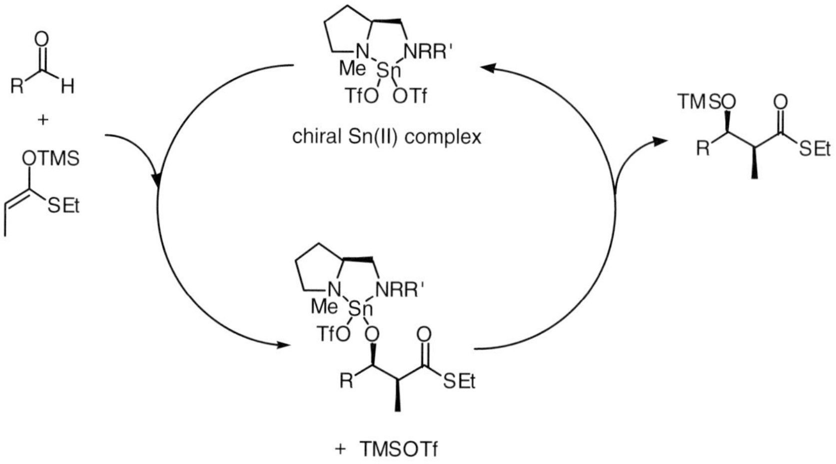

Scheme 4.5
Catalytic asymmetric aldol reaction using chiral Sn(II) complex.

Actually, an optically active trimethylsilyl ether of the aldol was obtained by the reaction of KSA derived from *S*-Et propanethioate with aldehydes by slow addition of a mixture of substrates to a solution including a catalytic amount of the chiral diamine–Sn(OTf)$_2$ complex (Eq. (79)) [61].

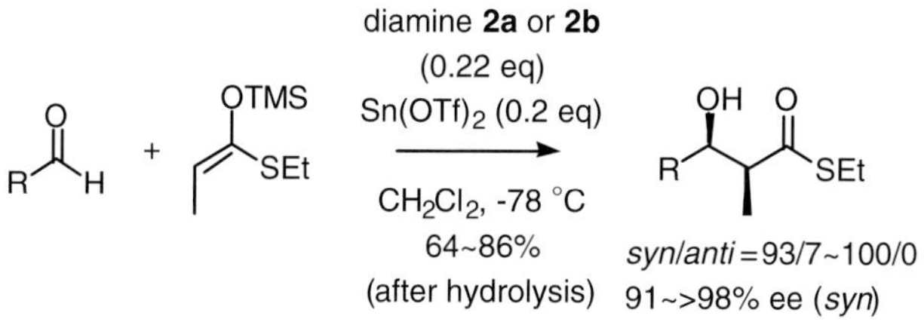

$$(79)$$

The rate of this transmetalation is affected by the conditions, particularly the solvent. Propionitrile was found to be a suitable reaction medium for the catalytic process, and a variety of the optically active aldol adducts were prepared with high ee when a solution of aldehydes and KSA derived from *S*-Et propane- or ethanethioate was added to the catalyst consisting of a chiral diamine and Sn(OTf)$_2$ in propionitrile (Eqs. (80) and (81)) [61, 62].

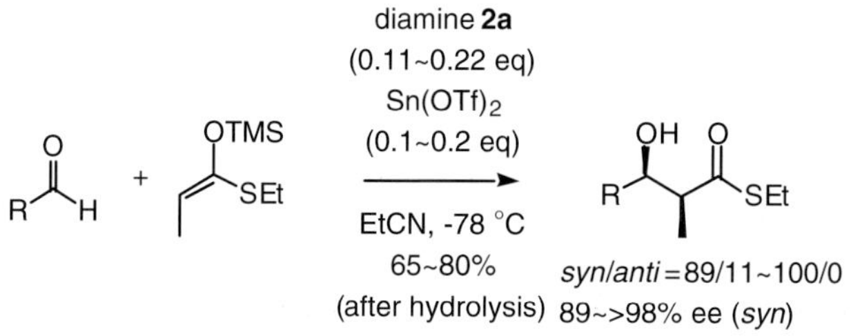

$$(80)$$

$$
\begin{array}{c}
\text{diamine } \mathbf{2a} \text{ or } \mathbf{2b} \\
(0.22 \text{ eq}) \\
\text{Sn(OTf)}_2 \ (0.2 \text{ eq}) \\
\hline
\text{EtCN, -78 °C} \\
48\text{~}90\% \\
\text{(after hydrolysis)}
\end{array}
\qquad (81)
$$

68~93% ee

A novel combined catalyst generated from SnO and TMSOTf was also developed for catalytic synthesis of the desired compounds, as shown in Eq. (82) [63]. Although the exact structure of the complex is unclear, SnO interacted with TMSOTf and formed an acidic species which functions as a chiral catalyst in this reaction. Kobayashi employed SnO as an effective additive for aldol reaction of KSA with aldehydes, promoted by the chiral diamine–Sn(OTf)$_2$ complex, that is, the optically active aldol adducts were synthesized with high stereoselectivity by using a new combination, a chiral diamine–Sn(OTf)$_2$–SnO (Eq. (83)) [64].

$$
\begin{array}{c}
\text{diamine } \mathbf{2a} \ (0.5 \text{ eq}) \\
\text{SnO} \ (1.0 \text{ eq}) \\
\text{TMSOTf} \ (0.65 \text{ eq}) \\
\hline
\text{CH}_2\text{Cl}_2, -78 \text{ °C} \\
58\text{~}82\% \\
\text{(after hydrolysis)}
\end{array}
\qquad (82)
$$

syn/anti = 91/9~98/2
67~94% ee (*syn*)

$$
\begin{array}{c}
\text{diamine } \mathbf{2a} \ (0.22 \text{ eq}) \\
\text{Sn(OTf)}_2 \ (0.2 \text{ eq}) \\
\text{SnO} \ (0.2\text{~}0.4 \text{ eq}) \\
\hline
\text{EtCN, -78 °C} \\
50\text{~}85\% \\
\text{(after hydrolysis)}
\end{array}
\qquad (83)
$$

R' = H, Me

syn/anti = 95/5~100/0
84~>98% ee (*syn*)

Evans recently designed an original chiral Sn(II) catalyst generated from bis(oxazoline) and Sn(OTf)$_2$ [65]. Bis(oxazoline) functions as a bidentate ligand to Sn(OTf)$_2$ and the complex formed might have a rigid C_2-symmetric structure creating an excellent asymmetric environment. This asymmetric aldol reaction proceeds with high diastereoselectivity and enantioselectivity to give the corresponding adducts when bis-functionalized electrophiles such as alkyl glyoxylates, α-ketoesters, and α-diketones are used, because of the formation of suitable complexes with bis(oxazoline) and Sn(OTf)$_2$. For example, the desired optically active *anti* aldols were obtained by reaction of KSA derived from thiol esters with ethyl glyoxylate using a catalytic amount of the Bn/Box–Sn(II) complex. *Anti*-(2*R*,3*S*) selectivity is also observed in the construction of β-hydroxy-α,β-dimethyl thioesters by aldol addition of the KSA of *S*-Et propanethioate to an α-ketoester in the presence of a Bn/Box–Sn(II) complex. A distinctive feature of this asymmetric aldol reaction is that an *anti*-(2*S*,3*R*) adduct, corresponding to the optical antipode of the above *anti*-(2*R*,3*S*) adduct, was produced on use of a Ph/PyBox–Sn(II) complex with the same chirality at the C4 position in the oxazoline moiety of Bn/Box (Eqs. (84) and (85)).

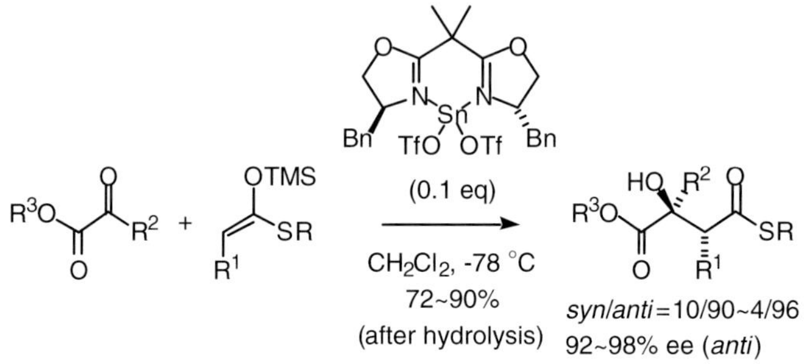

(84)

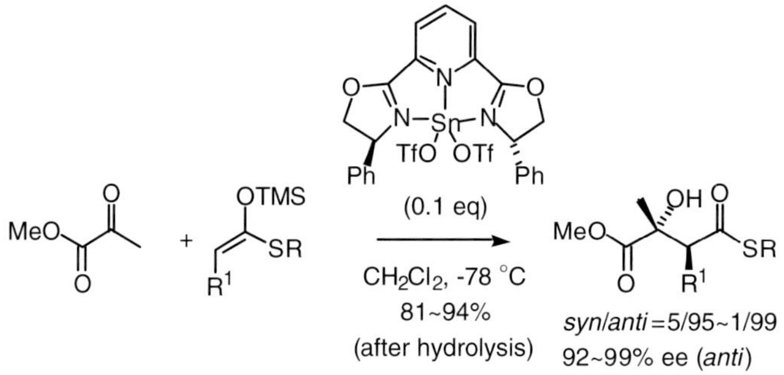

R³O—C(=O)—C(=O)—R² + (OTMS)C=C(R¹)—SR

Bn TfO OTf Bn
(0.1 eq)

CH₂Cl₂, –78 °C
72~90%
(after hydrolysis)

R³O—C(=O)—C(OH)(R²)—CH(R¹)—C(=O)—SR

syn/anti=10/90~4/96
92~98% ee (*anti*)

(85)

MeO—C(=O)—C(=O)—CH₃ + (OTMS)C=CH(R¹)—SR

Ph TfO OTf Ph
(0.1 eq)

CH₂Cl₂, –78 °C
81~94%
(after hydrolysis)

MeO—C(=O)—C(OH)—CH(R¹)—C(=O)—SR

syn/anti=5/95~1/99
92~99% ee (*anti*)

KSA prepared from acetic acid and α,α-dialkyl-substituted acetic acid derivatives also reacted with ethyl glyoxylate to produced the desired aldol adducts with high enantioselectivity, as shown in Eqs. (86) and (87) [65, 66].

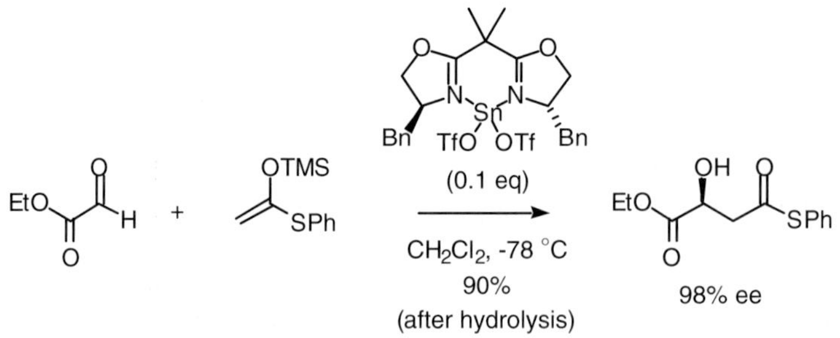

(86)

EtO—C(=O)—CHO + (OTMS)C=CH—SPh

Bn TfO OTf Bn
(0.1 eq)

CH₂Cl₂, –78 °C
90%
(after hydrolysis)

EtO—C(=O)—CH(OH)—CH₂—C(=O)—SPh

98% ee

$$\text{(87)}$$

4.4.5
Asymmetric Synthesis of *syn*- and *anti*-1,2-Diol Groups

Optically active 1,2-diol groups are often observed in nature as carbohydrates, macrolides, or polyethers, etc. Several excellent asymmetric dihydroxylation reactions of olefins using osmium tetroxide with chiral ligands have been developed to give the optically active 1,2-diol groups with high enantioselectivity. Some problems remain, however, for example, preparation of the optically active *anti*-1,2-diols, etc. The asymmetric aldol reaction of a KSA derived from an α-benzyloxy thioester with aldehydes has been developed by Mukaiyama et al. to introduce two hydroxyl groups simultaneously with stereoselective carbon–carbon bond-formation using the chiral Sn(II) Lewis acid.

First, a variety of optically active *anti*-α,β-dihydroxy thioester derivatives were obtained in good yield with excellent diastereoselectivity and enantioselectivity when the chiral diamine **1a** or **1b**, Sn(OTf)$_2$, and dibutyltin diacetate were employed together (Eq. (88)) [67]. By means of current aldol methodology, two hydroxyl groups can be stereoselectively introduced at the 1,2-position during formation of the new carbon–carbon bond.

$$\text{(88)}$$

On the other hand, several *syn*-aldol adducts are obtained under the same reaction conditions, i.e. in the presence of chiral diamine **1a**, Sn(OTf)$_2$, and dibutyltin diacetate. The reaction of a KSA with a *t*-butyldimethylsiloxy group at the 2-position with achiral aldehydes smoothly proceeds to give the corresponding *syn*-α,β-dihydroxy thioester derivatives in high yield with good stereoselectivity. When a chiral diamine **1c**, which is similar to **1a** in possessing a propyl group on the nitrogen of the pyrrolidine ring, is used the ee increases up to 94% (Eq. (89)) [68, 67b]. Now it becomes possible to control the enantiofacial selectivity of the KSA derived from α-hydroxy thioester derivatives just by choosing the appropriate protective groups of the hydroxy parts of the KSA, and the two diastereomers of the optically active α,β-dihydroxy thioesters can be synthesized.

$$
\begin{array}{c}
\text{diamine } \mathbf{1c} \\
\text{Sn(OTf)}_2 \\
\xrightarrow{\hspace{2cm}} \\
\text{Bu}_2\text{Sn(OAc)}_2 \\
\text{CH}_2\text{Cl}_2, -78\ ^\circ\text{C} \\
46\text{~}93\%
\end{array}
$$

syn/anti = 88/12~97/3
82~94% ee (syn)

(89)

Kobayashi also introduced several new types of chiral diamine, for example 4 and 5, to obtain rather higher selectivity for the synthesis of *syn*-α,β-dihydroxy thioester derivatives, as shown in Eq. (90) [69, 70].

$$
\begin{array}{c}
\text{diamine } \mathbf{4} \text{ or } \mathbf{5} \\
\text{Sn(OTf)}_2 \\
\xrightarrow{\hspace{2cm}} \\
\text{Bu}_2\text{Sn(OAc)}_2 \\
\text{CH}_2\text{Cl}_2, -78\ ^\circ\text{C} \\
\mathbf{4}; 69\text{~}89\% \\
\mathbf{5}; 61\text{~}86\%
\end{array}
$$

4; syn/anti = 94/6~99/1
86~96% ee (syn)
5; syn/anti = 98/2~>99/1
96~99% ee (syn)

(90)

Diastereoselective and enantioselective synthesis of both stereoisomers of α,β-dihydroxy-β-methyl thioester derivatives has also been achieved by reaction of KSA with a benzyloxy or *t*-butyldimethylsiloxy group at the 2-position, promoted by an Sn(II) Lewis acid including chiral diamine **1c** or **4** (Eqs. (91) and (92)) [69, 71].

R = Me, Ph

$$
\begin{array}{c}
\text{diamine } \mathbf{1c} \\
\text{Sn(OTf)}_2 \\
\xrightarrow{\hspace{2cm}} \\
\text{Bu}_3\text{SnF} \\
\text{CH}_2\text{Cl}_2, -78\ ^\circ\text{C} \\
66\text{~}93\%
\end{array}
$$

syn/anti = 13/87~7/93
91% ee (anti)

(91)

R = Me, Ph

$$
\begin{array}{c}
\text{diamine } \mathbf{4} \\
\text{Sn(OTf)}_2 \\
\xrightarrow{\hspace{2cm}} \\
\text{Bu}_3\text{SnF} \\
\text{CH}_2\text{Cl}_2, -78\ ^\circ\text{C} \\
76\text{~}89\%
\end{array}
$$

syn/anti = 84/16~94/6
87~88% ee (syn)

(92)

KSA derived from phenyl esters have a unique capacity to promote remarkable stereoselectivity in asymmetric aldol reactions involving a chiral diamine–Sn(II) complex [72–74]. For instance, Kobayashi found that (*E*)-KSA derived from *p*-methoxyphenyl (*t*-butyldimethylsiloxy)acetates reacts with aldehydes to afford the corresponding *anti*-1,2-diol derivatives with high diastereoselectivity and enantioselectivity when promoted by an Sn(II) Lewis acid complexed with chiral diamine **1a** (Eq. (93)) [74].

$$
\begin{array}{c}
\text{diamine } \mathbf{1a} \\
\text{Sn(OTf)}_2 \\
\hline
\text{Bu}_2\text{Sn(OAc)}_2 \\
\text{CH}_2\text{Cl}_2, -78\ ^\circ\text{C} \\
31\text{\textasciitilde}95\%
\end{array}
\qquad (93)
$$

syn/anti = 31/69~2/98
84~95% ee (*anti*)

Reaction of (*Z*)-KSA derived from phenyl benzyloxyacetate with aldehydes, using chiral diamine **3b**, also affords the optically active *anti*-1,2-aldols preferentially (Eq. (94)). However, the corresponding *syn* aldols were formed when the reaction was conducted in the presence of chiral diamine **6** (Eq. (95)) [75]. The latter reaction also proceeded when accelerated by a catalytic amount of chiral diamine **2a** or **2b** under Kobayashi conditions (Eq. (96)) [76].

$$
\begin{array}{c}
\text{diamine } \mathbf{3b} \\
\text{Sn(OTf)}_2 \\
\hline
\text{Bu}_2\text{Sn(OAc)}_2 \\
\text{CH}_2\text{Cl}_2, -78\ ^\circ\text{C} \\
51\text{\textasciitilde}80\%
\end{array}
\qquad (94)
$$

syn/anti = 12/88~7/93
90~94% ee (*anti*)

$$
\begin{array}{c}
\text{diamine } \mathbf{6} \\
\text{Sn(OTf)}_2 \\
\hline
\text{Bu}_2\text{Sn(OAc)}_2 \\
\text{CH}_2\text{Cl}_2, -78\ ^\circ\text{C} \\
85\text{\textasciitilde}90\%
\end{array}
\qquad (95)
$$

syn/anti = 94/6~>99/1
91~98% ee (*syn*)

$$
\begin{array}{c}
\text{diamine } \mathbf{2a} \text{ or } \mathbf{2b} \\
(0.2\ \text{eq}) \\
\text{Sn(OTf)}_2\ (0.2\ \text{eq}) \\
\hline
\text{SnO } (0.2\ \text{eq}) \\
\text{EtCN}, -78\ ^\circ\text{C} \\
68\text{\textasciitilde}87\% \\
(\text{after hydrolysis})
\end{array}
\qquad (96)
$$

syn/anti = 90/10~>98/2
80~96% ee (*syn*)

This method of producing chiral aldols is also applicable to the construction of an asymmetric quarternary carbon included in the 1,2-diol groups. In the presence of a chiral promoter consisting of the chiral diamine **1a**, $Sn(OTf)_2$, and dibutyltin diacetate, optically active *anti*-α,β-dihydroxy-α-methyl thioester and phenyl ester derivatives were synthesized in good yields with high stereoselectivity (Eqs. (97) and (98)) [72, 73].

diamine **1a**
$Sn(OTf)_2$

$Bu_2Sn(OAc)_2$
CH_2Cl_2, -78 °C
58%

E/Z=12/88

syn/anti = 2/98
97% ee (anti)

(97)

diamine **1a**
$Sn(OTf)_2$

$Bu_2Sn(OAc)_2$
CH_2Cl_2, -78 °C
44~72%

E/Z=58/42

syn/anti = 26/74~8/92
73~95% ee (anti)

(98)

Another interesting phenomenon in which the corresponding *syn*-α,β-dihydroxy-α-methyl ester derivatives were produced from similar KSA using a stoichiometric or catalytic amount of the chiral catalyst containing diamine **2a** as shown in Eqs. (99) and (100) [73].

diamine **2a**
$Sn(OTf)_2$

$Bu_2Sn(OAc)_2$
CH_2Cl_2, -78 °C
52%~quant.

R = Et, ^{i}Pr, Ph
E/Z=58/42~71/29

syn/anti = 81/19~98/2
80~97% ee (syn)

(99)

diamine **2a** (0.24 eq)
$Sn(OTf)_2$ (0.2 eq)

EtCN, -78 °C
60%
(after hydrolysis)

E/Z=71/29

syn/anti = 90/10
96% ee (syn)

(100)

Similarly, it was found that KSA (E/Z = 38 to 62) derived from *p*-methoxyphenyl α-benzyloxypropionate reacted with aldehydes in the presence of a $Sn(II)$ catalyst containing diamine **3a** to give the corresponding *anti*-aldol groups (Eq. (101)), whereas asymmetric aldol reaction of (E)-KSA, derived from *p*-methoxyphenyl α-benzyloxypropionate with a variety of aldehydes, promoted by Sn(OTf)$_2$ coordinated by chiral diamine **2a**, afforded the stereoisomeric *syn* compounds with high ee (Eq. (102)) [77]. Tetrasubstituted KSA with an alkylthio group also reacted with aldehydes to produce the *syn*-aldol compounds preferentially [78]; these were used as synthetic intermediates of *anti*-β-hydroxy-α-methyl groups in the total synthesis of octalactins, described in a later section (Eq. (103)).

$$ (101) $$

$$ (102) $$

$$ (103) $$

4.4.6
Enantioselective Synthesis of Both Enantiomers of Aldols Using Similar Diamines Derived from L-Proline

Kobayashi recently reported remarkable results in the synthesis of optically active aldol compounds using new chiral diamine–Sn(II) complexes as promoters. Reaction of KSA derived from *S*-Et (*t*-butyldimethylsiloxy)-ethanethioate with aldehydes using chiral diamine **6** mainly yielded the *syn*-(2*R*,3*S*) compounds which are optical antipodes of aldol adducts (*syn*-(2*S*,3*R*)) prepared by the reaction using chiral diamine **1c** (Eqs. (89) and (104)) [70, 79]. Optically active *syn*-(2*R*,3*R*) aldols were also prepared from propionic acid derivatives and promotion with an Sn(II) complex with chiral diamine **7**, whereas *syn*-(2*S*,3*S*) aldols were produced if chiral diamine **2a** was used in the same reaction (Eqs. (75) and (105)) [80]. Chiral diamines **1c**, **2a**, **6**, and **7** were all prepared starting from L-proline and have identical chirality at the C2 position. Artificial switching of the enantiofacial selectivity of the aldol reaction by using only one chiral source could therefore be achieved by use of these methods.

diamine **6**
Sn(OTf)$_2$

$$
\underset{R}{\overset{O}{\|}}H \;+\; \underset{\overset{|}{OTBS}}{\overset{OTMS}{\diagup}}SEt \;\xrightarrow[\substack{Bu_2Sn(OAc)_2 \\ CH_2Cl_2,\ -78\ ^\circ C \\ 63\sim 86\%}]{}\; R\underset{\overset{|}{OTBS}}{\overset{OH\quad O}{\diagup}}SEt
$$

syn/anti=>99/1
98~>99% ee (syn)

(104)

diamine **7**
Sn(OTf)$_2$

$$
\underset{R}{\overset{O}{\|}}H \;+\; \overset{OTMS}{\diagup}SEt \;\xrightarrow[\substack{Bu_2Sn(OAc)_2 \\ CH_2Cl_2,\ -78\ ^\circ C \\ 67\sim 83\%}]{}\; R\overset{OH\quad O}{\diagup}SEt
$$

syn/anti=>99/1
80~92% ee (syn)

(105)

4.5
Asymmetric Total Syntheses of Complex Molecules Using Chiral Diamine–Sn(II) Catalysts

Enantioselective aldol reactions can be powerful tools for the stereoselective synthesis of complex molecules, especially for construction of optically active 1,2-diol groups in the carbon backbones of the target compounds. Recent progress in this area will be illustrated by means of successful methods for stereoselective synthesis of natural and unnatural polyoxy compounds.

4.5.1
Monosaccharides

In the last decade chemical synthesis of monosaccharides has made a great advance as a result of stereoselective addition reactions of 2,3-*O*-isopropylidene-D- or -L-glyceraldehyde or 4-*O*-benzyl-2,3-*O*-isopropylidene-L-threose with enolate components or allyl nucleophiles, and many examples of the effective synthesis of sugars, both natural and unnatural, have been demonstrated [81]. In these syntheses one of the starting materials, glyceraldehyde or a threose derivative, is prepared from a natural chiral pool, mannitol and tartaric acid, respectively. In contrast, a general method has been developed for synthesis of a variety of sugars starting from both achiral KSA and α,β-unsaturated aldehydes (Scheme 4.6). Chiral induction can be accomplished by means of an asymmetric aldol reaction using a complex consisting of Sn(OTf)$_2$ and an appropriate chiral diamine. Subsequent dihydroxylation or epoxidation of the double bond in the aldol adducts affords several tetrahydroxy thioester derivatives which can be useful precursors for the syntheses of a variety of monosaccharides, including rare sugars.

072805E-2

42-5873   QD305   MARC

Modern aldol reactions: v.1: Enolates, organocatalysis, biocatalysis and natural product synthesis; v.2: Metal catalysis, ed. by Rainer Mahrwald. Wiley-VCH, 2004. 2v bibl index afp ISBN 3527307141, $405.00

This set is a comprehensive overview of the versatile and well-used aldol reaction in organic synthesis. This reaction is one of the most important methods of carbon-carbon bond formation in organic synthesis. The 22 distinguished researchers and contributors have thoroughly reviewed the development of the aldol reaction from the fundamentals to the use of enolates, metal and metal complexes, natural product synthesis, and enzymatic reactions. Each author examines thoroughly the stereoselectivity of the aldol reaction and its importance in organic chemistry, particularly in natural product synthesis. Illustrations are used abundantly to describe the reaction intermediates and products. Each author has provided a very extensive bibliography. This set will be a valuable addition to the library of researchers in the field and for graduate-level collections. **Summing Up:** Recommended. Graduate students through professionals.—*L. S. Smith, emeritus, Central State University (OH)*

3 44.25

Tim Peele
Chemistry

B

56328131

TJP

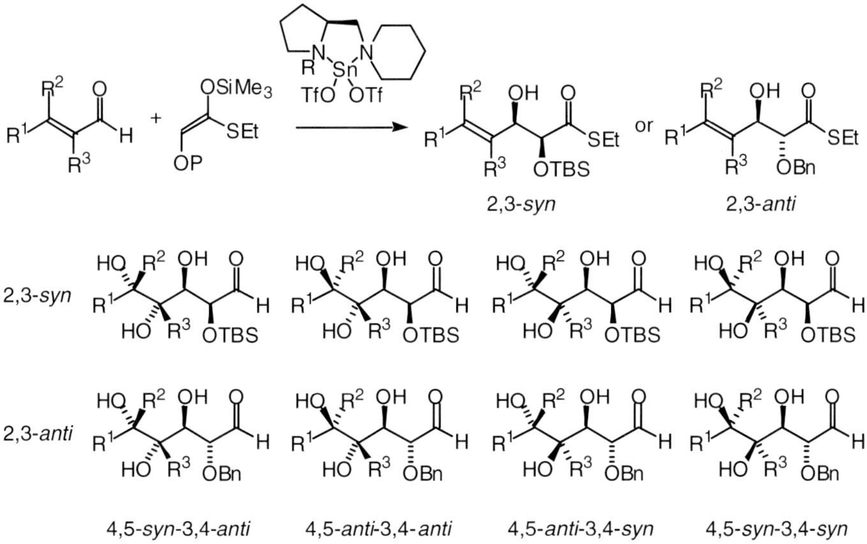

Scheme 4.6
Synthesis of monosaccharides by use of the asymmetric aldol reaction.

One example, synthesis of 6-deoxy-L-talose, is shown in Scheme 4.7 [82]. The asymmetric aldol reaction between crotonaldehyde and the KSA of α-benzyloxy thioester was carried out in the presence of Sn(OTf)$_2$, chiral diamine **1a**, and dibutyltin diacetate, and the corresponding aldol adduct was obtained in 85% yield with >97% enantiomeric excess. Dihydroxylation of this chiral synthon, subsequent reduction of the resulted lactone, and deprotection of the benzyl group gave the desired 6-deoxy-L-talose in good yield.

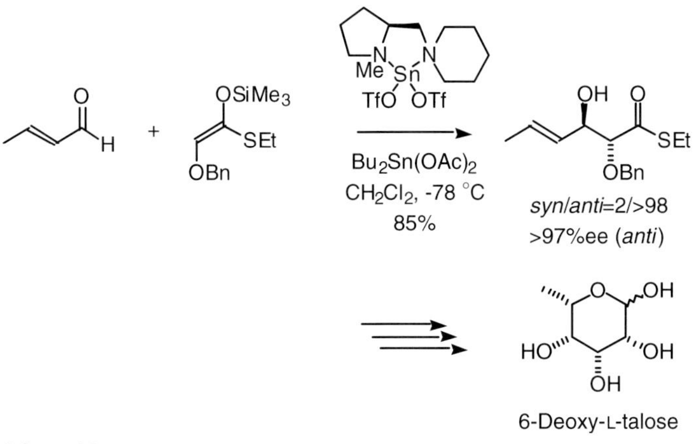

Scheme 4.7
Synthesis of 6-deoxy-L-talose.

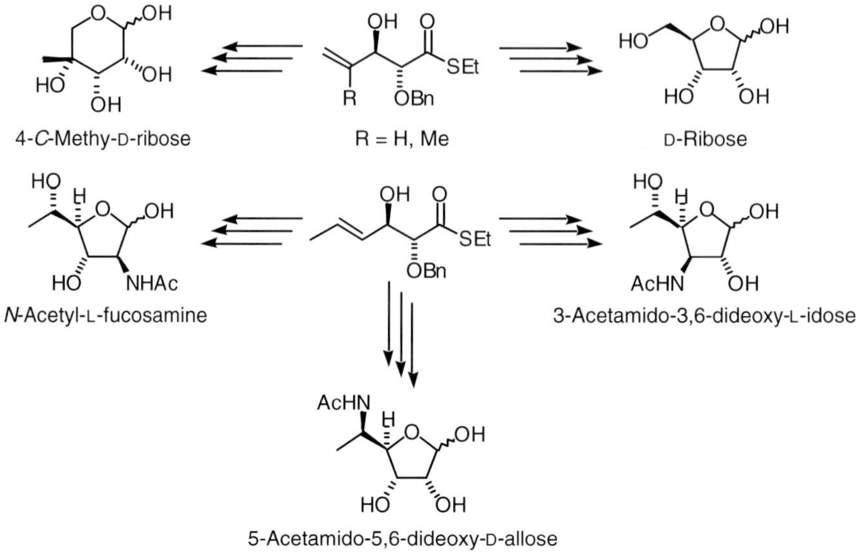

Scheme 4.8

Synthesis of D-ribose, 4-C-methyl-D-ribose and several amino sugars.

By use of this universal methodology, several monosaccharides including branched and amino sugars were synthesized as shown in Scheme 4.8 (D-ribose and 4-C-methyl-D-ribose (1990) [82], N-acetyl-L-fucosamine, 3-acetamide-3,6-dideoxy-L-idose and 5-acetamide-5,6-dideoxy-D-allose (1993) [83]).

Scheme 4.9 shows the syntheses of two stereoisomers of 6-deoxy-L-talose from the corresponding intermediates generated via asymmetric aldol reaction (6-deoxy-D-allose (1992) [84] and L-fucose (1993) [76]).

Several 2-branched saccharine acid γ-lactones, 2-C-methyl-D- or L-threono-1,4-lactones and 2-C-methyl-D-erythrono-1,4-lactone have been effectively prepared using this strategy, by enantioselective construction of asymmetric quaternary carbons developed in the former section (Schemes 4.10 and 4.11) [72, 73, 69].

Because the key asymmetric aldol reaction has wide flexibility in controlling newly created chiral centers, these methods are expected to provide useful routes to the synthesis of a variety of monosaccharides from achiral KSA and aldehydes.

4.5.2

Leinamycin and a Part of Rapamycin

Fukuyama used the asymmetric formation of a 1,2-diol group for the total synthesis of leinamycin in which it was shown that KSA with a p-methoxybenzyloxy group at the C2 position functions as a suitable nucleophile for the multifunctional aldehyde (Scheme 4.12) [85]. White also reported that

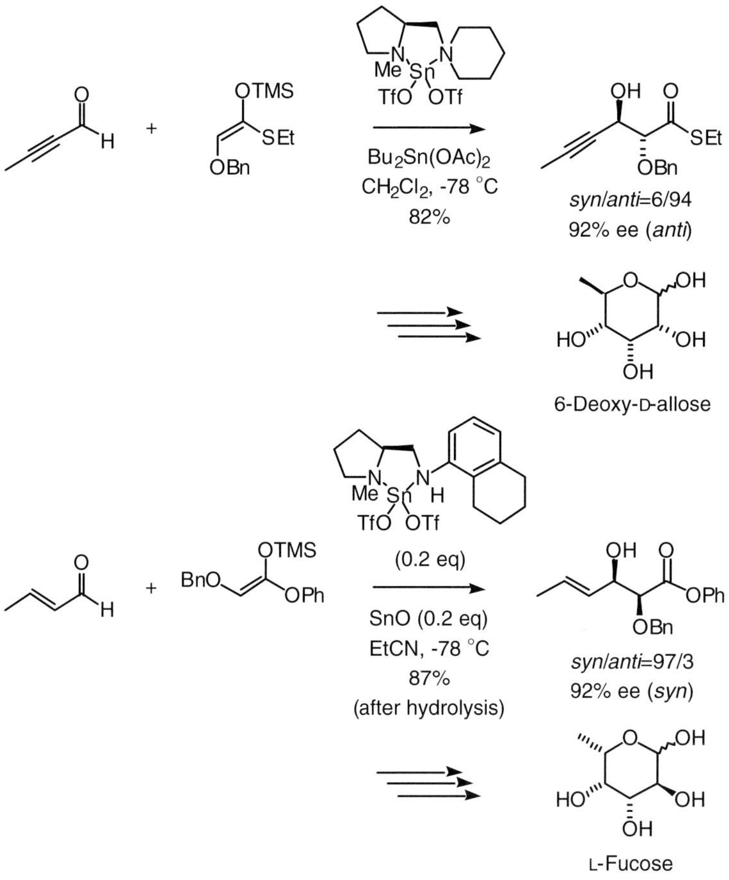

Scheme 4.9
Synthesis of 6-deoxy-D-allose and L-fucose.

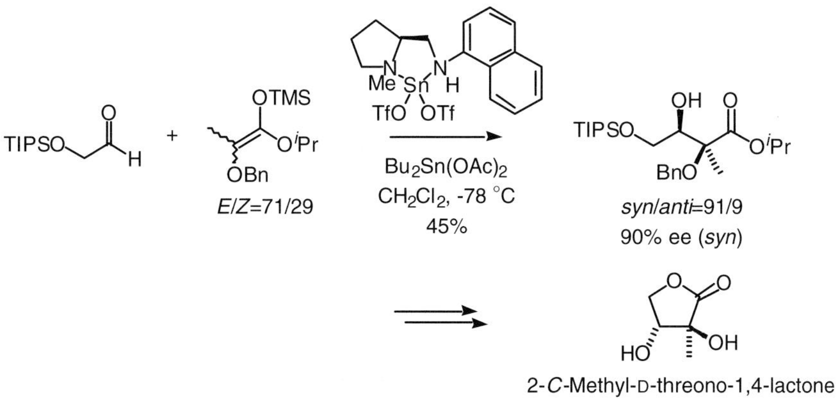

Scheme 4.10
Synthesis of 2-C-methyl-D-threono-1,4-lactone.

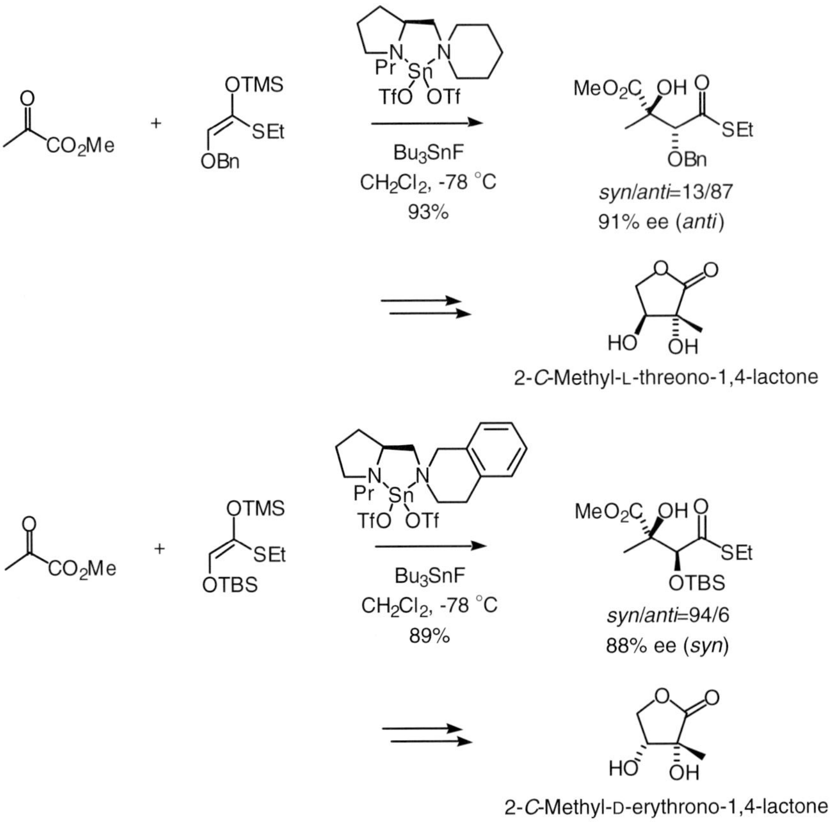

Scheme 4.11
Synthesis of 2-C-methyl-L-threono- and D-erythrono-1,4-lactones.

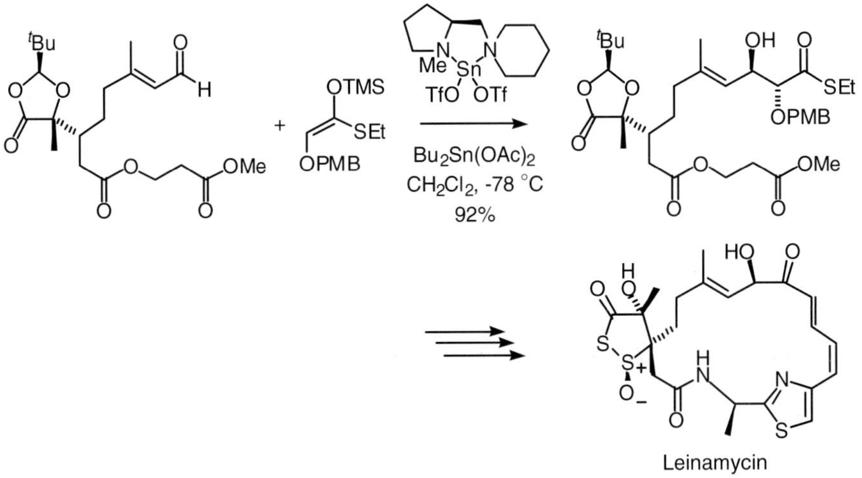

Scheme 4.12
Synthesis of leinamycin.

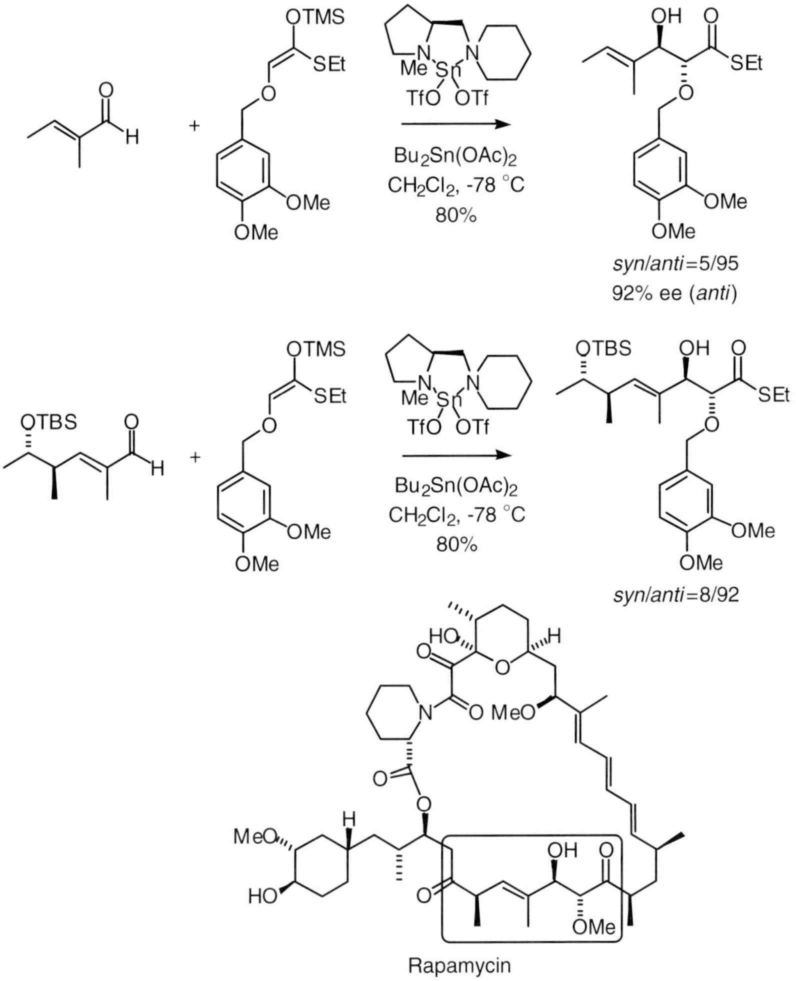

Scheme 4.13
Synthesis of a part of rapamycin.

reaction of the KSA generated from *S*-Et (3,4-dimethoxybenzyloxy)ethane-thioate with α,β-unsaturated aldehydes proceeded smoothly to afford the corresponding diol groups in high yields with excellent stereoselectivity, as shown in Scheme 4.13 [86].

4.5.3
Sphingosine, Sphingofungins, and Khafrefungin

Kobayashi used asymmetric reactions for stereoselective synthesis of a variety of polyoxygenated natural compounds. Initially, a new method for the preparation of sphingosine was developed using the catalytic asymmetric aldol reaction of KSA with α,β-eynal as a key step (Scheme 4.14) [87].

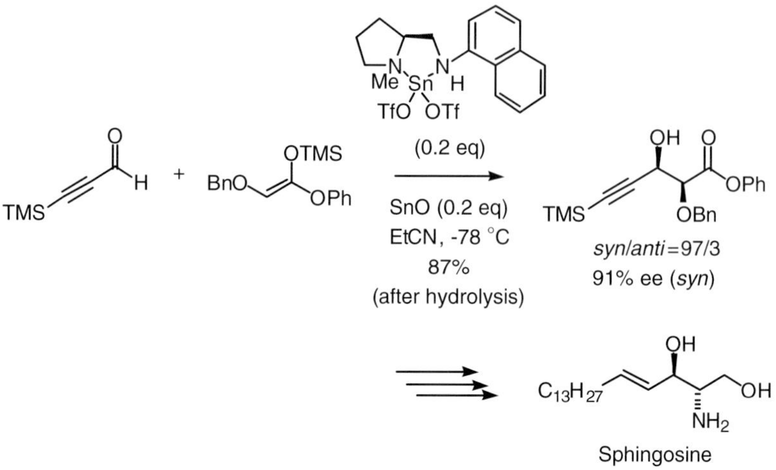

Scheme 4.14
Synthesis of sphingosine.

Sphingofungins B and F were also totally synthesized from small molecules by the asymmetric aldol strategy as shown in Scheme 4.15 [87b, 88]. Here the optically active polyol part was obtained by reaction of trisubstituted KSA using a chiral diamine *ent-2a*, and the sole asymmetric center in the side chain was synthesized by the reaction of KSA derived from an acetic acid derivative using the chiral diamine **2a**. These segments were coupled to form the basic skeleton of sphingofungins in the total synthesis.

A diastereoselective aldol reaction using an α-alkoxyaldehyde was also mentioned in this research (Eq. (106)) [88c].

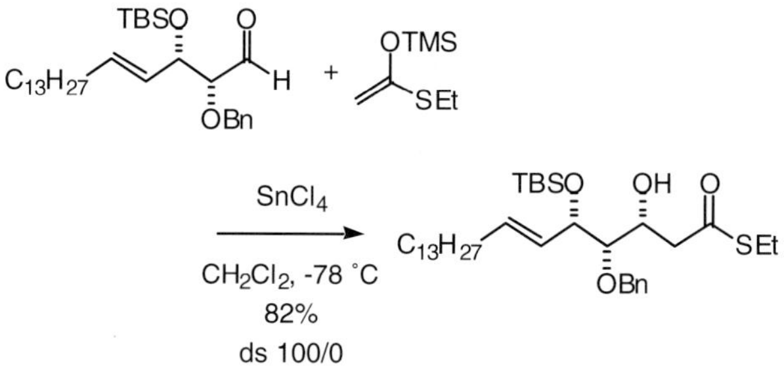

(106)

Total synthesis of khafrefungin and the determination of its stereochemistry was recently achieved by Kobayashi, who used chiral induction technology to give the optically active aldol compounds (Scheme 4.16) [89]. The asymmetric aldol reaction of KSA derived from *S*-Et propanethioate with aldehydes was applied not only to the first step to afford the corresponding thioester with high ee but also to the following stage to give the multifunctional linear thioester with excellent diastereoselectivity.

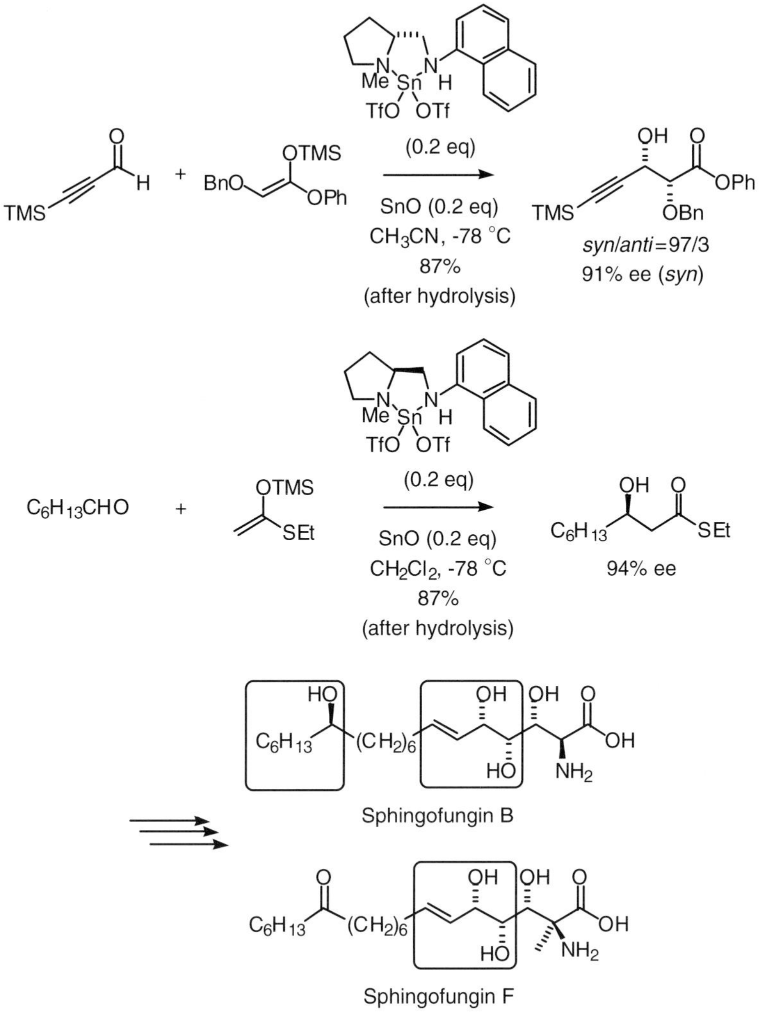

Scheme 4.15
Synthesis of sphingofungins B and F.

4.5.4
Febrifugine and Isofebrifugine

Kobayashi also reported the enantioselective total synthesis of febrifugine and isofebrifugine using the Sn(II)-mediated catalytic asymmetric aldol re-action giving the optically active diol groups (Scheme 4.17) [90]. The correct absolute stereochemistries of natural febrifugine and isofebrifugine were shown by comparison with spectral data and the sense of the optical rota-tions of four synthetic samples, including enantiomorphs.

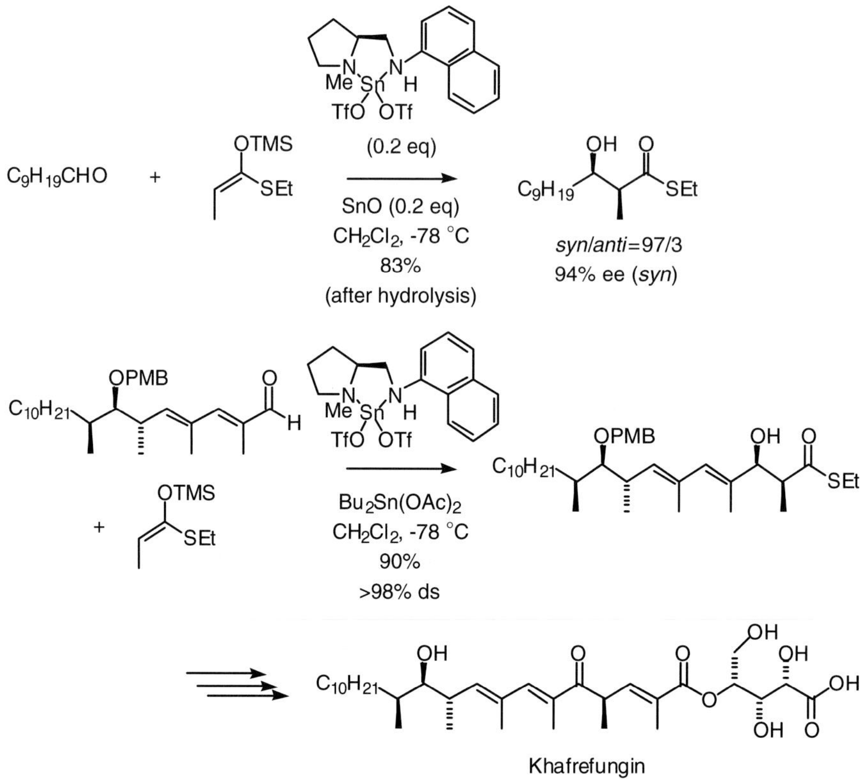

Scheme 4.16
Synthesis of khafrefungin.

4.5.5
Altohyrtin C (Spongistatin 2) and Phorboxazole B

Asymmetric aldol reaction accelerated by the chiral bis(oxazoline)–Sn(OTf)$_2$ complex also provides a powerful means of construction of poly-functionalized natural compounds. Evans succeeded in the total synthesis of altohyrtin C (spongistatin 2), a macrocyclic compound with many oxygenated functional groups (Scheme 4.18) [91]. Part of the tetrahydropyran segment (F ring) in altohyrtin C was stereoselectively obtained from the corresponding *anti-β*-hydroxy-*α*-methyl thioester generated by the asymmetric aldol reaction using the Ph/Box–Sn(II) complex catalyst.

This asymmetric aldol reaction is effective when using chelating electrophiles such as ethyl glyoxylate; therefore, *α*-oxazole aldehyde might be employed in the preparation of an optically active oxazole derivative. Indeed, the reaction of KSA generated from *S-^{t}*Bu ethanethioate with *α*-oxazole al-

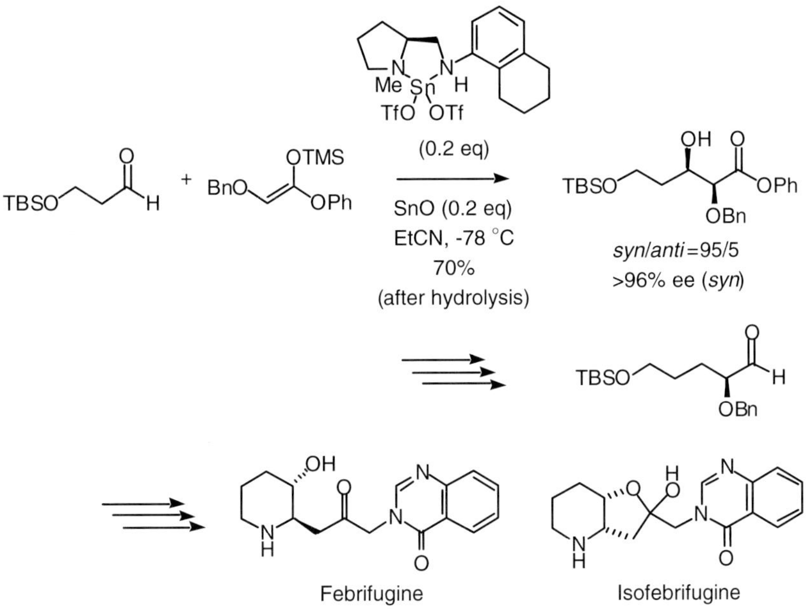

Scheme 4.17
Synthesis of febrifugine and isofebrifugine.

dehyde took place as expected to afford the corresponding aldol with high ee, and the total synthesis of phorboxazole B was successfully achieved using the adduct as a part of the complex structure (Scheme 4.19) [92].

4.5.6
Paclitaxel (Taxol)

Mukaiyama and Shiina accomplished the total synthesis of paclitaxel (taxol) by the strategy shown in Scheme 4.20, i.e. synthesis of the eight-membered B ring first, starting from an optically active polyoxy precursor generated by the highly controlled enantioselective aldol reaction and subsequent construction of the fused A and C ring systems on to the B ring [93].

The optically active diol unit **9** was prepared by the asymmetric aldol reaction of a KSA with a benzyloxy group at the C2 position with an achiral aldehyde **8** using a chiral diamine–Sn(II) complex (Scheme 4.21). Synthesis of the eight-membered ring aldols from an optically active polyoxy-group **10** containing all the functionality necessary for the construction of taxol was performed by the intramolecular aldol cyclization using SmI$_2$. Subsequent acetylation of this mixture of isomeric alcohols and treatment with DBU gave the desired eight-membered enone **11** in good yield.

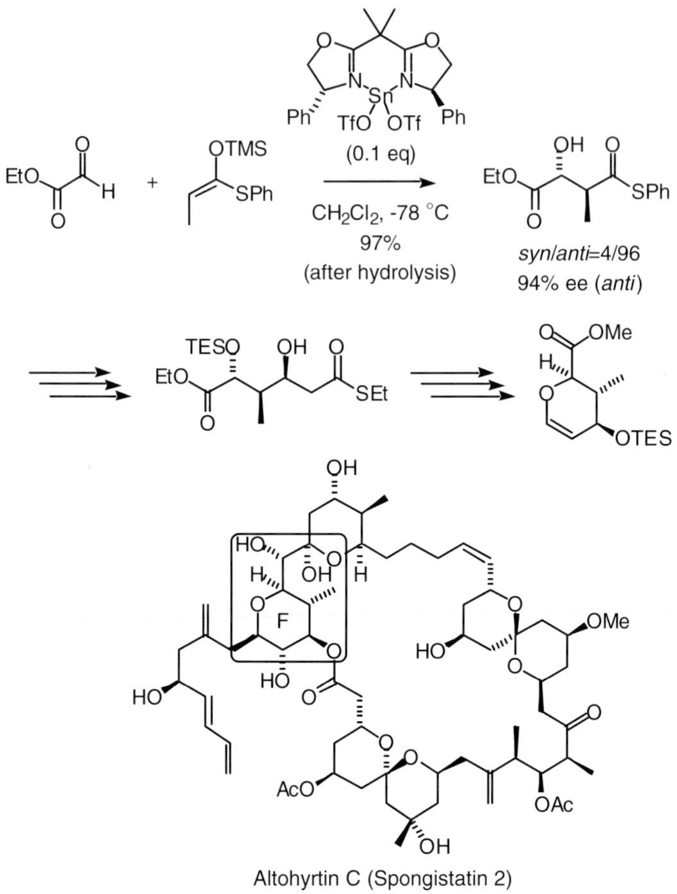

Scheme 4.18
Synthesis of altohyrtin C (spongistatin 2).

As shown in Scheme 4.22, fully functionalized BC ring system **12** was then synthesized from the optically active eight-membered ring compound **11** by successive Michael addition and intramolecular aldol cyclization of ketoaldehyde. Intramolecular pinacol coupling of the diketone derived from the above BC ring system using a low-valent titanium reagent resulted in the formation of ABC ring system **13**, a new taxoid, in good yield. 7-Triethylsilylbaccatin III was prepared from the above new taxoid **13** by oxygenation at the C13 position and construction of the oxetane ring.

It was also shown that the asymmetric aldol reaction is useful for preparation of the chiral side chains of taxol (Scheme 4.23). Because reaction of the KSA derived from *S*-Et benzyloxyethanethioate with benzaldehyde afforded the corresponding aldol adduct **14** in high yield with excellent selectivity, as shown in the last section, this adduct was successfully con-

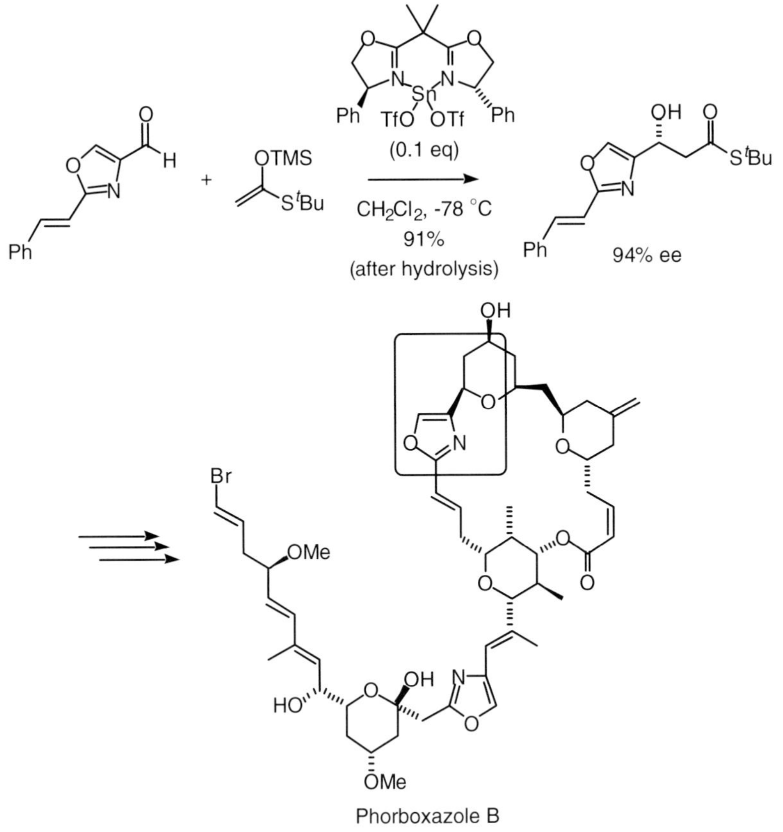

Scheme 4.19
Synthesis of phorboxazole B.

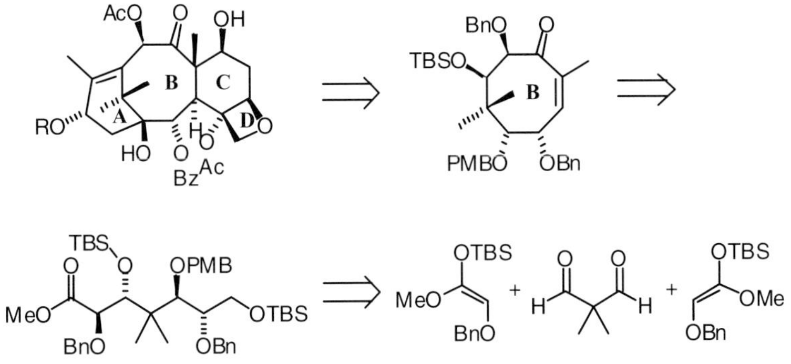

Scheme 4.20
Retrosynthesis of taxol.

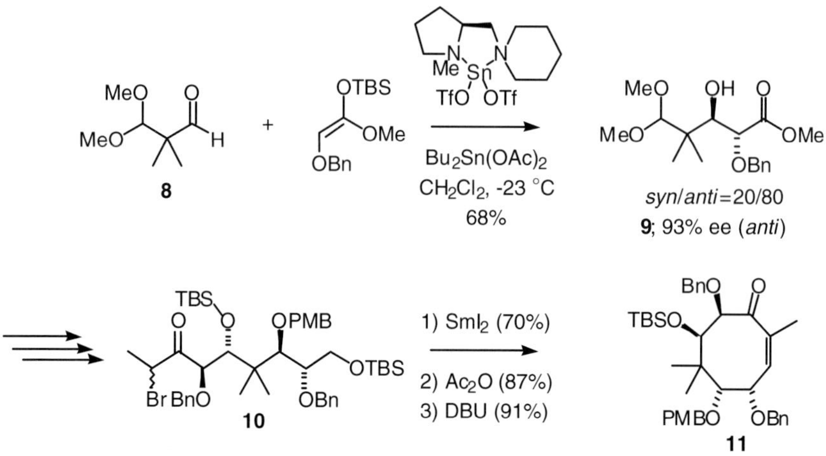

Scheme 4.21
Synthesis of the B ring of taxol.

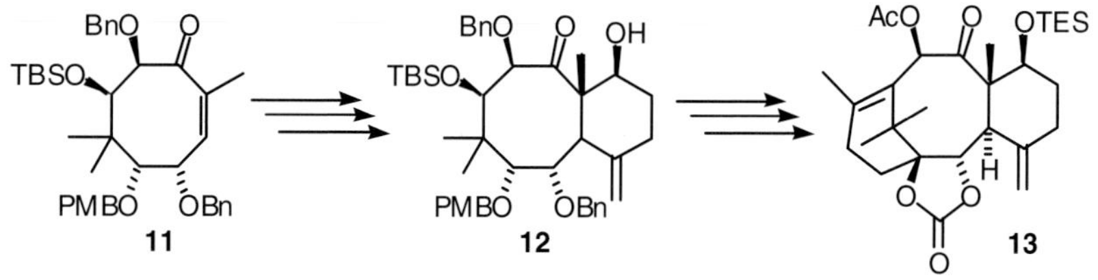

Scheme 4.22
Synthesis of the ABC ring system of taxol.

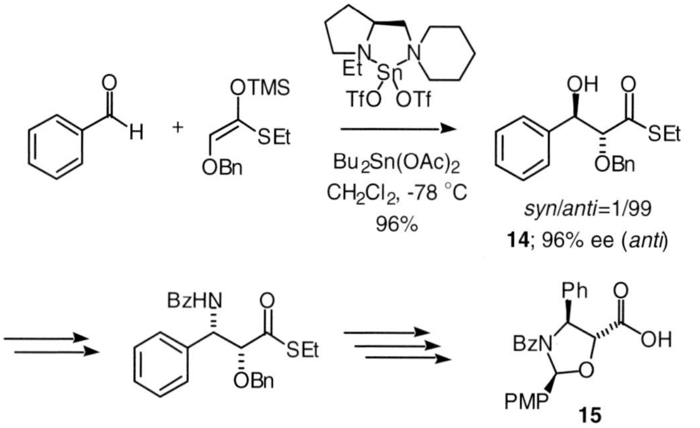

Scheme 4.23
Synthesis of the side chain of taxol.

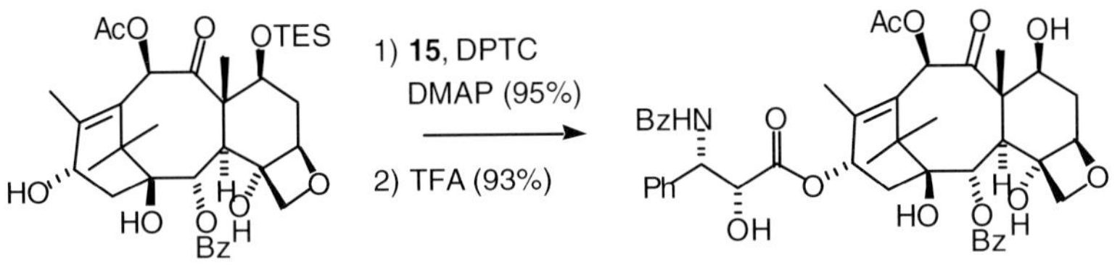

Scheme 4.24
Synthesis of taxol.

verted into the targeted β-amino acid **15** in good yield, with inversion of chirality at the β-position, by use of the Mitsunobu reaction. Introduction of N-benzoylphenylisoserine **15** to 7-triethylsilylbaccatin III was further studied, and dehydration condensation was found to proceed smoothly using DPTC (*O,O*-di(2-pyridyl)thiocarbonate) as a novel coupling reagent in the presence of DMAP to afford the desired ester in 95% yield at 93% conversion (Scheme 4.24) [93e, 94]. Finally, deprotection of the intermediate gave the final target molecule taxol in excellent yield.

This established a new method for asymmetric synthesis of baccatin III by way of B to BC to ABC to ABCD ring construction and completion of the total synthesis of taxol by preparation of the side chain by asymmetric aldol reaction and subsequent dehydration condensation with 7-TES baccatin III using DPTC. This synthetic route would be widely applicable to the preparation of a variety of derivatives of taxol and related taxoids.

4.5.7
Cephalosporolide D

Shiina developed a method for preparation of cephalosporolide D, a natural eight-membered ring lactone, and the exact stereochemistry of this compound was determined through the first total synthesis (Scheme 4.25) [95]. In this synthetic strategy two asymmetric carbon atoms were constructed by the asymmetric aldol reaction using the KSA derived from *S*-Et ethanethioate. It is also mentioned that the second diastereoselective aldol reaction afforded the desired compound in 3:97 ratio when using the chiral diamine *ent*-**2a**–Sn(II) complex and that the ratio ranges from 97:3 to 59:41 when the **2a**–Sn(II) complex or $SnCl_4$ was used as a catalyst. The desired eight-membered ring lactone moiety was constructed by cyclization of the seco acid via a novel mixed-anhydride method using (4-trifluoromethyl)benzoic anhydride (TFBA) with $Hf(OTf)_4$ [96].

4.5.8
Buergerinin F

The synthesis of buergerinin F, a natural compound consisting of a unique tricyclic skeleton, was achieved in the course of synthetic studies by Shiina

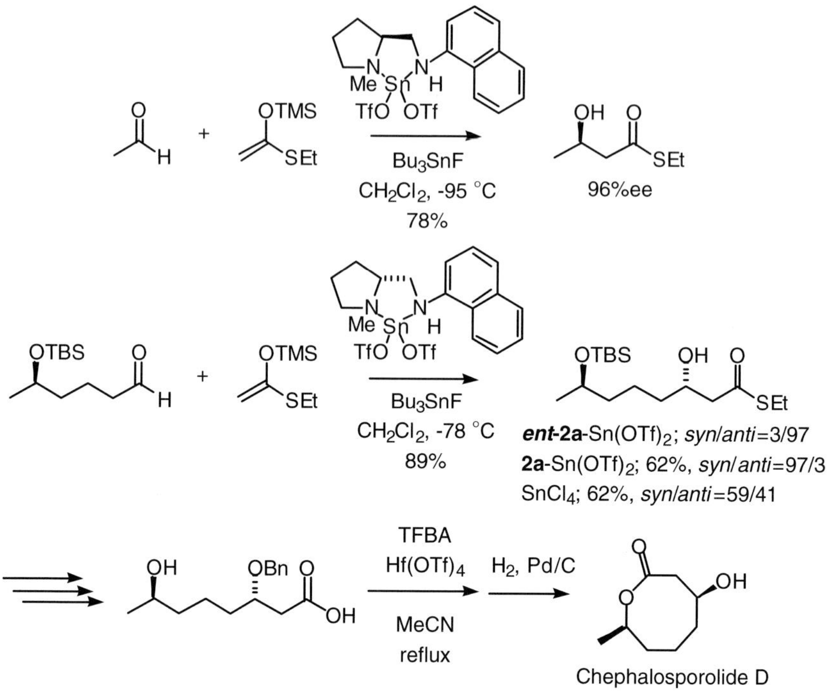

Scheme 4.25
Synthesis of cephalosporolide D.

on the utilization of the asymmetric aldol strategy [97]. The first key step is producing the optically active α,β,γ'-trioxy ester including an asymmetric quaternary carbon at the C2 position as shown in Scheme 4.26. It was also revealed that enantioselective aldol reaction of tetrasubstituted KSA with four oxygenated functional groups is quite effective for preparation of this complex synthetic intermediate. Successive intramolecular Wacker-type ketalization and one-carbon elongation of the intermediate afforded the optically active buergerinin F. On completion of the total synthesis using the asymmetric aldol reaction promoted by chiral diamine–Sn(OTf)$_2$ as catalyst, the absolute stereochemistry of natural buergerinin F was determined.

4.5.9
Octalactins A and B

Shiina recently developed a new method for synthesis of octalactin A, an antitumor agent consisting of an eight-membered ring lactone (Scheme 4.29) [98]. Because the lactone moiety includes two pairs of *anti*-β-hydroxy-α-methyl groups, enantioselective addition of the KSA derived from ethyl 2-methylthiopropanoate was efficiently used for construction of the required

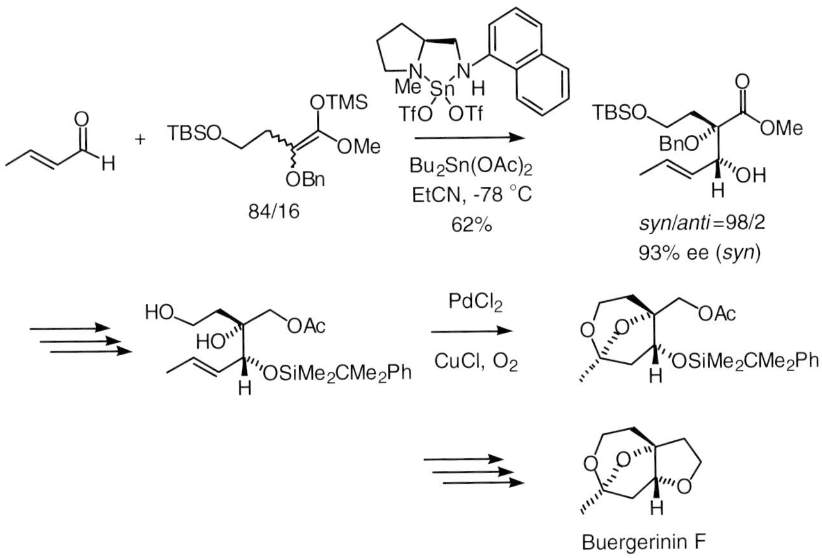

Scheme 4.26
Synthesis of buergerinin F.

components, i.e. asymmetric aldol reaction of the tetrasubstituted KSA with aldehydes [78] and subsequent treatment of the optically active adducts formed with Guidon's reduction [99] afforded the desired two chiral segments **16** and **17** (Scheme 4.27).

The optically active side chain **18** was also produced by means of the asymmetric aldol reaction of the KSA derived from *S*-Et ethanethioate with 2-methylpropionaldehyde (Scheme 4.28). A chiral linear precursor having repeated *anti*-β-hydroxy-α-methyl units was obtained by coupling segments **16** and **17**, and the resulting seco acid was eventually cyclized to form the eight-membered ring lactone by a new quite effective mixed-anhydride method using 2-methyl-6-nitrobenzoic anhydride (MNBA) with DMAP, as shown in Scheme 4.29 [100, 98b]. Finally, the side chain **18** was introduced to the eight-membered ring lactone moiety to afford the targeted multi-oxygenated compounds, octalactins A and B.

4.5.10
Oudemansin-antibiotic Analog

Uchiro and Kobayashi recently reported the synthesis of β-methoxyacrylate antibiotics (MOA) and their analogs (Scheme 4.30) [101]. In accordance with their strategy for preparation of the related compounds, asymmetric aldol reaction of the KSA generated from *S*-Et propanethioate with cinnamyl aldehyde was used for stereoselective synthesis of the intermediate of an MOA analog, as shown in Scheme 4.30.

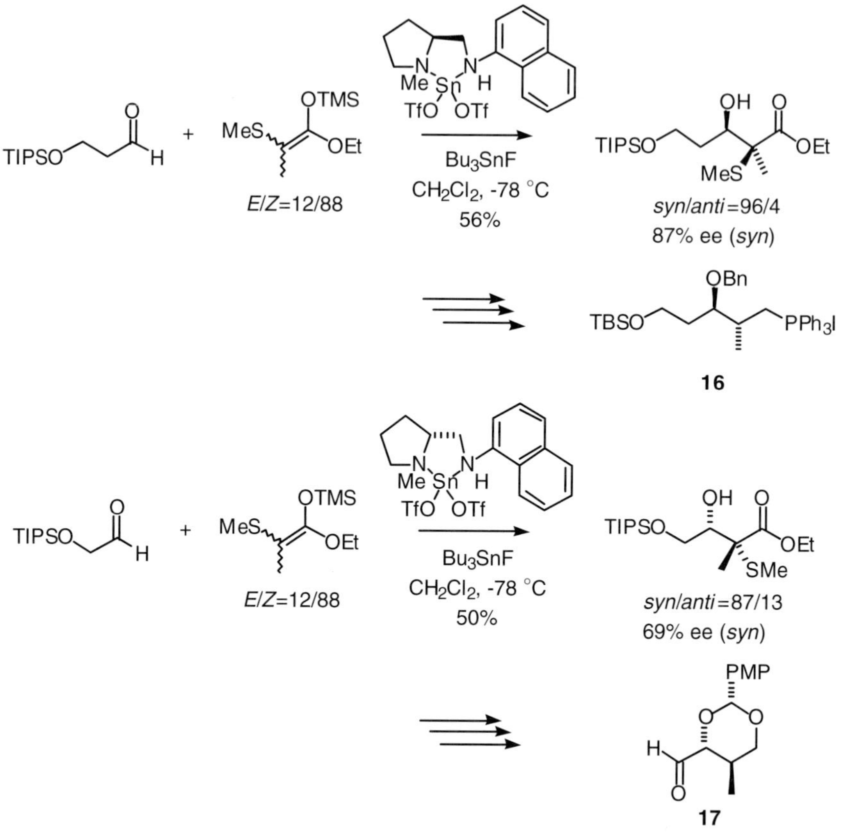

Scheme 4.27
Synthesis of two chiral segments of octalactins A and B.

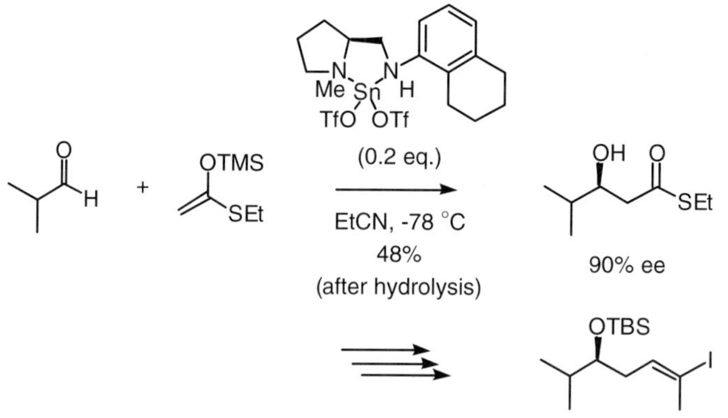

Scheme 4.28
Synthesis of the side chain of octalactins A and B.

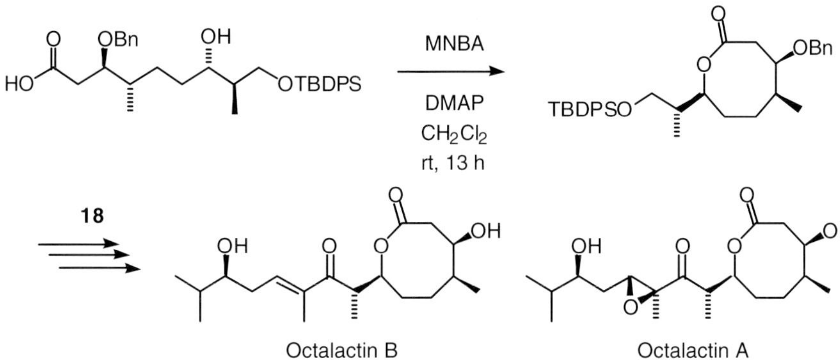

Scheme 4.29
Synthesis of octalactin A and B.

4.6
Conclusions

In this chapter a variety of Sn(IV) or Sn(II) metallic species-promoted aldol reactions have been presented, with their application in syntheses of complicated molecules with high stereoselectivity. Alkoxy aldehydes were effectively activated by SnCl$_4$, and reactions with particular ESE or KSA are highly applicable to the generation of 3,4-*syn* aldol compounds with high diastereoselectivity. Intramolecular reaction of ESE and KSA with an acetal moiety is also quite attractive for preparation of medium-sized compounds which are generally not available by other methods. Sn(II)-promoted asymmetric aldol reaction could be now used as a general and powerful method for the construction of not only optically active small molecules but highly

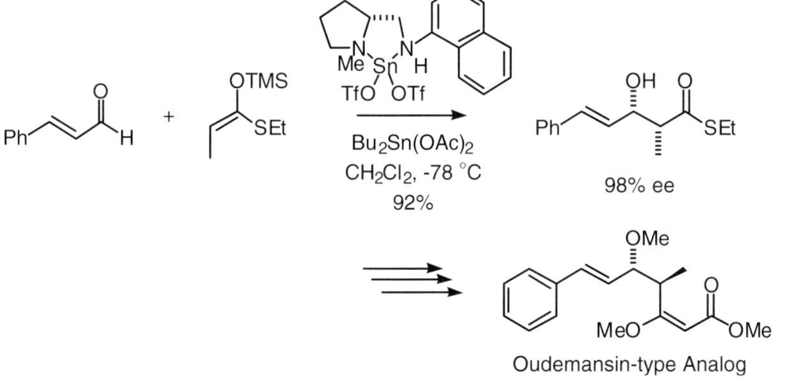

Scheme 4.30
Synthesis of oudemansin-antibiotic analog.

advanced multifunctional compounds. Progress in aldol reactions using tin reagents has contributed greatly to the syntheses of many useful substrates in the last decade, and this fruitful history might provide valuable information to organic and organometallic chemistry in the future.

4.7
Experimental

Typical Procedure for Catalytic Asymmetric Aldol Reaction of a KSA with Simple Achiral Aldehydes (Eqs. (80) and (81)) [61d]. A solution of **2a** (21.1 mg, 0.088 mmol) in EtCN (1 mL) was added to a solution of $Sn(OTf)_2$ (33.4 mg, 0.080 mmol, 20 mol%) in EtCN (1 mL). The mixture was cooled to $-78\ °C$ and a mixture of KSA (0.40 mmol) and an aldehyde (0.40 mmol) in EtCN (1.5 mL) was then added slowly over 3–4.5 h by means of a mechanical syringe. The mixture was further stirred for 2 h, and then quenched with saturated aqueous $NaHCO_3$. The organic layer was isolated and the aqueous layer was extracted with CH_2Cl_2 (three times). The organic solutions were combined, washed with H_2O and brine, then dried over Na_2SO_4. After evaporation of the solvent the crude product was purified by preparative TLC on silica gel to afford an aldol-type adduct as the corresponding trimethylsilyl ether. The trimethylsilyl ether was treated with THF–1 M HCl (20:1) at $0\ °C$ to give the corresponding alcohol.

Typical Procedure for Catalytic Asymmetric Aldol Reaction of a KSA with Ethyl Glyoxylate (Eqs. (84) and (86)) [65a]. (S,S)-bis(Benzyloxazoline) (19.9 mg, 0.055 mmol) and $Sn(OTf)_2$ (20.8 mg, 0.050 mmol) were placed, within an inert atmosphere box, in an oven-dried 8-mL vial containing a magnetic stirring bar. The flask was fitted with a serum cap, removed from the inert atmosphere box, and charged with CH_2Cl_2 (0.8 mL). The resulting suspension was stirred rapidly for 1 h to give a cloudy solution. The catalyst was cooled to $-78\ °C$ and the KSA (0.50 mmol) was added followed by distilled ethyl glyoxylate–toluene solution (8:2 mixture, 100 µL, 0.75 mmol). The resulting solution was stirred at $-78\ °C$ until the KSA was completely consumed (0.1–2 h), as determined by TLC. The reaction mixture was then filtered through a 0.3 cm × 5 cm plug of silica gel, which was then washed with Et_2O (8 mL). Concentration of the ether solution gave the crude silyl ether which was dissolved in THF (2 mL) and 1 M HCl (0.2 mL). After standing at room temperature for 0.5 h this solution was poured into a separatory funnel and diluted with Et_2O (20 mL) and H_2O (10 mL). After mixing the aqueous layer was discarded and the ether layer was washed with saturated aqueous $NaHCO_3$ (10 mL) and brine (10 mL). The resulting ether layer was dried over anhydrous Na_2SO_4, filtered, and concentrated to furnish the hydroxy esters. Purification by flash chromatography provided the desired aldols.

Typical Procedure for Synthesis of an Optically Active Diol Group by Asymmetric Aldol Reaction (Eq. (88)) [67b, 93f]. A solution of **1a** or **1b** (0.405 mmol) in CH_2Cl_2 (0.5 mL), then a solution of $Bu_2Sn(OAc)_2$ (131.1 mg, 0.373 mmol) in CH_2Cl_2 (0.5 mL), were added successively to a suspension of $Sn(OTf)_2$ (141.8 mg, 0.340 mmol) in CH_2Cl_2 (1 mL). The mixture was stirred for 30 min at room temperature, then cooled to −78 °C. A solution of 2-benzyloxy-1-ethylthio-1-(trimethylsiloxy)ethene (96.1 mg, 0.340 mmol) ($Z/E = 9$:1, the E isomer has no reactivity) in CH_2Cl_2 (0.5 mL), and a solution of aldehyde (0.228 mmol) in CH_2Cl_2 (0.5 mL) at −78 °C, were then added successively to the reaction mixture. The mixture was further stirred for 20 h then quenched with saturated aqueous $NaHCO_3$. The organic layer was isolated and the aqueous layer was extracted with CH_2Cl_2 (three times). The organic solutions were combined, washed with H_2O and brine, then dried over Na_2SO_4. After evaporation of the solvent the crude product was purified by preparative TLC on silica gel to afford the corresponding *anti-α,β*-dihydroxy thioester derivatives.

References

1 (a) MUKAIYAMA, T.; NARASAKA, K.; BANNO, K. *Chem. Lett.* **1973**, 1011. (b) MUKAIYAMA, T.; BANNO, K.; NARASAKA, K. *J. Am. Chem. Soc.* **1974**, *96*, 7503. (c) SAIGO, K.; OSAKI, M.; MUKAIYAMA, T. *Chem. Lett.* **1975**, 989. (d) MUKAIYAMA, T. *Org. React.* **1982**, *28*, 203. (e) MUKAIYAMA, T.; NARASAKA, K. *Org. Synth.* **1987**, *65*, 6.

2 WISSNER, A. *Synthesis* **1979**, 27.

3 (a) LOZZI, L.; RICCI, A.; TADDEI, M. *J. Org. Chem.* **1984**, *49*, 3408. (b) BROWNBRIDGE, P.; CHAN, T.-H. *Tetrahedron Lett.* **1980**, *21*, 3427.

4 (a) NAKAMURA, E.; KUWAJIMA, I. *Chem. Lett.* **1983**, 59. (b) NAKAMURA, E.; KUWAJIMA, I. *Tetrahedron Lett.* **1983**, *24*, 3347.

5 (a) KUWAJIMA, I.; INOUE, T.; SATO, T. *Tetrahedron Lett.* **1978**, 4887. (b) INOUE, T.; SATO, T.; KUWAJIMA, I. *J. Org. Chem.* **1984**, *49*, 4671.

6 (a) PONOMAREV, S. V.; BANKOV, Y. I.; DUDUKINA, O. V.; PETROSYAN, I. V.; PETROVSKAYA, L. I. *J. Gen. Chem. USSR* **1967**, 37, 2092. (b) NOLTES, J. G.; CREEMERS, H. M. J. C.; VAN DER KERK, G. J. M. *J. Organomet. Chem.* **1968**, *11*, 21. (c) PEREYRE, M.; BELLEGARDE, B.; MENDELSOHN, J.; VALADE, J. *J. Organomet. Chem.* **1968**, *11*, 97. (d) PEREYRE, M.; QUINTARD, J.-P.; RAHM, A. Tin in Organic Synthesis, Butterworths, London, 1987.

7 (a) YASUDA, M.; KATOH, Y.; SHIBATA, I.; BABA, A.; MATSUDA, H.; SONODA, N. *J. Org. Chem.* **1994**, *59*, 4386. (b) YASUDA, M.; HAYASHI, K.; KATOH, Y.; SHIBATA, I.; BABA, A. *J. Am. Chem. Soc.* **1998**, *120*, 715.

8 YAMAMOTO, Y.; YATAGAI, H.; MARUYAMA, K. *J. Chem. Soc. Chem. Commun.* **1981**, 162.

9 (a) SHENVI, S.; STILLE, J. K. *Tetrahedron Lett.* **1982**, *23*, 627. (b) LABADIE, S. S.; STILLE, J. K. *Tetrahedron* **1984**, *40*, 2329.

10 (a) Mukaiyama, T.; Stevens, R. W.; Iwasawa, N. *Chem. Lett.* **1982**, 353. (b) Mukaiyama, T.; Iwasawa, N. *Chem. Lett.* **1982**, 1903. (c) Mukaiyama, T.; Iwasawa, N.; Stevens, R. W.; Haga, T. *Tetrahedron* **1984**, *40*, 1381.

11 (a) Heathcock, C. H.; Hug, K. T.; Flippin, L. A. *Tetrahedron Lett.* **1984**, *25*, 5973. (b) Heathcock, C. H.; Davidsen, S. K.; Hug, K. T.; Flippin, L. A. *J. Org. Chem.* **1986**, *51*, 3027. (c) Heathcock, C. H.; Montgomery, S. H. *Tetrahedron Lett.* **1985**, *26*, 1001. (d) Montgomery, S. H.; Pirrung, M. C.; Heathcock, C. H. *Carbohydrate Res.* **1990**, *202*, 13.

12 (a) Reetz, M. T.; Kesseler, K.; Jung, A. *Tetrahedron Lett.* **1984**, *25*, 729. (b) Reetz, M. T.; Kesseler, K.; Jung, A. *Tetrahedron* **1984**, *40*, 4327. (c) Reetz, M. T.; Kesseler, K.; Schmidtberger, S.; Wenderoth, B.; Steinbach, R. *Angew. Chem. Int. Ed. Engl.* **1983**, *22*, 989. (d) Reetz, M. T. *Angew. Chem. Int. Ed. Engl.* **1984**, *23*, 556. (e) Reetz, M. T.; Kesseler, K. *J. Org. Chem.* **1985**, *50*, 5434. (f) Mahrwald, R. *Chem. Rev.* **1999**, *99*, 1095.

13 (a) Gennari, C.; Bernardi, A.; Poli, G.; Scolastico, C. *Tetrahedron Lett.* **1985**, *26*, 2373. (b) Gennari, C.; Beretta, M. G.; Bernardi, A.; Moro, G.; Scolastico, C.; Todeschini, R. *Tetrahedron* **1986**, *42*, 893. (c) Gennari, C.; Cozzi, P. G. *Tetrahedron* **1988**, *44*, 5965. (d) Gennari, C. Comprehensive Organic Synthesis, Pergamon Press, 1991, Vol. 2, 629.

14 Uenishi, J.; Tomozane, H.; Yamato, M. *Tetrahedron Lett.* **1985**, *26*, 3467.

15 (a) Bernardi, A.; Cardani, S.; Gennari, C.; Poli, G.; Scolastico, C. *Tetrahedron Lett.* **1985**, *26*, 6509. (b) Bernardi, A.; Cardani, S.; Colombo, L.; Poli, G.; Schimperna, G.; Scolastico, C. *J. Org. Chem.* **1987**, *52*, 888.

16 (a) Guanti, G.; Banfi, L.; Narisano, E.; Scolastico, C. *Tetrahedron Lett.* **1985**, *26*, 3517. (b) Banfi, L.; Cardani, S.; Potenza, D.; Scolastico, C. *Tetrahedron* **1987**, *43*, 2317.

17 Mikami, K.; Kaneko, M.; Loh, T.-P.; Terada, M.; Nakai, T. *Tetrahedron Lett.* **1990**, *31*, 3909.

18 (a) Annunziata, R.; Cinquini, M.; Cozzi, F.; Cozzi, P. G. *Tetrahedron Lett.* **1990**, *31*, 6733. (b) Annunziata, R.; Cinquini, M.; Cozzi, F.; Cozzi, P. G.; Consolandi, E. *J. Org. Chem.* **1992**, *57*, 456.

19 (a) Ewing, W. R.; Harris, B. D.; Li, W.-R.; Joullié, M. M. *Tetrahedron Lett.* **1989**, *30*, 3757. (b) Li, W.-R.; Ewing, W. R.; Harris, B. D.; Joullié, M. M. *J. Am. Chem. Soc.* **1990**, *112*, 7659.

20 (a) Danishefsky, S.; Barbachyn, M. *J. Am. Chem. Soc.* **1985**, *107*, 7761. (b) Danishefsky, S. J.; DeNinno, S. L.; Chen, S.; Boisvert, L.; Barbachyn, M. *J. Am. Chem. Soc.* **1989**, *111*, 5810.

21 (a) Cox, P.; Mahon, M. F.; Molloy, K. C.; Lister, S.; Gallagher, T. *Tetrahedron Lett.* **1989**, *30*, 2437. (b) Cox, P.; Lister, S.; Gallagher, T. *J. Chem. Soc. Perkin Trans. 1* **1990**, 3151.

22 Akiyama, T.; Ishikawa, K.; Ozaki, S. *Synlett* **1994**, 275.

23 Mukai, C.; Miyakawa, M.; Hanaoka, M. *J. Chem. Soc. Perkin Trans. 1* **1997**, 913.

24 (a) Mukai, C.; Kataoka, O.; Hanaoka, M. *Tetrahedron Lett.*

1994, *35*, 6899. (b) MUKAI, C.; KATAOKA, O.; HANAOKA, M. *J. Org. Chem.* **1995**, *60*, 5910.

25 BOECKMAN JR., R. K.; CLARK, T. J.; SHOOK, B. C. *Org. Lett.* **2002**, *4*, 2109.

26 MUKAI, C.; MOHARRAM, S. M.; KATAOKA, O.; HANAOKA, M. *J. Chem. Soc. Perkin Trans. 1* **1995**, 2849.

27 LIU, W.; SZEWCZYK, J. M.; WAYKOLE, L.; REPIC, O.; BLACKLOCK, T. J. *Tetrahedron Lett.* **2002**, *43*, 1373.

28 MUKAIYAMA, T.; HAYASHI, M. *Chem. Lett.* **1974**, 15.

29 MURATAKE, H.; NATSUME, M. *Heterocycles* **1990**, *31*, 691.

30 LINDERMAN, R. J.; GRAVES, D. M.; KWOCHKA, W. R.; GHANNAM, A. F.; ANKLEKAR, T. V. *J. Am. Chem. Soc.* **1990**, *112*, 7438.

31 PENG, S.; QING, F.-L. *J. Fluorine Chem.* **2000**, *103*, 135.

32 COSSY, J.; RAKOTOARISOA, H.; KAHN, P.; DESMURS, J.-R. *Tetrahedron Lett.* **2000**, *41*, 7203.

33 (a) PEARLMAN, B. A.; MCNAMARA, J. M.; HASAN, I.; HATAKEYAMA, S.; SEKIZAKI, H.; KISHI, Y. *J. Am. Chem. Soc.* **1981**, *103*, 4248. (b) MCNAMARA, J. M.; KISHI, Y. *J. Am. Chem. Soc.* **1982**, *104*, 7371. (c) SEKIZAKI, H.; JUNG, M.; MCNAMARA, J. M.; KISHI, Y. *J. Am. Chem. Soc.* **1982**, *104*, 7372.

34 CAMBIE, R. C.; LARSEN, D. S.; RICKARD, C. E. F.; RUTLEDGE, P. S.; WOODGATE, P. D. *Aust. J. Chem.* **1986**, *39*, 487.

35 CAMBIE, R. C.; RUTLEDGE, P. S.; WATSON, P. A.; WOODGATE, P. D. *Aust. J. Chem.* **1989**, *42*, 1939.

36 (a) ISHIHARA, K.; NAKAMURA, H.; YAMAMOTO, H. *J. Am. Chem. Soc.* **1999**, *121*, 7720. (b) NAKAMURA, H.; ISHIHARA, K.; YAMAMOTO, H. *J. Org. Chem.* **2002**, *67*, 5124.

37 (a) OTERA, J.; CHEN, J. *Synlett* **1996**, 321. (b) CHEN, J.; SAKAMOTO, K.; ORITA, A.; OTERA, J. *Synlett* **1996**, 877. (c) CHEN, J.; OTERA, J. *Synlett* **1997**, 29. (d) CHEN, J.; OTERA, J. *Tetrahedron* **1997**, *53*, 14275. (e) CHEN, J.; OTERA, J. *Angew. Chem. Int. Ed. Engl.* **1998**, *37*, 91. (f) CHEN, J.; OTERA, J. *Tetrahedron Lett.* **1998**, *39*, 1767. (g) CHEN, J.; SAKAMOTO, K.; ORITA, A.; OTERA, J. *Tetrahedron* **1998**, *54*, 8411. (h) SHIRAKAWA, S.; MARUOKA, K. *Tetrahedron Lett.* **2002**, *43*, 1469.

38 (a) ASAOKA, M.; SUGIMURA, N.; TAKEI, H. *Bull. Chem. Soc. Jpn.* **1979**, *52*, 1953. (b) ASAOKA, M.; YANAGIDA, N.; ISHIBASHI, K.; TAKEI, H. *Tetrahedron Lett.* **1981**, *22*, 4269.

39 (a) CASIRAGHI, G.; RASSU, G.; SPANU, P.; PINNA, L. *J. Org. Chem.* **1992**, *57*, 3760. (b) RASSU, G.; CASIRAGHI, G.; SPANU, P.; PINNA, L.; FAVA, G. G.; FERRARI, M. B.; PELOSI, G. *Tetrahedron: Asymmetry* **1992**, *3*, 1035. (c) CASIRAGHI, G.; RASSU, G.; SPANU, P.; PINNA, L. *Tetrahedron Lett.* **1994**, *35*, 2423. (d) RASSU, G.; ZANARDI, F.; CORNIA, M.; CASIRAGHI, G. *J. Chem. Soc. Perkin. Trans. 1* **1994**, 2431. (e) ZANARDI, F.; BATTISTINI, L.; RASSU, G.; CORNIA, M.; CASIRAGHI, G. *J. Chem. Soc. Perkin. Trans. 1* **1995**, 2471. (f) RASSU, G.; AUZZAS, L.; PINNA, L.; BATTISTINI, L.; ZANARDI, F. *J. Org. Chem.* **2000**, *65*, 6307. (g) RASSU, G.; AUZZAS, L.; PINNA, L.; ZAMBRANO, V.; ZANARDI, F.; BATTISTINI, L.; MARZOCCHI, L.; ACQUOTTI, D.; CASIRAGHI, G. *J. Org. Chem.* **2002**, *67*, 5338.

40 UNO, H.; BALDWIN, J. E.; RUSSELL, A. T. *J. Am. Chem. Soc.* **1994**, *116*, 2139.

41 (a) COCKERILL, G. S.; KOCIENSKI, P. *J. Chem. Soc. Chem. Commun.* **1983**, 705. (b) COCKERILL, G. S.; KOCIENSKI, P. *J. Chem. Soc. Perkin. Trans. 1* **1985**, 2093. (c) COCKERILL, G. S.; KOCIENSKI, P. *J. Chem. Soc. Perkin. Trans. 1* **1985**, 2101.

42 FRIEDRICH, D.; PAQUETTE, L. A. *J. Chem. Soc. Perkin. Trans. 1* **1991**, 1621.

43 TATSUTA, K.; TAKAHASHI, M.; TANAKA, N. *J. Antibiotics* **2000**, *53*, 88.

44 DIXON, D. J.; LEY, S. V.; TATE, E. W. *Synlett* **1998**, 1093.

45 HATANAKA, Y.; KUWAJIMA, I. *J. Org. Chem.* **1986**, *51*, 1932.

46 (a) HORIGUCHI, Y.; FURUKAWA, T.; KUWAJIMA, I. *J. Am. Chem. Soc.* **1989**, *111*, 8277. (b) FURUKAWA, T.; MORIHIRA, K.; HORIGUCHI, Y.; KUWAJIMA, I. *Tetrahedron* **1992**, *48*, 6975. (c) SETO, M.; MORIHIRA, K.; HORIGUCHI, Y.; KUWAJIMA, I. *J. Org. Chem.* **1994**, *59*, 3165. (d) NAKAMURA, T.; WAIZUMI, N.; TSURUTA, K.; HORIGUCHI, Y.; KUWAJIMA, I. *Synlett* **1994**, 584. (e) MORIHIRA, K.; HARA, R.; KAWAHARA, S.; NISHIMORI, T.; NAKAMURA, N.; KUSAMA, H.; KUWAJIMA, I. *J. Am. Chem. Soc.* **1998**, *120*, 12980. (f) KUSAMA, H.; HARA, R.; KAWAHARA, S.; NISHIMORI, T.; KASHIMA, H.; NAKAMURA, N.; MORIHIRA, K.; KUWAJIMA, I. *J. Am. Chem. Soc.* **2000**, *122*, 3811.

47 IWASAWA, N.; MUKAIYAMA, T. *Chem. Lett.* **1982**, 1441.

48 IWASAWA, N.; YURA, T.; MUKAIYAMA, T. *Tetrahedron* **1989**, *45*, 1197.

49 IWASAWA, N.; MUKAIYAMA, T. *Chem. Lett.* **1983**, 297.

50 STEVENS, R. W.; MUKAIYAMA, T. *Chem. Lett.* **1983**, 1799.

51 MUKAIYAMA, T.; IWASAWA, N. *Chem. Lett.* **1984**, 753.

52 MUKAIYAMA, T.; YAMASAKI, N.; STEVENS, R. W.; MURAKAMI, M. *Chem. Lett.* **1986**, 213.

53 YURA, T.; IWASAWA, N.; MUKAIYAMA, T. *Chem. Lett.* **1988**, 1021.

54 YURA, T.; IWASAWA, N.; NARASAKA, K.; MUKAIYAMA, T. *Chem. Lett.* **1988**, 1025.

55 MUKAIYAMA, T.; KOBAYASHI, S. *Org. React.* **1994**, *46*, 1.

56 (a) KOBAYASHI, S.; MUKAIYAMA, T. *Chem. Lett.* **1989**, 297. (b) KOBAYASHI, S.; UCHIRO, H.; FUJISHITA, Y.; SHIINA, I.; MUKAIYAMA, T. *J. Am. Chem. Soc.* **1991**, *113*, 4247. (c) MUKAIYAMA, T.; KOBAYASHI, S. Stereocontrolled Organic Synthesis, Blackwell Scientific Publications, 1994, 37.

57 (a) MUKAIYAMA, T.; UCHIRO, H.; KOBAYASHI, S. *Chem. Lett.* **1989**, 1001. (b) MUKAIYAMA, T.; UCHIRO, H.; KOBAYASHI, S. *Chem. Lett.* **1989**, 1757.

58 (a) KOBAYASHI, S.; SANO, T.; MUKAIYAMA, T. *Chem. Lett.* **1989**, 1319. (b) MUKAIYAMA, T.; KOBAYASHI, S.; SANO, T. *Tetrahedron* **1990**, *46*, 4653.

59 KOBAYASHI, S.; FUJISHITA, Y.; MUKAIYAMA, T. *Chem. Lett.* **1989**, 2069.

60 MUKAIYAMA, T.; ASANUMA, H.; HACHIYA, I.; HARADA, T.; KOBAYASHI, S. *Chem. Lett.* **1991**, 1209.

61 (a) MUKAIYAMA, T.; KOBAYASHI, S.; UCHIRO, H.; SHIINA, I. *Chem. Lett.* **1990**, 129. (b) KOBAYASHI, S.; OHTSUBO, A.; MUKAIYAMA, T. *Chem. Lett.* **1991**, 831. (c) MUKAIYAMA, T.; FURUYA, M.; OHTSUBO, A.; KOBAYASHI, S. *Chem. Lett.* **1991**, 989. (d) KOBAYASHI, S.; UCHIRO, H.; SHIINA, I.; MUKAIYAMA, T. *Tetrahedron* **1993**, *49*, 1761.

62 (a) KOBAYASHI, S.; FUJISHITA, Y.; MUKAIYAMA, T. *Chem. Lett.* **1990**, 1455. (b) KOBAYASHI, S.; FURUYA, M.; OHTSUBO, A.; MUKAIYAMA, T. *Tetrahedron: Asymmetry* **1991**, *2*, 635.

63 MUKAIYAMA, T.; UCHIRO, H.; KOBAYASHI, S. *Chem. Lett.* **1990**, 1147.

64 KOBAYASHI, S.; KAWASUJI, T.; MORI, N. *Chem. Lett.* **1994**, 217.

65 (a) EVANS, D. A.; MACMILLAN, D. W. C.; CAMPOS, K. R. *J. Am. Chem. Soc.* **1997**, *119*, 10859. (b) JOHNSON, J. S.; EVANS, D. A. *Acc. Chem. Res.* **2000**, *33*, 325.

66 EVANS, D. A.; WU, J.; MASSE, C. E.; MACMILLAN, D. W. C. *Org. Lett.* **2002**, *4*, 3379.

67 (a) MUKAIYAMA, T.; UCHIRO, H.; SHIINA, I.; KOBAYASHI, S. *Chem. Lett.* **1990**, 1019. (b) MUKAIYAMA, T.; SHIINA, I.; UCHIRO, H.; KOBAYASHI, S. *Bull. Chem. Soc. Jpn.* **1994**, *67*, 1708.

68 MUKAIYAMA, T.; SHIINA, I.; KOBAYASHI, S. *Chem. Lett.* **1991**, 1901.

69 KOBAYASHI, S.; HORIBE, M.; SAITO, Y. *Tetrahedron* **1994**, *50*, 9629.

70 KOBAYASHI, S.; HORIBE, M. *J. Am. Chem. Soc.* **1994**, *116*, 9805.

71 KOBAYASHI, S.; HORIBE, M. *Synlett* **1994**, 147.

72 KOBAYASHI, S.; SHIINA, I.; IZUMI, J.; MUKAIYAMA, T. *Chem. Lett.* **1992**, 373.

73 MUKAIYAMA, T.; SHIINA, I.; IZUMI, J.; KOBAYASHI, S. *Heterocycles* **1993**, *35*, 719.

74 (a) KOBAYASHI, S.; KAWASUJI, T. *Tetrahedron Lett.* **1994**, *35*, 3329. (b) KOBAYASHI, S.; HORIBE, M.; HACHIYA, I. *Tetrahedron Lett.* **1995**, *36*, 3173.

75 KOBAYASHI, S.; HAYASHI, T. *J. Org. Chem.* **1995**, *60*, 1098.

76 KOBAYASHI, S.; KAWASUJI, T. *Synlett* **1993**, 911.

77 KOBAYASHI, S.; HORIBE, M.; MATSUMURA, M. *Synlett* **1995**, 675.

78 SHIINA, I.; IBUKA, R. *Tetrahedron Lett.* **2001**, *42*, 6303.

79 (a) KOBAYASHI, S.; HORIBE, M. *Tetrahedron: Asymmetry* **1995**, *6*, 2565. (b) KOBAYASHI, S.; HORIBE, M. *Tetrahedron* **1996**, *52*, 7277. (c) KOBAYASHI, S.; HORIBE, M. *Chem. Eur. J.* **1997**, *3*, 1472.

80 KOBAYASHI, S.; HORIBE, M. *Chem. Lett.* **1995**, 1029.

81 (a) McGARVEY, G. J.; KIMURA, M.; OH, T.; WILLIAMS, J. M. *J. Carbohydr. Chem.* **1984**, *3*, 125. (b) MUKAIYAMA, T.; SUZUKI, K.; YAMADA, T.; TABUSA, F. *Tetrahedron* **1990**, *46*, 265.

82 MUKAIYAMA, T.; SHIINA, I.; KOBAYASHI, S. *Chem. Lett.* **1990**, 2201.

83 MUKAIYAMA, T.; ANAN, H.; SHIINA, I.; KOBAYASHI, S. *Bull. Soc. Chim. Fr.* **1993**, *130*, 388.

84 KOBAYASHI, S.; ONOZAWA, S.; MUKAIYAMA, T. *Chem. Lett.* **1992**, 2419.

85 (a) KANDA, Y.; FUKUYAMA, T. *J. Am. Chem. Soc.* **1993**, *115*, 8451. (b) FUKUYAMA, T.; KANDA, Y. *Synth. Org. Chem. Jpn.* **1994**, *52*, 888.

86 WHITE, J. D.; DEERBERG, J. *Chem. Commun.* **1997**, 1919.

87 (a) KOBAYASHI, S.; HAYASHI, T.; KAWASUJI, T. *Tetrahedron Lett.* **1994**, *35*, 9573. (b) KOBAYASHI, S.; FURUTA, T. *Tetrahedron* **1998**, *54*, 10275.

88 (a) KOBAYASHI, S.; HAYASHI, T.; IWAMOTO, S.; FURUTA, T.; MATSUMURA, M. *Synlett* **1996**, 672. (b) KOBAYASHI, S.; MATSUMURA, M.; FURUTA, T.; HAYASHI, T.; IWAMOTO, S. *Synlett* **1997**, 301. (c) KOBAYASHI, S.; FURUTA, T.; HAYASHI, T.; NISHIJIMA, M.; HANADA, K. *J. Am. Chem. Soc.* **1998**, *120*, 908.

89 WAKABAYASHI, T.; MORI, K.; KOBAYASHI, S. *J. Am. Chem. Soc.* **2001**, *123*, 1372.

90 (a) KOBAYASHI, S.; UENO, M.; SUZUKI, R.; ISHITANI, H. *Tetrahedron Lett.* **1999**, *40*, 2175. (b) KOBAYASHI, S.; UENO, M.; SUZUKI, R.; ISHITANI, H.; KIM, H.-S.; WATAYA, Y. *J. Org. Chem.* **1999**, *64*, 6833. (c) OKITSU, O.; SUZUKI, R.; KOBAYASHI, S. *J. Org. Chem.* **2000**, *66*, 809.

91 EVANS, D. A.; TROTTER, B. W.; COLEMAN, P. J.; CÔTÉ, B.; DIAS, L. C.; RAJAPAKSE, H. A.; TYLER, A. N. *Tetrahedron* **1999**, *55*, 8671.

92 (a) EVANS, D. A.; CEE, V. J.; SMITH, T. E.; FITCH, D. M.; CHO, P. S. *Angew. Chem. Int. Ed. Engl.* **2000**, *39*, 2533. (b) EVANS, D. A.; FITCH, D. M. *Angew. Chem. Int. Ed. Engl.* **2000**, *39*, 2536. (c) EVANS, D. A.; FITCH, D. M.; SMITH, T. E.; CEE, V. J. *J. Am. Chem. Soc.* **2000**, *122*, 10033.

93 (a) MUKAIYAMA, T.; SHIINA, I.; SAKATA, K.; EMURA, T.; SETO, K.; SAITOH, M. *Chem. Lett.* **1995**, 179. (b) SHIINA, I.; UOTO, K.; MORI, N.; KOSUGI, T.; MUKAIYAMA, T. *Chem. Lett.* **1995**, 181. (c) SHIINA, I.; IWADARE, H.; SAKOH, H.; TANI, Y.; HASEGAWA, M.; SAITOH, K.; MUKAIYAMA, T. *Chem. Lett.* **1997**, 1139. (d) SHIINA, I.; IWADARE, H.; SAKOH, H.; HASEGAWA, M.; TANI, Y.; MUKAIYAMA, T. *Chem. Lett.* **1998**, 1. (e) SHIINA, I.; SAITOH, K.; FRÉCHARD-ORTUNO, I.; MUKAIYAMA, T. *Chem. Lett.* **1998**, 3. (f) MUKAIYAMA, T.; SHIINA, I.; IWADARE, H.; SAITOH, M.; NISHIMURA, T.; OHKAWA, N.; SAKOH, H.; NISHIMURA, K.; TANI, Y.; HASEGAWA, M.; YAMADA, K.; SAITOH, K. *Chem. Eur. J.* **1999**, *5*, 121.

94 SAITOH, K.; SHIINA, I.; MUKAIYAMA, T. *Chem. Lett.* **1998**, 679.

95 (a) SHIINA, I.; FUKUDA, Y.; ISHII, T.; FUJISAWA, H.; MUKAIYAMA, T. *Chem. Lett.* **1998**, 831. (b) SHIINA, I.; FUJISAWA, H.; ISHII, T.; FUKUDA, Y. *Heterocycles* **2000**, *52*, 1105.

96 (a) SHIINA, I.; MIYOSHI, S.; MIYASHITA, M.; MUKAIYAMA, T. *Chem. Lett.* **1994**, 515. (b) SHIINA, I.; MIYASHITA, M.; MUKAIYAMA, T. *Chem. Lett.* **1994**, 677. (c) SHIINA, I. *Tetrahedron* **2004**, *59*, 1587.

97 SHIINA, I.; KAWAKITA, Y.; IBUKA, R. Abstracts of Papers, 83rd National Meeting of the Chemical Society of Japan, Tokyo, 2003, Vol. 2, 2C401; IBUKA, R. Ph.D. Thesis, Tokyo University of Science, Tokyo, Japan, 2003.

98 (a) SHIINA, I.; OSHIUMI, H.; HASHIZUME, M.; YAMAI, Y.; IBUKA, R. *Tetrahedron Lett.* **2004**, *45*, 543. (b) SHIINA, I.; KUBOTA, M.; OSHIUMI, H.; HASHIZUME, M. *J. Org. Chem.* **2004**, *69*, 1822.

99 (a) GUINDON, Y.; FAUCHER, A.-M.; BOURQUE, É.; CARON, V.; JUNG, G.; LANDRY, S. R. *J. Org. Chem.* **1997**, *63*, 9276. (b) GUINDON, Y.; JUNG, G.; GUÉRIN, B.; OGILVIE, W. W. *Synlett* **1998**, 213. (c) HENA, M. A.; TERAUCHI, S.; KIM, C.-S.;

Horiike, M.; Kiyooka, S. *Tetrahedron: Asymmetry* **1998**, *9*, 1883.

100 (a) Shiina, I.; Ibuka, R.; Kubota, M. *Chem. Lett.* **2002**, 286. (b) Shiina, I.; Kubota, M.; Ibuka, R. *Tetrahedron Lett.* **2002**, *43*, 7535.

101 Uchiro, H.; Nagasawa, K.; Kotake, T.; Hasegawa, D.; Tomita, A.; Kobayashi, S. *Bioorg. Med. Chem. Lett.* **2002**, *12*, 2821.

5
Zirconium Alkoxides as Lewis Acids

Yasuhiro Yamashita and Shū Kobayashi

5.1
Introduction

Zirconium is a group 4 element, and its low cost and low toxicity are advantageous compared with other metals used in industry. The usefulness of zirconium is well known in organic chemistry. Zirconium compounds promote some organic reactions efficiently and play important roles resulting in interesting selectivity [1]. In aldol reactions zirconium compounds have often been used to realize high and unique selectivity by forming zirconium enolates [2, 3]. Bis(cyclopentadienyl) zirconium compounds, which form zirconium enolates via metal exchange from lithium enolates, have been successfully used in stereoselective aldol reactions, and high *syn* selectivity was obtained. Both (*E*) and (*Z*) enolates gave the same *syn* adducts predominantly. This methodology was also applied to highly diastereoselective asymmetric aldol reactions to afford aldol adducts with excellent selectivity (Scheme 5.1) [3].

Zirconium alkoxides, especially zirconium tetra-*t*-butoxide, have been known to act as bases and directly deprotonate the α-hydrogen atoms of ketones to form zirconium enolates [4]. The enolates reacted with aldehydes to give aldol adducts (Scheme 5.2) [4c].

Following this report, asymmetric aldol and related reactions using chiral zirconium alkoxides as bases were investigated. Aldol–Tishchenko reactions are an efficient method for synthesizing 1,3-diol derivatives from aldehydes and enolates. It was recently shown that a zirconium enolate generated by retro-aldol reaction of a β-keto-*tert*-alcohol and a catalytic amount of a zirconium *t*-butoxide–TADDOL complex, reacted with an aldehyde to afford the corresponding 1,3-*anti*-diol via domino aldol–Tishchenko process in good yield with moderate enantioselectivity (Scheme 5.3) [5].

Similar to the aldol reaction of zirconium enolates, zirconium Lewis acid-mediated aldol reactions of silicon enolates with aldehydes (Mukaiyama aldol reaction) have also been well explored [6]. Zirconium Lewis acids are comparatively mild and have been employed in several stereoselective re-

Modern Aldol Reactions. Vol. 2: Metal Catalysis. Edited by Rainer Mahrwald
Copyright © 2004 WILEY-VCH Verlag GmbH & Co. KGaA, Weinheim
ISBN: 3-527-30714-1

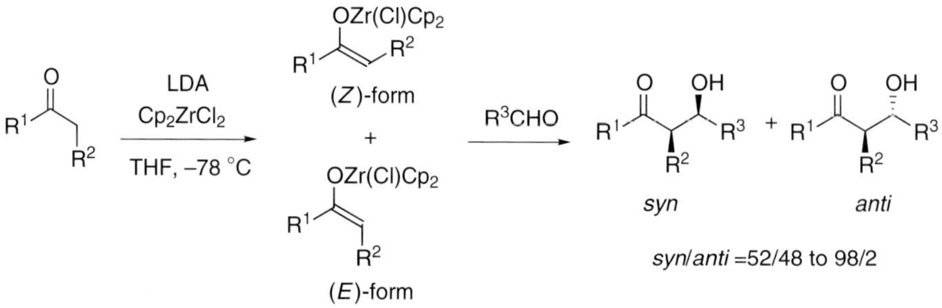

Scheme 5.1

Stereoselective aldol reactions using zirconium enolates.

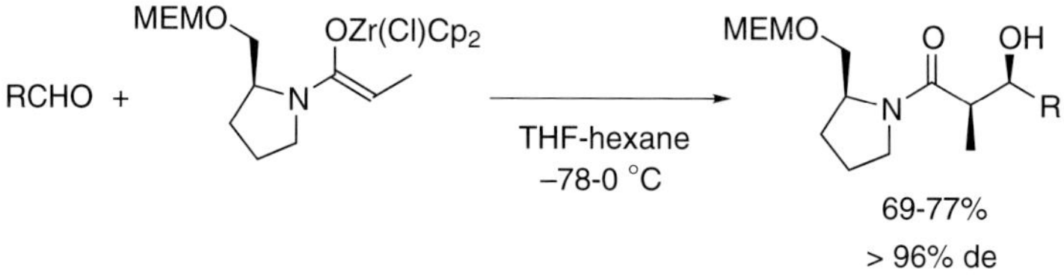

Scheme 5.2

Direct aldol reaction using a zirconium alkoxide.

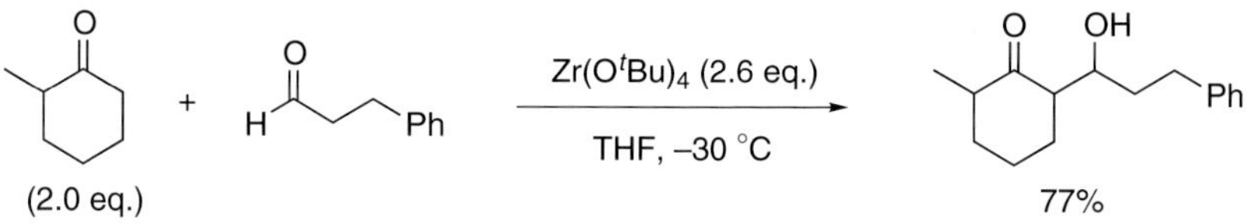

84%, 57% ee

Scheme 5.3

Asymmetric aldol-Tishchenko reaction using a chiral zirconium catalyst.

actions. Among these, zirconium alkoxides have served as mild and stereoselective Lewis acids, especially in asymmetric catalysis. Several catalytic asymmetric reactions using chiral zirconium alkoxides as Lewis acids have been developed [4c]. In chiral modification of zirconium catalysts, chiral alcohol derivatives or chiral phenol derivatives, especially 1,1'-binaphthalene-2,2'-diol (BINOL) derivatives, have often been employed. In this chapter, aldol and related reactions catalyzed by zirconium alkoxides as chiral Lewis acids are discussed.

5.2
The Asymmetric Mukaiyama Aldol Reaction

A chiral zirconium catalyst prepared from zirconium alkoxide and BINOL derivatives has been successfully applied to the catalytic asymmetric Mukaiyama aldol reaction [7]. Among several types of zirconium catalyst, a catalyst prepared from zirconium tetra-*t*-butoxide and 3,3'-dihalogeno-1,1'-bi-2-naphthol (3,3'-X$_2$BINOL) [8] was found to be effective. The aldol reactions of benzaldehyde **1a** with ketene silyl acetals proceeded smoothly in toluene at 0 °C in the presence of an additional alcohol. It was found that the additional alcohol played important roles in this reaction (vide infra) [9]. Among the catalysts screened, a Zr catalyst containing 3,3'-I$_2$BINOL was the most effective, and high level of enantiocontrol was achieved. The reproducibility of the reaction was not as good, however, and the enantioselectivity was sometimes poor. After several investigations to address this issue it was finally revealed that a small amount of water had an important effect on enantioselectivity [10]. The effect of water was significant. Under strictly anhydrous conditions enantioselectivity was occasionally quite low. It was revealed that the presence of a small amount of water was essential to realize high enantioselectivity (Table 5.1).

The effect of alcohol additives was investigated in the reaction of benzaldehyde **1a** with the ketene silyl acetal of *S*-ethyl ethanethioate (**2a**) (Table 5.2). In the presence of water reactions using normal primary alcohols such as ethanol (EtOH), propanol (PrOH), and butanol (BuOH) gave high yields and high enantioselectivity (entries 1–3). Other primary alcohols such as benzyl alcohol (BnOH) and 2,2,2-trifluoroethanol (CF$_3$CH$_2$OH) gave lower yields and selectivity (entries 4 and 5). Secondary and tertiary alcohols such as isopropanol (*i*PrOH) and *tert*-butyl alcohol (*t*BuOH) resulted in reduced yields and selectivity (entries 6 and 7). Phenol also resulted in lower yield and selectivity (entry 8). The best yields and enantioselectivity were obtained when 80–120 mol% PrOH was used (entries 9–12) and similar yields and selectivity were obtained when zirconium tetrapropoxide–propanol complex (Zr(OPr)$_4$–PrOH) was employed instead of Zr(O^{t}Bu)$_4$ (entry 14). The use of Zr(OPr)$_4$–PrOH is desirable economically.

Other substrates were then examined; the results are shown in Table 5.3. The ketene silyl acetal derived from methyl isobutyrate (**2b**) also worked well. For aldehydes, whereas aromatic and α,β-unsaturated aldehydes gave excellent yields and selectivity, aliphatic aldehydes resulted in high yields but somewhat lower selectivity.

Diastereoselective aldol reactions using this chiral zirconium catalyst were then examined (Table 5.4). First, the ketene silyl acetal derived from methyl

Tab. 5.1
Effect of water in asymmetric Mukaiyama aldol reactions using a chiral zirconium catalyst.

Entry	Silicon Enolate	Product	H_2O (mol%)	Yield (%)	ee (%)
1[a]	OSiMe₃	OH O	0	28	16
2			0	42	3
3	SEt	Ph SEt	10	86	83
4	**2a**	**4aa**	20	94	88
5	OSiMe₃ OMe	OH O Ph OMe	0	94	−10[b]
6	**2b**	**4ab**	20	88	94

[a] PrOH was not used.
[b] The product obtained was the opposite enantiomer.

(R)-3,3'-I₂BINOL (**3a**)

propionate (**2c**) was employed in the reaction with benzaldehyde. The reaction proceeded smoothly to afford the desired *anti*-aldol adduct in high yield with high diastereo- and enantioselectivity when ethanol was used as a primary alcohol. The selectivity was further improved by use of the ketene silyl acetal derived from phenyl propionate (**2d**). Other aldehydes such as *p*-anisaldehyde (**1b**), *p*-chlorobenzaldehyde (**1g**), cinnamaldehyde (**1d**), and 3-phenylpropionealdehyde (**1e**), etc., were tested, and all the reactions proceeded smoothly, and the desired *anti*-aldol adducts were obtained in high yield with high diastereo- and enantioselectivity. In the reactions of the ketene silyl acetals derived from propionate derivatives, most chiral Lewis acids led to *syn* diastereoselectivity. Few catalyst systems giving *anti*-aldol adducts with high selectivity are known, so the general *anti* selectivity was a remarkable feature of the zirconium aldol reaction [11].

Although the high *anti* selectivity observed in these reactions is remarkable, examination of the effect of the geometry of the ketene silyl acetals revealed further important information on the selectivity – when the (E) and (Z) ketene silyl acetals (**2e** and **2f**) derived from methyl propionate were employed in reactions with benzaldehyde, high *anti* selectivity was obtained for both, and it was confirmed that selectivity was independent of the ge-

Tab. 5.2
Effect of alcohol (additive).

$$Ph\underset{1a}{\overset{O}{\downarrow}}H + \underset{2a}{\overset{OSiMe_3}{\diagdown}}SEt \xrightarrow[\text{toluene, 0°C, 14 h}]{\substack{Zr(O^tBu)_4 \ (10 \ mol\%) \\ (R)\text{-}3,3'\text{-}I_2BINOL \ (\mathbf{3a}) \ (12 \ mol\%) \\ PrOH, H_2O \ (20 \ mol\%)}} \underset{4aa}{Ph\overset{OH}{\diagup}\overset{O}{\diagdown}SEt}$$

Entry	ROH (mol%)	Yield (%)	ee (%)
1	EtOH (50)	85	87
2	PrOH (50)	94	88
3	BuOH (50)	92	86
4	BnOH (50)	76	76
5	CF_3CH_2OH (50)	47	62
6	^{i}PrOH (50)	87	85
7	^{t}BuOH (50)	39	44
8	PhOH (50)	36	54
9	PrOH (30)	68	74
10	PrOH (80)	91	95
11	PrOH (100)	95	95
12	PrOH (120)	94	95
13	PrOH (160)	95	91
14[a]	PrOH (60)	98	92

[a] $Zr(OPr)_4$-PrOH was used instead of $Zr(O^tBu)_4$.

Tab. 5.3
Asymmetric aldol reactions (1).

$$R^1\overset{O}{\underset{H}{\diagup}} + \underset{R^3}{\overset{OSiMe_3}{\diagup}}\overset{R^3}{\diagdown}XR^2 \xrightarrow[\text{toluene, 0°C, 18 h}]{\substack{Zr(O^tBu)_4 \ (10 \ mol\%) \\ (R)\text{-}3,3'\text{-}I_2BINOL \ (\mathbf{3a}) \ (12 \ mol\%) \\ PrOH \ (80 \ mol\%), H_2O \ (20 \ mol\%)}} R^1\underset{R^3 \ R^3}{\overset{OH \quad O}{\diagup \diagdown}}XR^2$$

Entry	Aldehyde (R¹)	Silicon Enolate	Yield (%)	ee (%)
1	Ph (1a)	2a	91	95
2	Ph (1a)	2b	95	98
3	p-MeOC$_6$H$_4$ (1b)	2b	92	96
4	(E)-CH$_3$CH=CH (1c)	2b	76	97
5	(E)-PhCH=CH (1d)	2b	94	95
6	PhCH$_2$CH$_2$ (1e)	2a	92	80
7	CH$_3$(CH$_2$)$_4$(1f)	2a	93 (93)[a]	84 (87)[a]

[a] The reaction was performed at -20 °C.

$$\underset{2a}{\overset{OSiMe_3}{\diagup}SEt} \qquad \underset{2b}{\overset{OSiMe_3}{\diagup}OMe}$$

Tab. 5.4
Asymmetric aldol reactions (2).

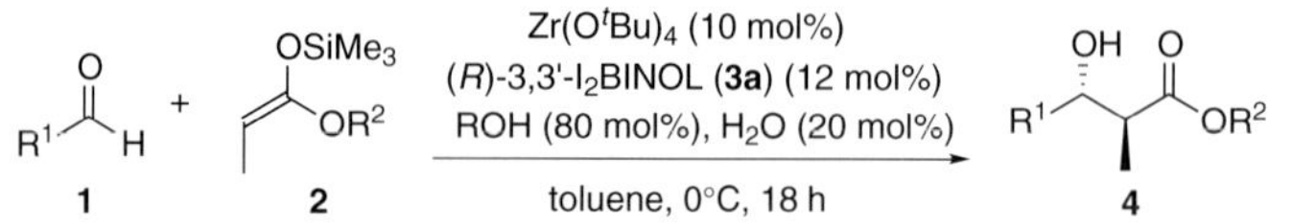

Entry	Aldehyde (R^1)	Silicon Enolate	ROH	Yield (%)	syn/ anti	ee (%)
1	Ph (**1a**)	**2c**	PrOH	79	7/93	96
2	Ph (**1a**)	**2c**	EtOH	87	7/93	97
3	Ph (**1a**)	**2d**	PrOH	94	5/95	99
4	Ph (**1a**)	**2d**	EtOH	90	4/96	99
5	p-MeOC$_6$H$_4$ (**1b**)	**2d**	PrOH	89	7/93	98
6	p-ClC$_6$H$_4$ (**1g**)	**2d**	PrOH	96	9/91	96
7	(E)-CH$_3$CH=CH (**1c**)	**2d**	PrOH	65	11/89	92
8	(E)-PhCH=CH (**1d**)	**2d**	PrOH	92	15/85	98
9	PhCH$_2$CH$_2$ (**1e**)	**2d**	PrOH	61	14/86	89

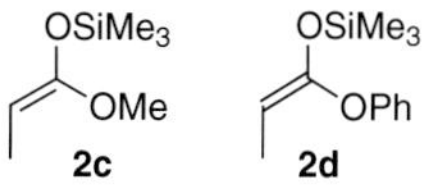

ometry of the ketene silyl acetals (Scheme 5.4). For the transition states of these reactions, acyclic pathways are assumed (details are discussed below).

Further investigation of the effect of the aldehyde structure was conducted (Table 5.5). Reactions of other aliphatic aldehydes were examined

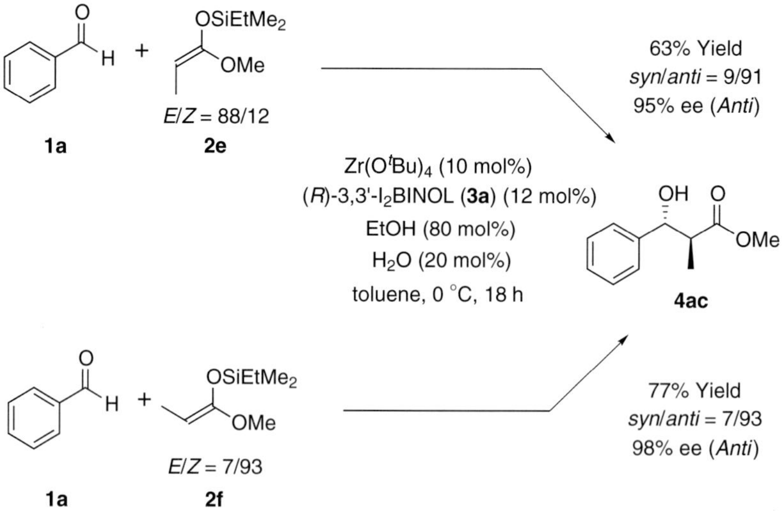

Scheme 5.4
Effect of geometry of the silicon enolates.

Tab. 5.5
Effect of structures of aliphatic aldehydes.

$$R \text{–CHO (1)} + \text{(2d: OSiMe}_3/\text{OPh)} \xrightarrow[\text{toluene, 0°C, 18 h}]{\substack{\text{Zr(O}^t\text{Bu)}_4 \text{ (20 mol\%)} \\ (R)\text{-3,3'-I}_2\text{BINOL (3a) (24 mol\%)} \\ \text{PrOH (160 mol\%), H}_2\text{O (20 mol\%)}}} \text{4 (OH / O / OPh)}$$

Entry	Aldehyde		Yield (%)	syn/anti	ee (%) (anti)
1	⌒⌒⌒CHO	(1f)	64	12/88	85
2	Ph⌒⌒CHO	(1e)	71	10/90	82
3	⌒⌒CHO	(1h)	71	15/85	81
4	(CH₃)₂CH⌒CHO	(1i)	56	12/88	89
5	cyclohexyl-CH₂⌒CHO	(1j)	52	14/86	78
6	(CH₃)₂CH–CH(CH₃)CHO	(1k)	16	17/83	28
7	cyclohexyl–CH₂CHO	(1l)	14	21/79	31
8	cyclohexyl–CHO	(1m)	Trace	–	–

using a chiral zirconium catalyst consisting of Zr(O^{t}Bu)$_4$, (R)-3,3'-I$_2$BINOL (**3a**), PrOH, and water. For normal linear aliphatic aldehydes, for example hexanal and butanal, the reactions proceeded with high selectivity. γ-Branched aldehydes also reacted smoothly to afford the desired *anti* adducts in good yield and with high diastereo- and enantioselectivity. The catalyst did not, however, work well in reactions of α-branched and β-branched aliphatic aldehydes, possibly because of by steric interaction between the BINOL parts (especially large di-iodo atoms at the 3,3'-positions) of the catalysts and the alkyl moieties of the aldehydes. These results indicated that the environment around the zirconium of the catalyst was crowded and that the catalyst recognized the structure of the aldehydes strictly.

To create more effective catalyst systems, improvement of catalyst activity is important. It has recently been revealed that introduction of stronger electron-withdrawing groups at the 6,6'-positions of the binaphthyl rings effectively improved the Lewis acidity of the zirconium catalyst system [12]. The effect of stronger electron-withdrawing groups at the 6,6'-positions of (R)-3,3'-I$_2$BINOL in this aldol system was therefore examined. Bromo, iodo, and pentafluoroethyl groups were selected as the electron-withdrawing

Tab. 5.6

Improvement of catalyst activity in the aldol reaction.

Entry	BINOL Derivatives	Yield (%)	syn/anti	ee (%) (anti)
1	(R)-3,3'-I$_2$BINOL (**3a**) (6 mol%)	38	5/95	96
2	(R)-3,3'-I$_2$-6,6'-Br$_2$BINOL (**3b**) (7.5 mol%)	61	4/96	98
3	(R)-3,3'-I$_2$-6,6'-(C$_2$F$_5$)$_2$BINOL (**3c**) (7.5 mol%)	71	7/93	96
4	(R)-3,3',6,6'-I$_4$BINOL (**3d**) (7.5 mol%)	70	4/96	98

X: H, Br, I, C$_2$F$_5$

(R)-3,3'-I$_2$-6,6'-X$_2$BINOL (**3**)

groups. In the aldol reaction of benzaldehyde with the ketene silyl acetal derived from *S*-ethyl propanethioate (**2g**), the new catalysts prepared from 6,6'-disubstituted-3,3'-I$_2$BINOL (**3b–d**) had greater activity and the reaction proceeded much faster than with the catalyst prepared from 3,3'-I$_2$BINOL. In particular, iodo and pentafluoroethyl groups at the 6,6'-positions led to better results, giving the desired *anti* adducts in high yields with high diastereo- and enantioselectivity (Table 5.6). The new catalyst system was successfully applied to the reactions of aliphatic aldehydes. In the reactions of hexanealdehyde with the ketene silyl acetals derived from phenyl propionate (**2d**) and *S*-ethyl propanethioate (**2g**), the best results were obtained when (R)-3,3',6,6'-I$_4$BINOL (**3d**) was employed (Table 5.7). The electron-withdrawing substituents at the 3,3'- and 6,6'-positions of the BINOL derivatives were assumed to increase the Lewis acidity of the zirconium catalysts. By changing the chiral ligands, chemical yields were improved (38% to 71% in Table 5.6, entries 1 and 3; 9% to 92% in Table 5.7, entries 4 and 6) much more than enantioselectivity (80% ee to 87% ee in Table 5.7, entries 1 and 3).

A chiral zirconium complex prepared from zirconium tetra-*t*-butoxide, 2.2 equiv. 6,6'-dibromo-1,1'-binaphthalene-2,2'-diol (6,6'-Br$_2$BINOL), and a small amount of water, which was originally developed in asymmetric Mannich-type reactions [12], was found to be an effective catalyst in asymmetric aldol-type reactions using ethyl diazoacetate [13]. The chiral zirco-

Tab. 5.7
Effect of new BINOLs in the aldol reactions of hexanaldehyde **1f**.

Entry	X	Silicon Enolate	Yield (%)	syn/anti	ee (%) (anti)
1	H (**3a**)	OSiMe₃	53	16/84	80
2	C₂F₅ (**3c**)	OPh	39	11/89	84
3	I (**3d**)	**2d**	66	12/88	87
4	H (**3a**)	OSiMe₃	9	12/88	93
5	C₂F₅ (**3c**)	SEt	80	17/83	93
6	I (**3d**)	**2g**	92	11/89	93

(R)-3,3'-I₂-6,6'-X₂BINOL (**3**)

nium complex deprotonates the α-hydrogen atom of the diazo ester directly (Scheme 5.5).

Scheme 5.5
Asymmetric aldol reaction of an azoester.

5.3
Asymmetric Hetero Diels–Alder Reaction

Hetero Diels–Alder (HDA) reactions of aldehydes with 1-methoxy-3-trimethylsiloxy-1,3-butadiene (Danishefsky's diene) also proceeded in the presence of the chiral zirconium catalyst [14]. HDA reactions of aldehydes with Danishefsky's dienes [15] mediated by Lewis acids, which provide 2,3-dihydro-4H-pyran-4-one derivatives, are promising tools for construction of pyran ring systems [16]. Because Danishefsky's dienes contain a silicon enolate moiety, an aldol-type reaction and subsequent cyclization process

Tab. 5.8

Asymmetric hetero Diels–Alder reactions of benzaldehyde (**1a**) using a chiral zirconium catalyst.

Zr(O^{t}Bu)$_4$ (10 mol%)
(R)-3,3'-I$_2$BINOL (**3a**) (12 mol%)
PrOH, H$_2$O (20 mol%)   TFA
toluene, 18 h

Entry	Diene	PrOH (mol%)	Temp. (°C)	Yield (%)	ee (%)
1	2h	50	0	39	22
2	2h	80	0	35	62
3	2i	50	0	50	91
4	2i	80	0	65	94
5	2i	120	0	44	94
6	2i	80	−20	70	97
7	2i	80	−45	Trace	–
8	2j	80	−20	80	97
9	2k	80	−20	24	89
10[a]	2j	80	−20	Quant.	97

[a] Toluene/tBuOMe (2:1) was used as solvent.

OSiMe$_3$

OMe

2h

OSiR$_3$

O^{t}Bu

2i: SiR$_3$ = SiMe$_3$
2j: SiR$_3$ = SiEtMe$_2$
2k: SiR$_3$ = Si^{t}BuMe$_2$

are a possible pathway in HDA reactions using Danishefsky's dienes [17]. It was therefore assumed that the chiral zirconium catalyst effective in aldol reactions might work well in HDA reactions of aldehydes with Danishefsky's dienes. On the basis of this assumption a model HDA reaction of benzaldehyde with 1-methoxy-3-trimethylsiloxy-1,3-butadiene (**2h**) using a chiral zirconium catalyst prepared from Zr(O^{t}Bu)$_4$, (R)-3,3'-I$_2$BINOL, PrOH, and water was examined (Table 5.8). The initial result was rather disappointing, however, and the corresponding pyranone derivative was obtained in lower yield and selectivity (Table 5.8, entry 1). The selectivity was improved to 62% ee by increasing the amount of PrOH, but the yield was even lower (entry 2). According to the reaction pathway the product was formed with elimination of methanol. It was suspected that this methanol might decompose the diene, reducing the yield, and that the selectivity might be reduced by interaction of the methanol with the zirconium catalyst. To prevent production of methanol, 1-*tert*-butoxy-3-trimethylsiloxy-1,3-butadiene (**2i**) was used instead of **2h**. As expected, yield and selectivity were improved, and the desired adduct was obtained in 50% yield with 91% ee (entry 3). Yield and selectivity were also found to be affected by the amount of PrOH and the reaction temperature, and the desired product was ob-

Tab. 5.9

Asymmetric hetero Diels–Alder reactions (1).

$$
\begin{array}{c}
\text{Zr(O}^t\text{Bu)}_4 \text{ (10 mol\%)} \\
\text{(R)-3,3'-I}_2\text{BINOL (3a) (12 mol\%)} \\
\text{PrOH (80 mol\%), H}_2\text{O (20 mol\%)} \quad \text{TFA} \\
\hline
\text{solvent, } -20°\text{C, 18 h}
\end{array}
$$

Entry	Aldehyde; R	Solvent	Yield (%)	ee (%)
1	Ph (**1a**)	Toluene	80	97
2	Ph (**1a**)	Toluene-^{t}BuOMe[a]	Quant.	97
3	p-MeC$_6$H$_4$ (**1n**)	Toluene	63	95
4	p-MeC$_6$H$_4$ (**1n**)	Toluene-^{t}BuOMe[a]	95	95
5	p-ClC$_6$H$_4$ (**1g**)	Toluene	65	84
6	p-ClC$_6$H$_4$ (**1g**)	Toluene-^{t}BuOMe[a]	90	84
7	PhCH$_2$CH$_2$ (**1e**)	Toluene	84	90
8	PhCH$_2$CH$_2$ (**1e**)	Toluene-^{t}BuOMe[a]	Quant.	90
9	CH$_3$(CH$_2$)$_4$ (**1f**)	Toluene	69	91
10	CH$_3$(CH$_2$)$_4$ (**1f**)	Toluene-^{t}BuOMe[a]	98	93
11[b]	(E)-PhCH=CH (**1d**)	Toluene-^{t}BuOMe[a]	97	90

[a] Toluene/^{t}BuOMe (2:1) was used.
[b] Sc(OTf)$_3$ (10 mol%) was used instead of TFA.

tained in 70% yield with 97% ee when the reaction was performed using 80 mol% PrOH at −20 °C (entry 6). The effects of substituents on the silicon atoms of Danishefsky's dienes and of the solvents in this HDA reaction were further examined. When diene **2j**, with an ethyldimethylsilyloxy group, was employed the desired product was obtained in 80% yield with 97% ee (entry 8), although the more stable diene **2k**, with a *tert*-butyldimethylsilyloxy group, did not work well under these conditions (entry 9). Finally, the desired pyranone derivative was obtained quantitatively with 97% ee by use of the mixed solvent system toluene–^{t}BuOMe, 2:1 (entry 10) [18].

The reactions of different aldehydes with diene **2j** in toluene or toluene–^{t}BuOMe, 2:1, were then tested (Table 5.9). Aromatic, α,β-unsaturated, and even aliphatic aldehydes reacted with the Danishefsky diene to afford the desired HDA adducts in good to high yields with high enantioselectivity. It was noted that high stereocontrol was achieved even in reactions of aliphatic aldehydes. With regard to solvents, use of the toluene–^{t}BuOMe system always resulted in higher yields than use of toluene. In these HDA pathways the reaction was quenched by adding saturated aqueous NaHCO$_3$, and after usual work-up the crude adduct was treated with trifluoroacetic acid (TFA) to accelerate formation of the pyranone derivative. In the reaction with cinnamaldehyde, however, the desired cyclic product was not obtained in good yield after treatment with TFA, because side-reactions occurred. When other work-up conditions were investigated it was found that treatment with a

Tab. 5.10

Asymmetric hetero Diels–Alder reactions (2).

Entry	Aldehyde; R	BINOL	Conditions	Yield (%)	cis/ trans	ee (%) (trans)
1	Ph (**1a**)	**3a**	Toluene, 0 °C, 18 h	Trace	–	–
2	Ph (**1a**)	**3c**	Toluene, −20 °C, 18 h	Quant.	8/92	98
3	Ph (**1a**)	**3c**	Toluene, −40 °C, 24 h	99	4/96	97
4	p-MeC$_6$H$_4$ (**1n**)	**3c**	Toluene, −20 °C, 18 h	93	13/87	90
5	p-MeC$_6$H$_4$ (**1n**)	**3c**	Toluene, −40 °C, 24 h	99	6/94	93
6	p-ClC$_6$H$_4$ (**1g**)	**3c**	Toluene, −20 °C, 18 h	99	10/90	97
7	p-ClC$_6$H$_4$ (**1g**)	**3c**	Toluene, −40 °C, 24 h	99	4/96	98
8[a]	p-ClC$_6$H$_4$ (**1g**)	**3c**	Toluene, −40 °C, 168 h	90	6/94	97
9[b]	(E)-PhCH=CH (**1d**)	**3c**	Toluene, −10 °C, 18 h	78	12/88	87
10[b]	(E)-PhCH=CH (**1d**)	**3c**	Toluene, −20 °C, 60 h	96	10/90	90
11[c]	PhCH$_2$CH$_2$ (**1e**)	**3c**	Toluene, 0 °C, 48 h	23	15/85	79
12[c]	PhCH$_2$CH$_2$ (**1e**)	**3d**	Toluene, −20 °C, 48 h	68	10/90	87
13[c]	PhCH$_2$CH$_2$ (**1e**)	**3d**	Toluene/^{t}BuOMe (2:1), −20 °C, 72 h	97	10/90	90
14[c]	CH$_3$(CH$_2$)$_4$ (**1f**)	**3d**	Toluene, −20 °C, 48 h	63	9/91	88
15[c]	CH$_3$(CH$_2$)$_4$ (**1f**)	**3d**	Toluene/^{t}BuOMe (2:1), −20 °C, 72 h	94	9/91	95

[a] 2 mol% Zr catalyst. PrOH (80 mol%). [b] Sc(OTf)$_3$ (10 mol%) was used instead of TFA. [c] PrOH (120 mol%).

(R)-**3a**: X=H
(R)-**3c**: X=C$_2$F$_5$
(R)-**3d**: X=I

catalytic amount of scandium triflate (Sc(OTf)$_3$) [19] gave a high yield of the desired product.

The HDA reactions of 4-methyl-substituted Danishefsky's diene, which include diastereo- and enantiofacial selectivity issues, were then investigated. 1-*tert*-Butoxy-2-methyl-3-trimethylsiloxy-1,3-pentadiene (**2l**) was selected as a model, and was reacted with benzaldehyde using a chiral zirconium catalyst with an (R)-3,3'-I$_2$BINOL moiety under the conditions shown in Table 5.9 (toluene was used as solvent). Unexpectedly, the reaction proceeded sluggishly (Table 5.10, entry 1), a result which clearly showed that 4-substituted diene **2l** was less reactive than 4-unsubstituted dienes **2h–2k**. To increase the Lewis acidity of the zirconium catalyst, introduction of electron-withdrawing groups at the 6,6'-positions of the BINOL derivatives was investigated. (R)-3,3′-I$_2$-6,6′-(C$_2$F$_5$)$_2$BINOL (**3c**) was chosen as chiral ligand and a chiral zirconium catalyst was prepared from Zr(O^{t}Bu)$_4$, (R)-**3c**, PrOH, and water. It was remarkable that in the presence of 10 mol% of this chiral zirconium catalyst the reaction of benzaldehyde (**1a**) with **2l** proceeded smoothly in toluene at −20 °C to afford the desired HDA adduct quantitatively. It was also of interest that the stereochemistry of the product was 2,3-*trans*, and that the enantiomeric excess of the *trans* adduct was proved to be 98% (Table 5.10, entry 2). The *trans* selectivity was further improved when the reaction was performed at −40 °C (entry 3). The reactions of other aldehydes, including aromatic, α,β-unsaturated, and aliphatic aldehydes, using this new chiral zirconium catalyst were examined. With aromatic and α,β-unsaturated aldehydes the reactions proceeded smoothly to give the desired pyranone derivatives in high yield with high *trans* selectivity; the enantiomeric excesses of the *trans* adducts were also high. Yield and selectivity were, however, lower in the reaction of an aliphatic aldehyde (entry 11). They were finally improved when (R)-3,3′,6,6′-I$_4$BINOL ((R)-**3d**) was used instead of (R)-**3c** and the reaction was conducted in toluene–^{t}BuOMe, 2:1, as a solvent; the desired products were obtained in high yield with high diastereo- and enantioselectivity. It should be noted this was the first example of catalytic asymmetric *trans*-selective HDA reactions of aldehydes, and that a wide variety of aldehydes react with high yields and selectivity.

HDA reactions using more functionalized Danishefsky's dienes were also investigated. 3-Oxygenated 2-alkyl-2,3-dihydro-4*H*-pyran-4-one derivatives are important synthetic intermediates, affording hexose derivatives [20]. As a new approach to hexose derivatives, an HDA reaction of an aldehyde with a Danishefsky's diene having an oxy-substituent at the 4-position has already been developed [21]. As a catalytic asymmetric HDA reaction using this type of diene the reaction of benzaldehyde (**1a**) with 1-*tert*-butyldimethylsilyloxy-4-*tert*-butoxy-2-trimethylsilyloxy-1,3-butadiene (**2m**) was conducted using the Zr-3,3′,6,6′-I$_4$BINOL catalyst system. The reaction proceeded sluggishly and it was speculated that the bulky substituent at the 4-position prevented the smooth progress of the reaction. Next, the reaction employing 1-benzyloxy-4-*tert*-butoxy-2-trimethylsilyloxy-1,3-butadiene (**2n**) as diene component was investigated.

The HDA reactions of aldehydes with diene **2n** using the zirconium catalyst were conducted under optimized conditions (Table 5.11). The reactions proceeded smoothly in toluene–*tert*-butyl methyl ether, 2:1, to afford the desired cycloadducts in high yield with high diastereo- and enantioselectivity. It should be noted that the stereochemistry of the adduct obtained was 2,3-*cis*, completely opposite to the 2,3-*trans* selectivity obtained in the reaction with diene **2l** (details of the selectivity are discussed in Section 5.4). In the reaction of other aldehydes, aromatic aldehydes and α,β-unsaturated and aliphatic aldehydes reacted with the diene smoothly to afford the desired products in high yield with high *cis* selectivity; the enantiomeric excess of the *cis* adducts was also high.

Tab. 5.11

Asymmetric hetero Diels–Alder reactions (3).

Zr(O^{t}Bu)$_4$ (10 mol%)
(*R*)-3,3',6,6'-I$_4$BINOL (**3d**) (12 mol%)
PrOH (160 mol%), H$_2$O (20 mol%)   Sc(OTf)$_3$

toluene/^{t}BuOMe (2:1)
−20 °C, 96 h

Entry	Aldehyde; R	Yield (%)	cis/trans	ee (%) (cis)
1	Ph (**1a**)	95	97/3	97
2	*p*-MeC$_6$H$_4$ (**1n**)	90	95/5	94
3	*p*-ClC$_6$H$_4$ (**1g**)	Quant.	97/3	97
4	*p*-NO$_2$C$_6$H$_4$ (**1o**)	85	93/7	90
5	(*E*)-PhCH=CH (**1d**)	Quant.	85/15	92
6	PhCH$_2$CH$_2$ (**1e**)	54	92/8	81

OSiMe$_3$

^{t}BuMe$_2$SiO   O^{t}Bu

2m

5.4
Reaction Mechanism

In asymmetric aldol reactions the zirconium catalyst had *anti* selectivity irrespective of enolate geometry. This remarkable feature was in contrast with most *syn*-selective aldol reactions mediated by known chiral Lewis acids [22]. In the usual reactions affording *syn*-aldol adducts the selectivity was explained in terms of steric repulsion between the alkyl groups of the aldehydes and the α-methyl groups of enolates in acyclic transition state models. In the zirconium-catalyzed reactions, *anti*-aldol adducts were obtained from both (*E*) and (*Z*) enolates, showing that acyclic transition states were most likely. It was speculated that the origin of the *anti* selectivity was steric interaction between the α-methyl groups of enolates and not the alkyl groups of aldehydes but chiral Lewis acids coordinated to carbonyl oxygen atoms (Figure 5.1) [23]. The asymmetric environment around the zirconium center seemed to be very crowded, because of the bulky iodo groups at the 3,3'-positions of the BINOL derivatives. Experiments in which this zirconium complex strictly recognized the structures of aliphatic aldehydes also seemed to be indicative of highly steric hindrance around the active site of the catalyst.

Two mechanistic pathways have, on the other hand, been considered for HDA reactions of carbonyl compounds with Danishefsky's diene catalyzed by Lewis acids (Scheme 5.6) [17]. One is a concerted [4+2] cycloaddition pathway, the other a stepwise cycloaddition pathway (Mukaiyama-aldol reaction and cyclization). In most reports of HDA reactions of aldehydes with 4-substituted Danishefsky's dienes catalyzed by chiral Lewis acids, fa-

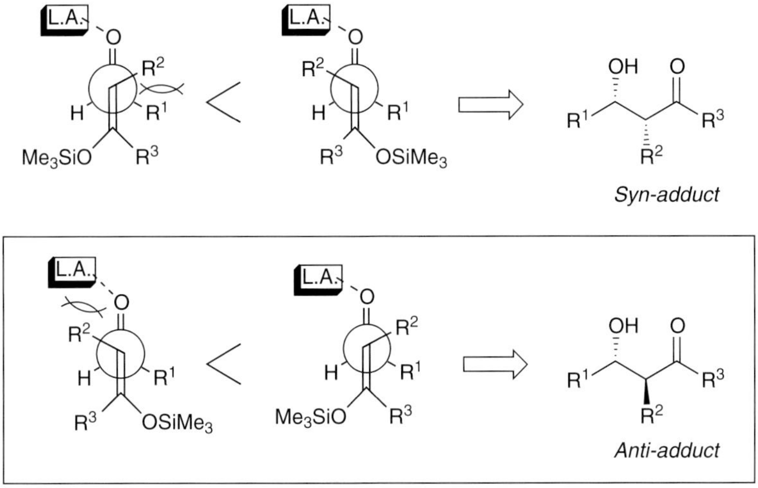

Fig. 5.1
Origin of the *anti*-selectivity.

vored products were 2,3-*cis*-disubstituted pyranones [24]. In reactions with 4-methyl Danishefsky's diene using the chiral zirconium complex, however, remarkable 2,3-*trans* selectivity was observed. This unique selectivity is difficult to explain on the basis of the concerted [4+2] cycloaddition mechanism, because of the disadvantage of the *exo*-type transition state. In almost

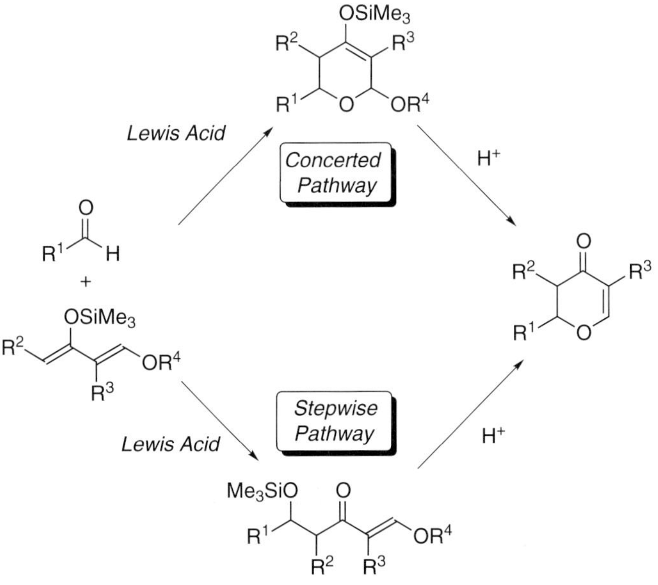

Scheme 5.6
Proposed reaction pathways in HDA reaction.

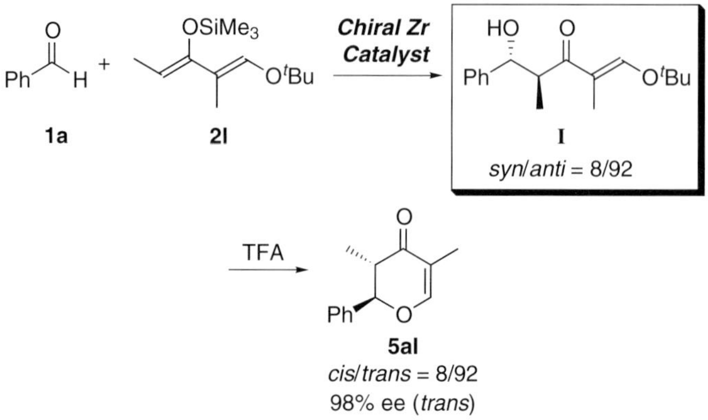

Scheme 5.7
Isolation of an intermediate of the HDA reaction.

all HDA reactions using the zirconium catalyst, it has been reported that the reaction mixture was simple, and that only one new product was observed by TLC analysis. In the reaction of benzaldehyde (**1a**) with 1-*tert*-butoxy-2-methyl-3-trimethylsiloxy-1,3-pentadiene (**2l**), the product was carefully isolated by use of deactivated silica gel column chromatography before treatment with TFA. It was revealed that the product isolated was the corresponding *anti*-aldol adduct (**I**) as a hydroxy-free form, and that high *anti* selectivity was observed (*syn/anti* = 8:92), as shown in Scheme 5.7. In addition, the aldol adduct (**I**) readily cyclized quantitatively under acidic conditions to afford the product with high selectivity (*cis/trans* = 8:92, 98% ee (*trans*)). These facts, the observed 2,3-*trans* selectivity, and the isolation of the *anti*-aldol intermediate, indicate that the HDA reaction catalyzed by the chiral zirconium complex proceeds via a stepwise (Mukaiyama-aldol reaction and cyclization) pathway. This unique 2,3-*trans* selectivity can therefore be explained by the remarkable *anti*-selective Mukaiyama aldol reactions using the chiral zirconium catalyst system, which proceeded with *anti* preference irrespective of the *E* and *Z* geometry of the silicon enolates, as already mentioned. On this basis the lower reactivity of diene **2k** would be understood by considering the greater stability of the *tert*-butyldimethylsilyloxy group than that of the trimethylsilyloxy or ethyldimethylsilyloxy group. The effect of the *tert*-butoxy group of the dienes on the enantioselectivity could, moreover, be explained in terms of steric hindrance effectively preventing the [4+2] cycloaddition pathway. Remarkable *cis* selectivity obtained in the reaction with 1-benzyloxy-4-*tert*-butoxy-2-trimethylsilyloxy-1,3-butadiene (**2n**) could also be accounted for by interaction of the benzyloxy group with the zirconium center of the catalyst. In the mechanism of zirconium-catalyzed aldol reactions steric repulsion between the methyl

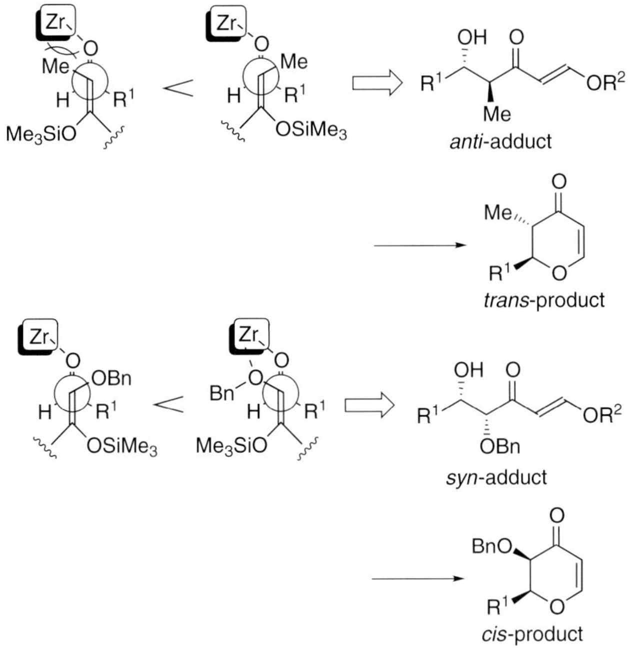

Fig. 5.2
origin of the *trans*- and *cis*-selectivity.

group of the enolate and the zirconium catalyst seemed to be an important factor explaining *anti* selectivity in an open-chain transition-state model. In reactions of the diene **2n** coordination of the oxygen atom of the benzyloxy group would be more favored than steric repulsion, and the stereochemical outcome results in *cis* selectivity (Figure 5.2) [25].

An assumed catalytic cycle for this aldol and HDA reaction is shown in Scheme 5.8. First, the zirconium catalyst **A** is produced by mixing $Zr(O^tBu)_4$, (*R*)-3,3'-I$_2$BINOL, a primary alcohol, and H_2O. At this stage, the remaining *t*-butoxide groups are exchanged for the primary alcohols or H_2O. An aldehyde coordinated to this catalyst and a silicon enolate attack the carbonyl carbon of the aldehyde to generate intermediate **B**. The silyl group on the carbonyl oxygen is then removed by the primary alcohol, directly generating the aldol product and the original catalyst again, or moves to the most anionic atom in the same complex, the oxygen of the binaphthol, to form intermediate **C**. The Si–O and Zr–O bonds of the intermediate **C** are also cleaved by the primary alcohols to form the aldol adduct and the catalyst again. This mechanism is supported by the observation that aldol adducts are obtained with free hydroxyl groups and that the trimethylsilyl ether of the alcohol and the mono trimethylsilyl ether of 3,3'-I$_2$BINOL are observed in the reaction system.

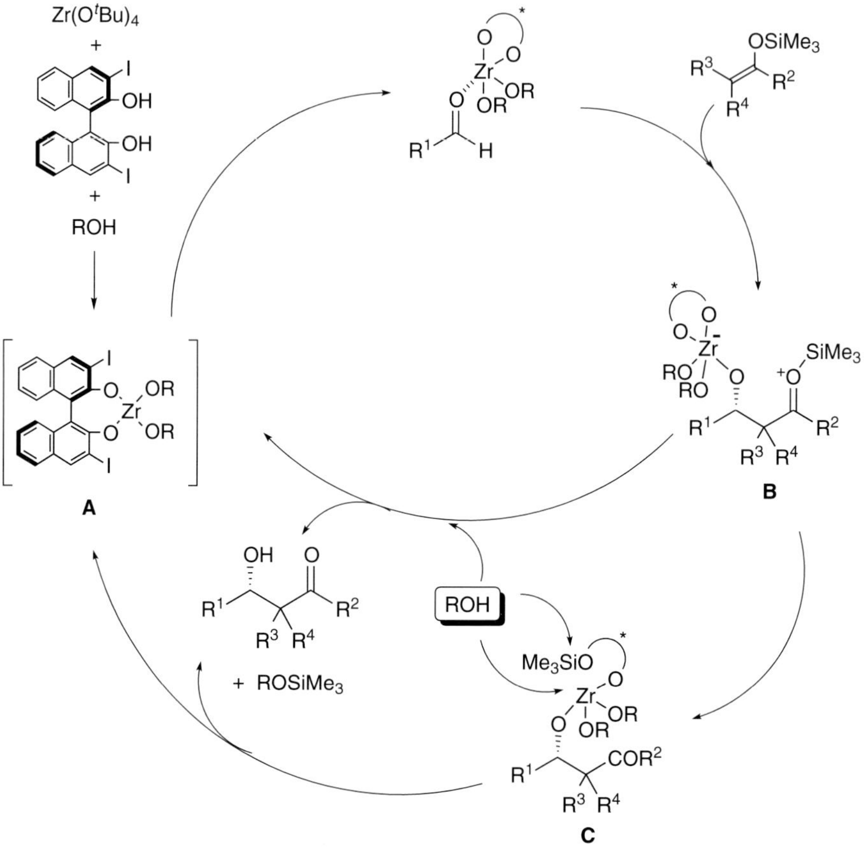

Scheme 5.8
Assumed catalytic cycle of the aldol reaction.

5.5
Structure of the Chiral Zirconium Catalyst

NMR experiments were performed to clarify the structure of the chiral zirconium catalyst. The catalyst was prepared from 1 equiv. $Zr(OPr)_4$–PrOH, 1 equiv. $3,3'-I_2BINOL$, and 1 equiv. H_2O in toluene-d_8. ^1H and ^{13}C NMR spectra were acquired at room temperature, and clear and simple signals were observed (Figure 5.3). It was revealed that this catalyst was stable in the presence of excess PrOH at room temperature and that almost the same spectra were obtained after one day. In the ^{13}C NMR spectrum two new kinds of signal corresponding to the naphthyl rings and two kinds of signal corresponding to the propoxide groups were observed in addition to the signals corresponding to the free BINOL. The presence of these two kinds of sharp signal suggested that the catalyst formed a dimeric structure. We

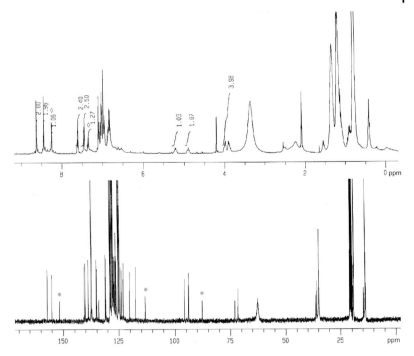

a The complex was prepared from $Zr(OPr)_4$-PrOH (1.0 equiv.),
(R)-3,3'-I$_2$BINOL (**3a**) (1.0 equiv.), and H$_2$O (1.0 equiv.).
* : free 3,3'-I$_2$BINOL

Fig. 5.3
^{1}H and ^{13}C NMR spectra of the zirconium complex[a].

also observed characteristic signals of propoxide protons connected directly to the carbon atoms attached to the oxygen atoms at 3.8, 4.0, 4.8, and 5.2 ppm in the ^{1}H NMR spectrum (O–CH$_2$–). Integration of the proton signals indicated the presence of two kinds of propoxide moiety (one pair observed at 3.8 and 4.0 ppm and the other at 4.8 and 5.2 ppm) in the catalyst; the ratio was 2:1.

The role of a small amount of water in this catalyst system was also revealed by NMR analysis [10, 26]. In the absence of PrOH and water a clear ^{13}C NMR spectrum was obtained from the combination of Zr(O^{t}Bu)$_4$ and 3,3'-I$_2$BINOL (Figure 5.4a). When PrOH was added to this system, rather complicated signals were observed (Figure 5.4b). Clear signals appeared once again when water was added to the catalyst system consisting of Zr(O^{t}Bu)$_4$, 3,3'-I$_2$BINOL, and PrOH (Figure 5.4c). From these results, it was assumed that the role of water in this catalyst system was to arrange the structure of the catalyst, i.e. the desired structure was formed from the oligomeric structure by adding water.

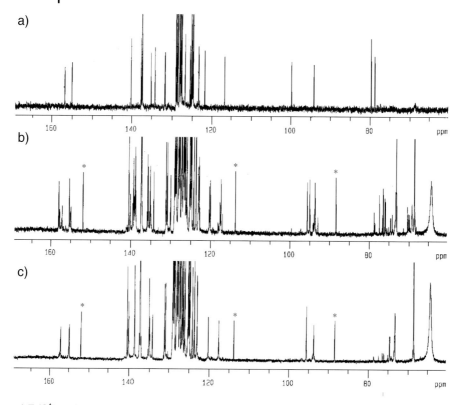

a) Zr(O^{t}Bu)$_4$ (1.0 equiv.) + (R)-3,3'-I$_2$BINOL (**3a**) (1.0 equiv.)

b) Zr(O^{t}Bu)$_4$ (1.0 equiv.) + (R)-3,3'-I$_2$BINOL (**3a**) (1.0 equiv.) + PrOH (5.0 equiv.)

c) Sample b) +PrOH (2.0 equiv.) + H$_2$O (1.0 equiv.)

* : free 3,3'-I$_2$BINOL

Fig. 5.4
Effect of water.

Because of the dimeric structure of the catalyst, the possibility of a non-linear effect in the asymmetric aldol reaction was examined [27]. The reaction of benzaldehyde with the ketene silyl acetal derived from S-ethyl etha-nethioate (**2a**) was chosen as a model, and the chiral Zr catalysts prepared from 3,3'-I$_2$BINOLs with lower enantiomeric excess were employed. It was found that a remarkable positive non-linear effect was observed, as illustrated in Figure 5.5. After preparation of the chiral Zr catalysts from (R)-3,3'-I$_2$BINOL and (S)-3,3'-I$_2$BINOL, respectively, they were combined and correlation between the ee of the zirconium catalyst and the ee of the product was investigated. A linear correlation between them was observed (Figure 5.6) [28]. These results also supported the dimeric structure of the cata-

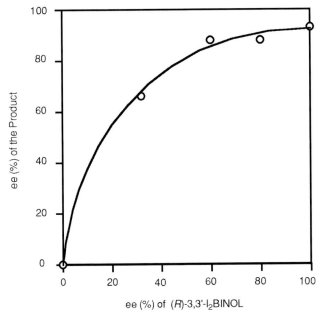

Fig. 5.5
Correlation between the ee of the product
and the ee of (*R*)-3,3'-I$_2$BINOL (**3a**) in the
aldol reaction using the catalyst prepared
from (*R*)-3,3'-I$_2$BINOLs with low ee.

lyst; on the basis of these experiments it was assumed the catalyst structure
was as shown in Figure 5.7.

5.6
Air-stable and Storable Chiral Zirconium Catalyst

Although fruitful results have been obtained by use of chiral Lewis acids as
catalysts in asymmetric synthesis, it has been known that Lewis acid catalysts are often sensitive to moisture and/or oxygen, even in air, and decompose rapidly in the presence of a small amount of water. Accordingly, most
chiral Lewis acids must be prepared in situ under strictly anhydrous conditions just before use, often with tedious handling, and they cannot be
stored for extended periods. This is also true for chiral zirconium catalysts.
Development of air-stable and storable chiral Lewis acid catalysts is therefore desirable [29].

To address this issue, an air-stable, storable chiral zirconium catalyst
(ZrMS) has been developed in catalytic asymmetric Mannich-type reactions
[30]. This catalyst is stable in air at room temperature and is easy to handle

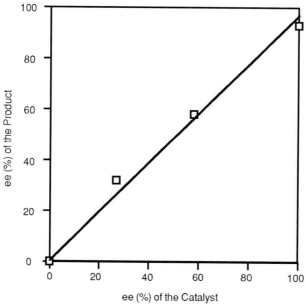

Fig. 5.6
Correlation between the ee of the product
and the ee of the catalyst in the aldol
reaction using the catalyst prepared by
mixing (R)- and (S)-catalyst.

A

Fig. 5.7
Assumed catalyst structure.

without requiring strict attention to levels of moisture and oxygen, etc. The
key to this stability was combination of the catalyst and zeolite (5-Å molecular sieves (MS 5A)). This method should be generally applicable to zirconium catalysts.

The zirconium catalyst with MS (3I-ZrMS) was prepared simply by combining the already prepared zirconium catalyst and MS 5A [31]. In the
course of development of 3I-ZrMS, the amount of water in MS 5A was
found to be important to achieving high enantioselectivity in the reaction

Tab. 5.12
Effect of storage time of 3I-ZrMS.

Entry	Storage Time (weeks)	Yield (%)	syn/anti	ee (%) (anti)
1	0	Quant.	5/95	99
2	2	Quant.	5/95	99
3	6	Quant.	5/95	99
4	13	Quant.	5/95	99

[10]. In practice 3I-ZrMS was prepared by first combining a zirconium propoxide propanol complex (Zr(OPr)$_4$–PrOH) and 3,3'-I$_2$BINOL in toluene at room temperature for 3 h. MS 5A (2.5 g mmol^{-1}) containing 10% (w/w) H$_2$O was then added and the mixture was stirred for 5 min. After removal of the solvents under reduced pressure at room temperature for 1 h the 3I-ZrMS catalyst was formed. Compared with zirconium *tert*-butoxide Zr(OPr)$_4$–PrOH is an economical source of zirconium alkoxide. The aldol reaction of benzaldehyde (**1a**) with ketene silyl acetal **2d** was then conducted using 5 mol% 3I-ZrMS catalyst in toluene at 0 °C in the presence of PrOH (80 mol%). The reaction proceeded smoothly to afford the desired product in high yield with high selectivity (quantitative yield, *syn/anti* = 5/95, *anti* 99% ee). The result obtained by use of 3I-ZrMS was almost comparable with that obtained by use of the zirconium catalyst prepared in situ. It is worthy of note that this 3I-ZrMS catalyst was remarkably stable in air and moisture, and that the catalyst could be stored for at least 13 weeks at room temperature without loss of reactivity and selectivity (Table 5.12).

The 3I-ZrMS catalyst was successfully applied to asymmetric aldol reactions of a variety of substrates; the results are summarized in Table 5.13. In reactions of benzaldehyde (**1a**) with other silicon enolates (**2a** and **2b**) the 3I-ZrMS catalyst worked well and excellent yields and enantioselectivity were obtained (entries 1–3). In reactions with the ketene silyl acetal derived from phenyl propionate (**2d**), *anti*-aldol adducts were obtained with high diastereo- and enantioselectivity (entry 4). The reactions of *p*-methoxy- and *p*-chlorobenzaldehyde (**1b**, **1n**) and α,β-unsaturated aldehyde with **2d** also occurred with high diastereo- and enantioselectivity (entries 5–7). With 3-phenylpropionaldehyde (**1e**), slight decreases of yield and selectivity were observed (entry 8). It is noted that high stereocontrol was achieved in reactions of several aldehydes and that *anti*-aldol adducts were obtained with excellent diastereo- and enantioselectivity.

It was also found that the hetero Diels–Alder reaction of benzaldehyde (**1a**) with Danishefsky's diene **2j** proceeded smoothly in the presence of

Tab. 5.13
Asymmetric aldol reactions using 3I-ZrMS.

Entry	Aldehyde; R^1	Silicon Enolate	Yield (%)	syn/anti	ee (%) (anti)
1	Ph (**1a**)	**2a**	Quant.	–	92
2[a]	Ph (**1a**)	**2a**	97	–	94
3	Ph (**1a**)	**2b**	92	–	94
4	Ph (**1a**)	**2d**	Quant.	5/95	99
5	p-MeOC$_6$H$_4$ (**1b**)	**2d**	80	5/95	94
6	p-ClC$_6$H$_4$ (**1n**)	**2d**	Quant.	8/92	95
7	PhCH=CH (**1d**)	**2d**	94	16/84	98
8	PhCH$_2$CH$_2$ (**1e**)	**2d**	65	15/85	87

[a] 10 mol% catalyst.

Scheme 5.9
Asymmetric hetero Diels–Alder reaction using 3I-ZrMS.

3I-ZrMS to afford the desired product in high yield with high enantiose-lectivity (Scheme 5.9).

5.7
Conclusion

Asymmetric aldol reactions and hetero Diels–Alder reactions via the Mu-kaiyama aldol process using chiral zirconium alkoxides as Lewis acids have been discussed. Both reactions proceeded under milder conditions to afford β-hydroxy esters and 2,3-dihydro-4*H*-pyran-4-one derivatives in high yield with high diastereo- and enantioselectivity. Addition of primary alcohols played an important role in catalyst turnover. It was discovered that a small amount of water affected the structure of the catalyst and that the presence of water was essential for high enantioselectivity. The remarkable stereoselectivity obtained would be ascribed to the unique steric features of the zirconium complex. Zirconium catalysts can be stored for a long time without loss of activity after stabilization on molecular sieves 5A. This methodology will contribute to practical asymmetric aldol chemistry.

5.8
Experimental

Typical Experimental Procedure for Asymmetric Aldol Reactions Using Chiral Zirconium Catalyst Prepared from 3,3′-I$_2$BINOL (3a). Zr(O^{t}Bu)$_4$ (0.040 mmol) in toluene (1.0 mL) was added at room temperature to a suspension of (R)-3,3′-diiodo-1,1′-binaphthalene-2,2′-diol ((R)-3,3′-I$_2$BINOL, 0.048 mmol) in toluene (1.0 mL) and the solution was stirred for 30 min. Propanol (0.32 mmol) and H$_2$O (0.080 mmol) in toluene (0.5 mL) were then added and the mixture was stirred for 3 h at room temperature. After cooling to 0 °C aldehyde **1** (0.40 mmol) in toluene (0.75 mL) and silicon enolate **2** (0.48 mmol) in toluene (0.75 mL) were successively added. The mixture was stirred for 18 h and saturated aqueous NaHCO$_3$ (10 mL) was added to quench the reaction. After addition of dichloromethane (10 mL) the organic layer was isolated and the aqueous layer was extracted with dichloromethane (2 × 10 mL). The organic layers were combined and dried over anhydrous Na$_2$SO$_4$. After filtration and concentration under reduced pressure the residue was treated with THF–1 M HCl (20:1) for 1 h at 0 °C. The solution was then made alkaline with saturated aqueous NaHCO$_3$ and extracted with dichloromethane. The organic layers were combined and dried over anhydrous Na$_2$SO$_4$. After filtration and concentration under reduced pressure the crude product was purified by preparative thin-layer chromatography (benzene–ethyl acetate, 20:1) to afford the desired aldol adduct **4**. The optical purity was determined by HPLC analysis on a chiral column. For some compounds optical purity was determined after acetylation or benzoylation of the hydroxy group.

Typical Experimental Procedure for Asymmetric Hetero Diels–Alder Reactions Using a Chiral Zirconium Catalyst Prepared from (R)-3,3′-I$_2$-6,6′-X$_2$BINOL (X: C$_2$F$_5$ (3c), I (3d)). Zr(O^{t}Bu)$_4$ (0.040 mmol) in toluene (0.5 mL) was added at room temperature to a suspension of (R)-3,3′-diiodo-6,6′-disubstituted-1,1′-binaphthalene-2,2′-diol ((R)-3,3′-I$_2$-6,6′-X$_2$BINOL (X: I, C$_2$F$_5$), 0.048 mmol) in toluene (0.5 mL) and the solution was stirred for 3 h. A mixture of propanol (0.32 mmol) and H$_2$O (0.080 mmol) in toluene (0.3 mL) was added and the mixture was stirred for 30 min at room temperature. After cooling to −78 °C, aldehyde **1** (0.40 mmol) in toluene (0.35 mL) and diene **2** (0.48 mmol) in toluene (0.35 mL) were successively added. The mixture was warmed to −20 °C and stirred for 18 h. Saturated aqueous NaHCO$_3$ (10 mL) was then added to quench the reaction. After addition of CH$_2$Cl$_2$ (10 mL) the organic layer was isolated and the aqueous layer was extracted with CH_2Cl_2 (2×10 mL). The organic layers were combined and dried over anhydrous Na_2SO_4. After filtration and concentration under reduced pressure the residue was treated with TFA (0.5 mL) in CH_2Cl_2 (8 mL) for 1 h at 0 °C. For reactions of α,β-unsaturated aldehydes and reactions with diene **2n**, scandium triflate ($Sc(OTf)_3$, 0.040 mmol, 10 mol% relative to the employed aldehyde) was used instead of TFA in CH_2Cl_2 at room temperature for 12 h. The solution was made alkaline with saturated aqueous $NaHCO_3$ (20 mL), the organic layer was isolated, and the aqueous layer was extracted with CH_2Cl_2 (2×10 mL). The organic layers were combined and dried over anhydrous Na_2SO_4. After filtration and concentration under reduced pressure, the *trans* and *cis* isomers were separated and purified by preparative thin-layer chromatography (benzene–ethyl acetate, 20:1). The optical purity of *trans* and *cis* isomers were determined by HPLC analysis on a chiral column (see following analytical data). For reactions with diene **2n** diastereoselectivity was determined by ^1H NMR analysis of the diastereomixtures and enantioselectivity was determined by HPLC analysis of the diastereomixtures.

Typical Experimental Procedure for Asymmetric Aldol Reactions Using 3l-ZrMS. PrOH (19.2 mg, 0.32 mmol) in toluene (0.3 mL) was added at room temperature to a suspension of 3l-ZrMS (74.5 mg, 5 mol%) in toluene (0.9 mL) and the mixture was stirred for 1 h at the same temperature. After cooling to 0 °C aldehyde **1** (42.5 mg, 0.4 mmol) in toluene (0.4 mL) and silicon enolate **2** (107 mg, 0.48 mmol) in toluene (0.4 mL) were successively added and the mixture was stirred for 18 h at the same temperature. The reaction was quenched with saturated aqueous $NaHCO_3$, and dichloromethane (CH_2Cl_2) was added. The organic layer was isolated, and the aqueous layer was extracted with CH_2Cl_2. The organic extracts were combined and dried over anhydrous sodium sulfate. After filtration and concentration under reduced pressure the crude mixture was purified by preparative thin-layer chromatography (SiO_2, benzene–ethylacetate) to afford the desired aldol adduct. The diastereomer ratio was determined by ^1H NMR analysis, and the optical purity was determined by HPLC analysis on a chiral column directly or after acetylation.

References

1 (a) CARDIN, D. J.; LAPPERT, M. F.; RASTON, C. L.; RILEY, P. I. In *Comprehensive Organometallic Chemistry*; WILKINSON, G.; STONE, F. G. A.; ABEL, E. W. Eds.; Pergamon, New York, **1982**, Vol. 3, 549; (b) NEGISHI, E.; TAKAHASHI, T. *Acc. Chem. Res.* **1994**, *72*, 2591; (c) TAKAHASHI, T.; KOTORA, M.; HARA, R.; XI, Z. *Bull. Chem. Soc. Jpn.* **1999**, 72; for review of zirconium alkoxide in catalysis see: (d) YAMASAKI, S.; KANAI, M.;

SHIBASAKI, M. *Chem. Eur. J.* **2001**, *7*, 4066; see also: (e) KROHN, K. *Synthesis*, **1997**, 1115.

2 (a) EVANS, D. A.; MCGEE, L. R. *Tetrahedron Lett.* **1980**, *21*, 3975; (b) YAMAMOTO, Y.; MARUYAMA, K. *Tetrahedron Lett.* **1980**, *21*, 4607; (c) SAUVÉ, G.; SHWARTZ, D. A.; RUEST, L.; DESLONGCHAMPS, P. *Can. J. Chem.* **1984**, *62*, 2929; (d) BROWN, D. W.; CAMPBELL, M. M.; TAYLOR, A. P.; ZHANG, X.-a. *Tetrahedron Lett.* **1987**, *28*, 985; (e) PANEK, J. S.; BULA, O. A. *Tetrahedron Lett.* **1988**, *29*, 1661; (f) CURRAN, D. P.; CHAO, J.-C. *Tetrahedron* **1990**, *46*, 7325; (g) YAMAGO, S.; MACHII, D.; NAKAMURA, E. *J. Org. Chem.* **1991**, *56*, 2098; (h) WIPF, P.; XU, W.; SMITROVICH, J. H. *Tetrahedron* **1994**, *50*, 1935.

3 Asymmetric reactions see: (a) EVANS, D. A.; MCGEE, L. R. *J. Am. Chem. Soc.* **1981**, *103*, 2876; (b) D'ANGELO, J.; PECQUET-DUMAS, F. *Tetrahedron Lett.* **1983**, *24*, 1403; (c) BERNARDI, A.; COLOMBO, L.; GENNARI, C.; PRATI, L. *Tetrahedron* **1984**, *40*, 3769; (d) KATSUKI, T.; YAMAGUCHI, M. *Tetrahedron Lett.* **1985**, *26*, 5807; (e) BRAUN, M.; SACHA, H. *Angew. Chem. Int. Ed.* **1991**, *30*, 1318; (f) SACHA, H.; WALDMÜLLER, D.; BRAUN, M. *Chem. Ber.* **1994**, *127*, 1959; (g) VICARIO, J. L.; BADIA, D.; DOMINGUEZ, E.; RODRIGUEZ, M.; CARRILLO, L. *J. Org. Chem.* **2000**, *65*, 3754; (h) KUROSU, M.; LORCA, M. *J. Org. Chem.* **2001**, *66*, 1205.

4 (a) STORK, G.; SHINER, C. S.; WINKLER, J. D. *J. Am. Chem. Soc.* **1982**, *104*, 310; (b) STORK, G.; WINKLER, J. D.; SHINER, C. S. *J. Am. Chem. Soc.* **1982**, *104*, 3767; (c) SASAI, H.; KIRIO, Y.; SHIBASAKI, M. *J. Org. Chem.* **1990**, *55*, 5306.

5 (a) MASCARENHAS, C. M.; DUFFEY, M. O.; LIU, S.-Y.; MORKEN, J. P. *Org. Lett.* **1999**, *1*, 1427; (b) SCHNEIDER, C.; HANSCH, M. *Chem. Commun.* **2001**, 1218; (c) SCHNEIDER, C.; HANSCH, M. *Synlett* **2003**, 837.

6 (a) HOLLIS, T. K.; ROBINSON, N. P.; BOSNICH, B. *Tetrahedron Lett.* **1992**, *33*, 6423; (b) HOLLIS, T. K.; ODENKIRK, W.; ROBINSON, N. P.; WHELAN, J.; BOSNICH, B. *Tetrahedron* **1993**, *49*, 5415; (c) COZZI, P. G.; FLORIANI, C.; CHIESI-VILLA, A.; RIZZOLI, C. *Synlett* **1994**, 857; (d) COZZI, P. G.; FLORIANI, C. *J. Chem. Soc. Perkin Trans. 1* **1995**, 2557.

7 (a) ISHITANI, H.; YAMASHITA, Y.; SHIMIZU, H.; KOBAYASHI, S. *J. Am. Chem. Soc.* **2000**, *122*, 5403; (b) YAMASHITA, Y.; ISHITANI, H.; SHIMIZU, H.; KOBAYASHI, S. *J. Am. Chem. Soc.* **2002**, *124*, 3292.

8 COX, P. J.; WANG, W.; SNIECKUS, V. *Tetrahedron Lett.* **1992**, *33*, 2253.

9 Additional alcohol effect in catalysis see: (a) KAWARA, A.; TAGUCHI, T. *Tetrahedron Lett.* **1994**, *35*, 8805; (b) KITAJIMA, H.; KATSUKI, T. *Synlett* **1997**, 568; (c) KITAJIMA, H.; ITO, K.; KATSUKI, T. *Tetrahedron* **1997**, *53*, 17015; (d) YUN, J.; BUCHWALD, S. L. *J. Am. Chem. Soc.* **1999**, *121*, 5640; (e) EVANS, D. A.; JOHNSON, D. S. *Org. Lett.* **1999**, *1*, 595; (f) TAKAMURA, M.; HAMASHIMA, Y.; USUDA, H.; KANAI, M.; SHIBASAKI, M. *Angew. Chem. Int. Ed.* **2000**, *39*, 1650; (g) EVANS, D. A.; SCHEIDT, K. A.; JOHNSTON, J. N.; WILLIS, M. C. *J. Am. Chem. Soc.* **2001**, *123*, 4480; (h) ONITSUKA, S.; MATSUOKA, Y.; IRIE, R.; KATSUKI, T. *Chem. Lett.* **2003**, *32*, 974.

10 In some metal catalyzed asymmetric reactions, water affected the yields and selectivities, RIBE, S.; WIPF, P. *Chem. Commun.* **2001**, 299. POSNER et al. and MIKAMI et al. also reported that a small amount of water affected catalytic enantioselective ene reactions using a Ti–BINOL complex. See, (a) POSNER, G. H.; DAI, H.; BULL, D. S.; LEE, J.-K.; EYDOUX, F.; ISHIHARA, Y.; WELSH, W.; PRYOR, N.; PETR JR., S. *J. Org. Chem.* **1996**, *61*, 671; (b) TERADA, M.; MATSUMOTO, Y.; NAKAMURA, Y.; MIKAMI, K. *Chem. Commun.* **1997**, 281. (c) TERADA, M.; MATSUMOTO, Y.; NAKAMURA, Y.; MIKAMI, K. *J. Molecular Catalysis A: Chemical* **1998**, *132*, 165. (d) MIKAMI, K.; TERADA, M.; MATSUMOTO, Y.; TANAKA, M.; NAKAMURA, Y. *Microporous and Mesoporous Materials* **1998**, *21*, 461. (e) TERADA, M.; MATSUMOTO, Y.; NAKAMURA, Y.; MIKAMI, K. *Inorg. Chim. Acta* **1999**, *296*, 267.

11 Recent examples of *anti*-selective aldol reactions, (a) PARMEE, E. R.; HONG, Y.; TEMPKIN, O.; MASAMUNE, S. *Tetrahedron Lett.* **1992**, *33*, 1729; (b) MIKAMI, K.; MATSUKAWA, S. *J. Am. Chem. Soc.* **1994**, *116*, 4077; (c) EVANS, D. A.; MACMILLAN, W. C.; CAMPOS, K. R. *J. Am. Chem. Soc.* **1997**, *119*, 10859; (d) YANAGISAWA, A.; MATSUMOTO, Y., NAKASHIMA, H.; ASAKAWA, K.; YAMAMOTO, H. *J. Am. Chem. Soc.* **1997**, *119*, 9319; (e) DENMARK, S. E.; WONG, K.-T.; STAVENGER, R. A. *J. Am. Chem. Soc.* **1997**, *119*, 2333; (f) NORTHRUP, A. B.; MACMILLAN, D. W. C. *J. Am. Chem. Soc.* **2002**, *124*, 6798; (g) YANAGISAWA, A.; MATSUMOTO, Y.; ASAKAWA, K.; YAMAMOTO, H. *Tetrahedron* **2002**, *58*, 8331; (h) DENMARK, S. E.; WYNN, T.; BEUTNER, G. L. *J. Am. Chem. Soc.* **2002**, *124*, 13405; (i) WADAMOTO, M.; OZAWA, N.; YANAGISAWA, A.; YAMAMOTO, H. *J. Org. Chem.* **2003**, *68*, 5593.

12 (a) ISHITANI, H.; UENO, M.; KOBAYASHI, S. *J. Am. Chem. Soc.* **1997**, *119*, 7153; (b) ISHITANI, H.; UENO, M.; KOBAYASHI, S. *J. Am. Chem. Soc.* **2000**, *122*, 8180.

13 YAO, W.; WANG, J. *Org. Lett.* **2003**, *5*, 1527.

14 (a) YAMASHITA, Y.; SAITO, S.; ISHITANI, H.; KOBAYASHI, S. *Org. Lett.* **2002**, *4*, 1221; (b) YAMASHITA, Y.; SAITO, S.; ISHITANI, H.; KOBAYASHI, S. *J. Am. Chem. Soc.* **2003**, *125*, 3793.

15 DANISHEFSKY, S.; KITAHARA, T. *J. Am. Chem. Soc.* **1974**, *96*, 7807.

16 (a) DANISHEFSKY, S. J. *Chemtracts: Org. Chem.* **1989**, 273; (b) DANISHEFSKY, S. J. *Aldrichimica Acta* **1986**, *19*, 59; (c) BOGER, D. L. in Comprehensive Organic Synthesis; TROST, B. M. Ed.; Pergamon Press: Oxford, **1991**; Vol. 5, 451; (d) WALDMANN, H. *Synthesis* **1994**, 535.

17 DANISHEFSKY, S. J.; LARSON, E.; ASKIN, D.; KATO, N. *J. Am. Chem. Soc.* **1985**, *107*, 1246.

18 *tert*-Butyl methyl ether was used as an efficient solvent in asymmetric HDA reactions previously. See: SCHAUS, S. E.; BRÅNALT, J.; JACOBSEN, E. N. *J. Org. Chem.* **1998**, *63*, 403.

19 (a) KOBAYASHI, S. *Synlett* **1994**, 689; (b) KOBAYASHI, S. *Eur. J. Org. Chem.* **1999**, 15; (c) KOBAYASHI, S.; SUGIURA, M.; KITAGAWA, H.; LAM, W. W.-L. *Chem. Rev.* **2002**, *102*, 2227.

20 DANISHEFSKY, S. J.; MARING, C. J. *J. Am. Chem. Soc.* **1985**, *107*, 1269.

21 (a) DANISHEFSKY, S. J.; WEBB II, R. R. *J. Org. Chem.* **1984**, *49*, 1955; (b) DANISHEFSKY, S. J.; MARING, C. J. *J. Am. Chem. Soc.* **1985**, *107*, 1269.

22 Reviews of catalytic asymmetric aldol reactions, (a) BACH, T. *Angew. Chem. Int. Ed. Engl.* **1994**, *33*, 417; (b) NELSON, S. G. *Tetrahedron: Asymmetry.* **1998**, *9*, 357; (c) GROGER, H.; VOGEL, E. M.; SHIBASAKI, M. *Chem. Eur. J.* **1998**, *4*, 1137; (d) MAHRWALD, R. *Chem. Rev.* **1999**, *99*, 1095; (e) JOHNSON, J. S.; EVANS, D. A. *Acc. Chem. Res.* **2000**, *33*, 325; (f) MACHAJEWSKI, T. D.; WONG, C.-H. *Angew. Chem. Int. Ed.* **2000**, *39*, 1352; (g) LIST, B. *Tetrahedron* **2002**, *58*, 5573; (h) ALCAIDE, B.; ALMENDROS, P. *Eur. J. Org. Chem.* **2002**, 1595; (i) PALOMO, C.; OIARBIDE, M.; GARCIA, J. M. *Chem. Eur. J.* **2002**, *8*, 37; (j) CARREIRA, E. M. in *Comprehensive Asymmetric Catalysis*; JACOBSEN, E. N.; PFALTZ, A.; YAMAMOTO, H. Eds.; Springer: Heidelberg, **1999**. Vol. 3, p 998; (k) CARREIRA, E. M. in *Catalytic Asymmetric Synthesis 2nd Edition*, I. OJIMA Ed.; Wiley–VCH, New York, **2000**, p 513; (l) SAWAMURA, M.; ITO, Y. in *Catalytic Asymmetric Synthesis 2nd Edition*, I. OJIMA Ed.; Wiley–VCH, New York, **2000**, p 493.

23 (a) MUKAIYAMA, T.; KOBAYASHI, S.; MURAKAMI, M. *Chem. Lett.* **1985**, 447. (b) GENNARI, C.; BERETTA, M. G.; BERNARDI, A.; MORO, G.; SCOLASTICO, C.; TODESCHINI, R. *Tetrahedron* **1986**, *42*, 893.

24 Reviews of catalytic asymmetric hetero Diels–Alder reactions, (a) OOI, T.; MARUOKA, K. in *Comprehensive Asymmetric Catalysis*; JACOBSEN, E. N.; PFALTZ, A.; YAMAMOTO, H. Eds.; Springer: Heidelberg, **1999**. Vol. 3, p 1237; (b) JØRGENSEN, K. A. *Angew. Chem. Int. Ed. Engl.* **2000**, *39*, 3558. See also references in ref. 14b.

25 KOBAYASHI, S.; UENO, M.; ISHITANI, H. *J. Am. Chem. Soc.* **1998**, *120*, 431.

26 (a) HANAWA, H.; HASHIMOTO, T.; MARUOKA, K. *J. Am. Chem. Soc.* **2003**, *125*, 1708; (b) HANAWA, H.; URAGUCHI, D.; KONISHI, S.; HASHIMOTO, T.; MARUOKA, K. *Chem. Eur. J.* **2003**, *9*, 4405.

27 GIRARD, C.; KAGAN, H. B. *Angew. Chem. Int. Ed.* **1998**, *37*, 2922.

28 MIKAMI, K.; MOTOYAMA, T.; TERADA, M. *J. Am. Chem. Soc.* **1994**, *116*, 2812.

29 A stable chiral La catalyst has been reported: Y. S. KIM, S. MATSUNAGA, J. DAS, A. SEKINE, T. OHSHIMA and M. SHIBASAKI, *J. Am. Chem. Soc.* **2000**, *122*, 6506.

30 M. UENO, H. ISHITANI and S. KOBAYASHI, *Org. Lett.*, **2002**, *4*, 3395.

31 KOBAYASHI, S.; SAITO, S.; UENO, M.; YAMASHITA, Y. *Chem. Commun.* **2003**, 2016.

6
Direct Catalytic Asymmetric Aldol Reaction Using Chiral Metal Complexes

Masakatsu Shibasaki*, Shigeki Matsunaga, and Naoya Kumagai

6.1
Introduction

The aldol reaction has established a position in organic chemistry as a re-markably useful synthetic tool which provides access to β-hydroxy carbonyl compounds and related building blocks. Intensive efforts have raised this classic process to a highly enantioselective transformation employing only catalytic amounts of chiral promoters, as reviewed in this handbook [1]. Although many effective applications have been reported, most methods necessarily involve the preformation of latent enolates such as ketene silyl acetals, by use of less than stoichiometric amounts of base and silylating reagents (Scheme 6.1, top). Because of an increasing demand for environ-mentally benign and atom-efficient processes, such stoichiometric amounts of reagents, which inevitably result in waste, for example salts, should be excluded from the procedure. Thus, the development of a *direct* catalytic asymmetric aldol reaction (Scheme 6.1, bottom), which employs unmodified ketone as a donor, is desired.

   The clue for success in achieving the direct enolization of unmodified ketone with a catalytic amount of reagent is found in enzymatic reactions.

(a) Mukaiyama-type reactions

(b) Direct reactions

Scheme 6.1
(a) Mukaiyama-type reactions and (b) direct reactions.

Modern Aldol Reactions. Vol. 2: Metal Catalysis. Edited by Rainer Mahrwald
Copyright © 2004 WILEY-VCH Verlag GmbH & Co. KGaA, Weinheim
ISBN: 3-527-30714-1

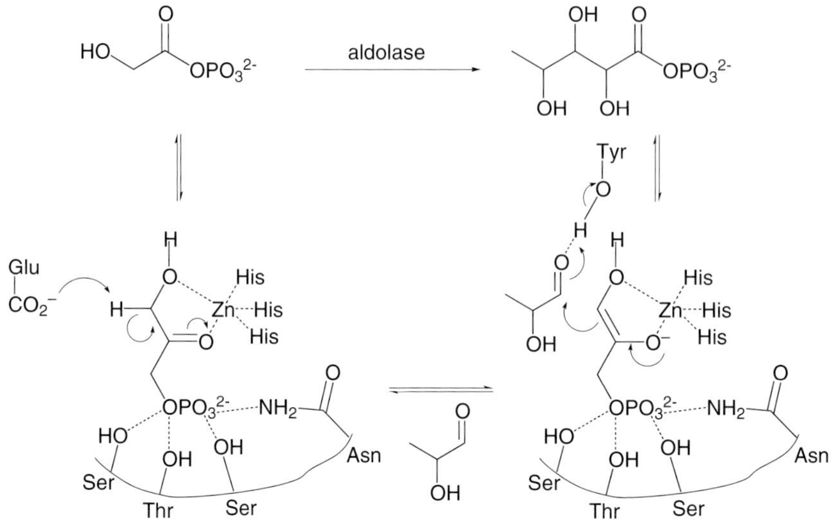

Fig. 6.1
Mechanism of action of the class II aldolase fructose-1-phosphate aldolase.

Aldolases efficiently promote the direct aldol reaction under mild in-vivo conditions – fructose-1,6-bisphosphate and dihydroxyacetone phosphate aldolases are typical examples. Such enzymes function as a bifunctional catalyst, activating a donor (ketone) with Brønsted basic functionality and an acceptor (aldehyde) with an acidic functionality. As shown in Figure 6.1, the mechanism of class II aldolases (e.g. fructose-1-phosphate aldolase) is thought to involve co-catalysis by a Zn^{2+} cation and a basic functional group in the enzyme's active site [2]. The Zn^{2+} functions as a Lewis acid to activate a carbonyl group, and the basic part abstracts an α-proton to form a Zn-enolate. Synthetic organic chemists regarded this bifunctional mechanism of the aldolases as a very promising strategy for achieving direct catalytic asymmetric aldol reactions with an artificial small molecular catalyst. This chapter focuses on notable advances recently achieved by use of metallic catalysis. This catalysis mimics that of the class II aldolases [3]. Important early contributions on direct catalytic asymmetric aldol reactions using Au as catalyst by Ito, Hayashi, and their coworkers [4] are described in Chapter 1 of Part II of this book. The other type of the direct aldol reaction, the organocatalysis as mimics of class I aldolases (amino acid based mechanism) [5] is discussed in Chapter 4 of Part I of this book.

6.2
Direct Aldol Reactions with Methyl Ketones

A possible catalytic cycle for achieving direct catalytic asymmetric aldol reaction by means of bifunctional metallic catalysis is shown in Scheme 6.2,

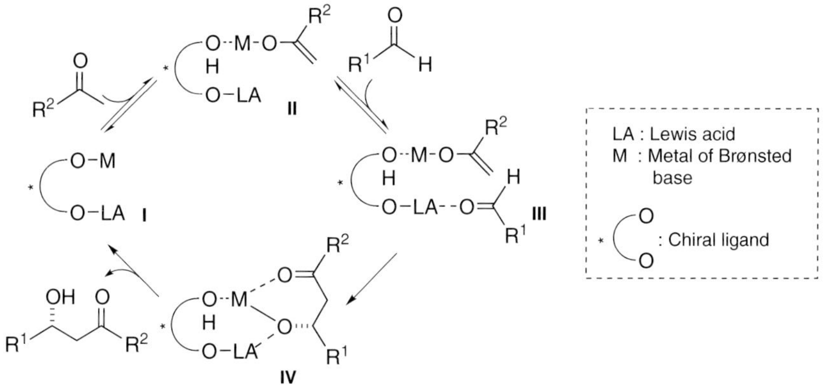

Scheme 6.2
Possible catalytic cycle for direct catalytic asymmetric aldol reactions.

which involves synergistic action of the Brønsted basic and the Lewis acidic moieties in the catalyst. The Brønsted base functionality (OM) in the catalyst I deprotonates an α-proton of a ketone to generate the metal enolate II. Lewis acid functionality activates an aldehyde to give III. The activated aldehyde then reacts with the metal enolate in a chelation-controlled asymmetric environment to afford a β-keto metal alkoxide IV. Proton exchange between the metal alkoxide moiety and an aromatic hydroxy proton or an

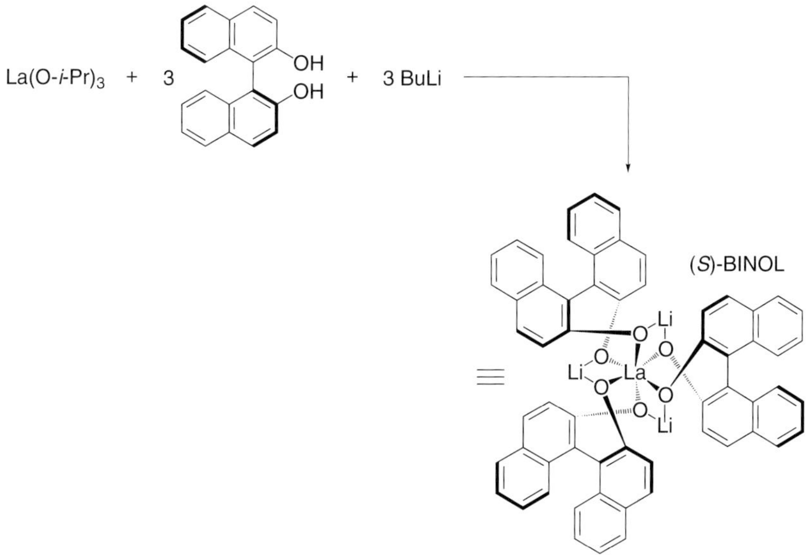

Fig. 6.2
The structure of LaLi$_3$tris((S)-binaphthoxide) ((S)-LLB).

Scheme 6.3
Direct catalytic asymmetric aldol reactions of methyl ketone **2** promoted by (S)-LLB.

α-proton of a ketone leads to the optically active aldol adduct, and at the same time leads to regeneration of the catalyst.

In 1997, Shibasaki reported that a heterobimetallic (S)-LaLi$_3$tris(binaphthoxide) complex (Figure 6.2, (S)-LLB), prepared from La(O-i-Pr)$_3$, BINOL and BuLi, was effective in the direct catalytic asymmetric aldol reaction of unmodified ketones (Scheme 6.3) [6]. As shown in Table 6.1, the LLB catalyst was effective for methyl aryl ketones (**2a** and **2b**) and methyl alkyl ketones

Tab. 6.1
Direct catalytic asymmetric aldol reactions of methyl ketone **2** promoted by (S)-LLB[a].

Entry	Aldehyde		Ketone (equiv.)		Product	Time (h)	Yield (%)	ee (%)
1	CHO	1a	-Ph	2a (5)	3aa	88	76	88
2		1a	-Ph	2a (1.5)	3aa	135	43	87
3		1a	-Ph	2a (10)	3aa	91	81	91
4		1a		2b (8)	3ab	253	55	76
5		1a	-CH$_3$	2c (10)	3ac	100	53	73
6	Ph CHO	1b	-Ph	2a (7.4)	3ba	87	90	69
7[b]		1b	-CH$_3$	2c (10)	3bc	185	82	74
8		1b	-CH$_2$CH$_3$	2d (50)	3bd	185	71	94
9	CHO	1c	-Ph	2a (8)	3ca	169	72	44
10	CHO	1d	-Ph	2a (8)	3da	277	59	54
11	Ph CHO	1e	-Ph	2a (10)	3ea	72	28	52

[a] Reaction conditions: (S)-LLB (20 mol%), THF, $-20\,°C$.
[b] The reaction was carried out at $-30\,°C$.

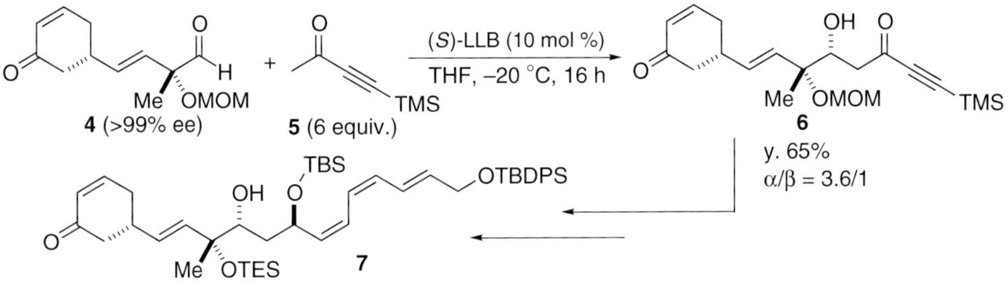

Scheme 6.4
Application of (S)-LLB to the formal total synthesis of fostriecin.

(**2c** and **2d**). Excess ketone was necessary to achieve good yield (entries 1–3). Although the reactivity was moderate even using 20 mol% catalyst, the enantiomeric excess reached 94% ee. The results demonstrated that the concept shown in Scheme 6.2 was possible.

As shown in Scheme 6.4, LLB was applicable to optically active functionalized aldehyde **4** and acetylenic ketone **5**. The aldol reaction of **4** and **5** proceeded smoothly with 10 mol% (S)-LLB at −20 °C to afford **6** in 65% yield. Diastereoselectivity was 3.6:1 and the desired α-OH **6** was obtained as a major product. Compound **6** was successfully converted into known intermediate, **7**, of fostriecin [7]. It is worthy of note that the aldol reaction did not proceed when a standard base, for example LDA, was used.

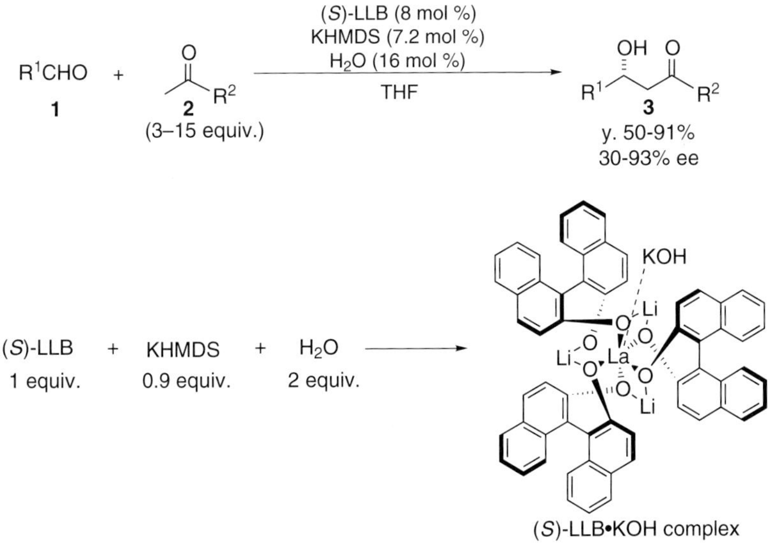

Scheme 6.5
Direct catalytic asymmetric aldol reactions of methyl ketone **2** promoted by (S)-LLB–KOH.

Tab. 6.2

Direct catalytic asymmetric aldol reactions of methyl ketone **2** promoted by (S)-LLB–KOH[a].

Entry	Aldehyde		Ketone (equiv.)		Product	Temp (°C)	Time (h)	Yield (%)	ee (%)
1	(CHO)	1a	-Ph	2a (5)	3aa	−20	15	75	88
2		1b	-Ph	2a (5)	3ba	−20	28	85	89
3		1b	-CH₃	2c (10)	3bc	−20	20	62	76
4		1b	-CH₂CH₃	2d (15)	3bd	−20		95	72	88
5	Ph (CHO)	1b	-Ph	2a (5)	3ba	−20	18	83	85
6[b]		1b	-Ph	2a (5)	3ba	−20	33	71	85
7	BnO (CHO)	1f	-Ph	2a (5)	3fa	−20	36	91	90
8		1f	-Ph	2a (5)	3fa	−20	24	70	93
9	(CHO)	1d	-Ph	2a (5)	3da	−30	15	90	33
10		1d	-m-NO₂-C₆H₄	2e (3)	3de	−50	70	68	70
11	(CHO)	1g	-m-NO₂-C₆H₄	2e (3)	3ge	−45	96	60	80
12	(CHO)	1h	-m-NO₂-C₆H₄	2e (5)	3he	−50	96	55	42
13	Ph (CHO)	1e	-m-NO₂-C₆H₄	2e (3)	3ee	−40	31	50	30

[a] Reaction conditions: (S)-LLB (8 mol%), KHMDS (7.2 mol%), H₂O (16 mol%), THF, unless otherwise noted.
[b] (S)-LLB (3 mol%), KHMDS (2.7 mol%), H₂O (6 mol%), THF.

Substantial acceleration of this reaction was achieved using a (S)-LLB–KOH catalyst prepared from (S)-LLB, KHMDS, and H₂O; this enabled reduction of the catalyst loading from 20 to 3–8 mol% with a shorter reaction time (Scheme 6.5) [8]. The results are summarized in Table 6.2. Aldol adducts were obtained in good yield and in moderate to good enantioselectivity (30–93% ee). The LLB–KOH complex was also applicable to the reaction between cyclopentanone and aldehyde **1b**, affording the aldol adduct **3bf** syn selectively (syn/anti = 93:7, syn:76% ee, anti:88% ee) in 95% yield (Scheme 6.6). Kinetic studies, including initial rate kinetics and isotope effects, indicated that the rate-determining step was deprotonation of the ketone. Additional KOH is supposed to accelerate the rate-determining enolization step. Because high ee was achieved in the presence of additional KOH, self-assembly of the LLB complex and KOH was assumed. On the basis of mechanistic studies a working model was proposed for the direct

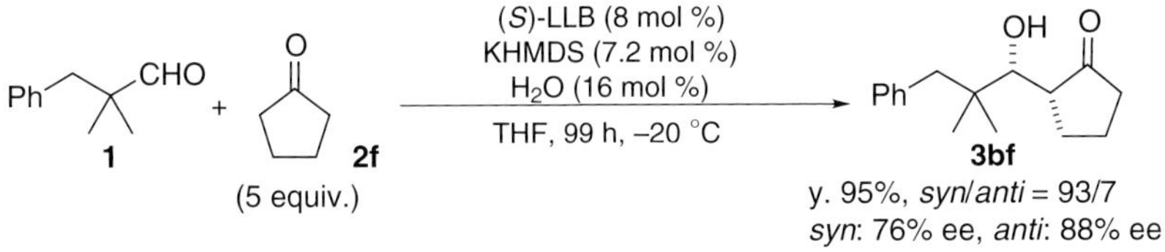

Scheme 6.6
Direct catalytic asymmetric aldol reactions
of cyclopentanone (**2f**) promoted by
(*S*)-LLB–KOH.

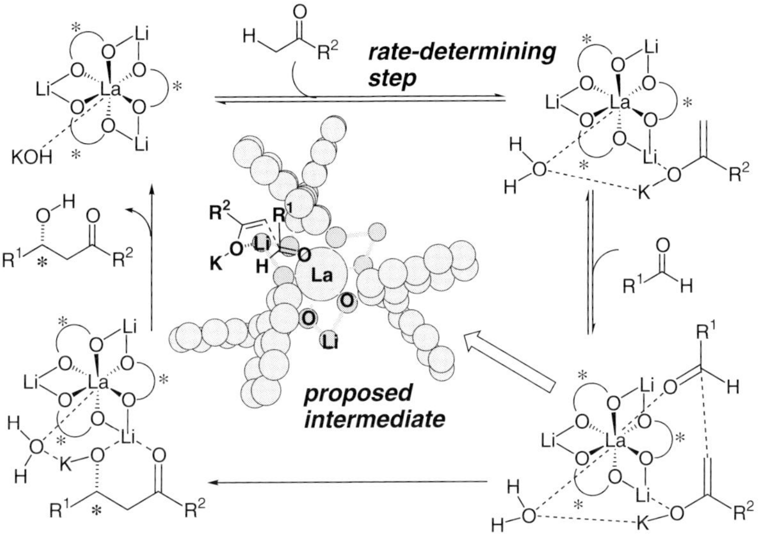

Fig. 6.3
Working model for direct catalytic
asymmetric aldol reactions promoted by the
(*S*)-LLB–KOH complex.

catalytic asymmetric aldol reaction promoted by the LLB–KOH complex (Figure 6.3). KOH functions as a Brønsted base, generating an enolate from the ketone (rate-determining step); the lanthanum ion acts as a Lewis acid to activate the aldehyde. 1,2-Addition and the protonation of an alkoxide leads to the aldol adduct and regenerates the catalyst. As shown in Scheme 6.7, a key-intermediate of bryostatin 7 was synthesized by using the LLB–KOH complex twice. **3fa** was prepared by the aldol reaction using (*R*)-LLB–KOH. Baeyer–Villiger oxidation, followed by functional group manipulation, afforded aldehyde **9**. The aldol reaction of **9** with (*S*)-LLB–KOH proceeded in a catalyst-controlled manner and the *anti* adduct was obtained (*anti*/*syn* = 7:1). The LLB–KOH complex was also applied to the total synthesis of epothilones; the LLB–KOH complex was effectively applied for

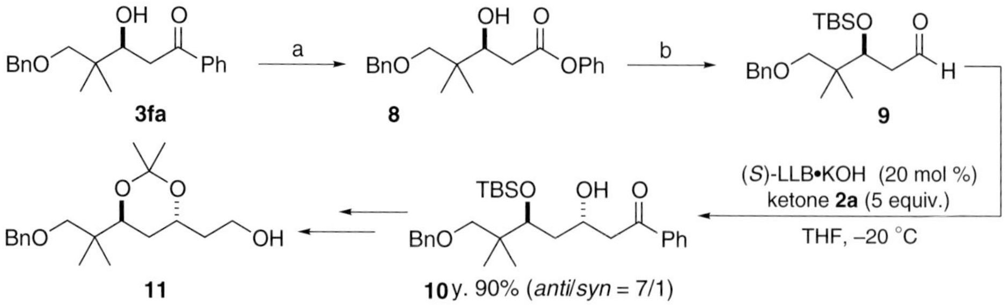

(a) mCPBA, NaH$_2$PO$_4$, ClCH$_2$CH$_2$Cl, y. 73%; (b) i)TBSOTf, Hünig base; ii) DIBAL; iii) PCC y. 87% (3 steps).

Scheme 6.7
Application to the synthesis of intermediate to bryostatin 7.

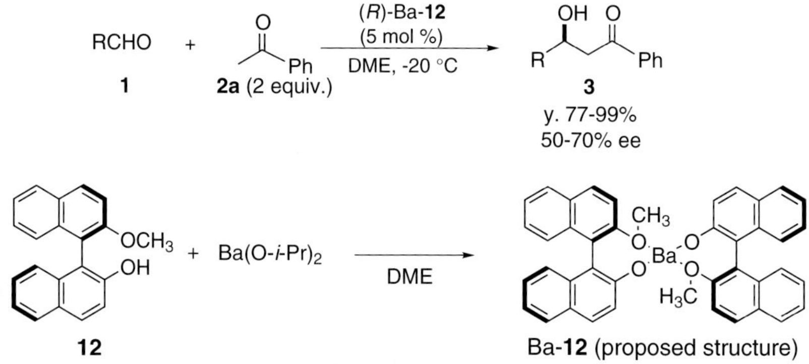

Scheme 6.8
Direct catalytic asymmetric aldol reactions of acetophenone (**2a**) promoted by (R)-Ba-**12**.

resolution of the racemic aldehyde [9]. Further details are given in Chapter 7 in Vol. 1.

To reduce the amount of ketone Shibasaki prepared a Ba-**12** complex from Ba(O-i-Pr)$_2$ and **12** [10]. As shown in Scheme 6.8, the Ba-**12** complex afforded aldol adducts in good yield (77–99%) from as little as 2 equiv. ketone **2a**, although ee was modest (50–70% ee).

In 2000 Trost reported a dinuclear Zn complex prepared from Et$_2$Zn and **13** [11]. **13** was easily synthesized from p-cresol in four steps. On the basis of ethane gas emission measurement and ESI-MS analysis the dinuclear complex Zn$_2$-**13** was proposed (Scheme 6.9). As shown in Table 6.3, the Zn complex was effective in direct catalytic asymmetric aldol reactions with a variety of methyl aryl ketones (**2a**, **2g–2j**; 10 equiv.). Excellent enantioselectivity (up to 99% ee) was achieved by use of 5 mol% **13**, although excess ketone was used and yields were occasionally moderate. It is worthy of note that high ee was achieved at relatively high reaction temperature

Tab. 6.3

Direct catalytic asymmetric aldol reactions of methyl ketone **2** promoted by dinuclear Zn_2-**13** complex[a].

Entry	Aldehyde			Ketone (Equiv.)	Product	Temp (°C)	Time (h)	Yield (%)	ee (%)
1	∼CHO	1i	-Ph	2a (10)	3ia	−5	48	33	56
2		1i	-Ph	2a (10)	3ia	−15	48	24	74
3	⤷CHO	1j	-Ph	2a (10)	3ja	−5	48	49	68
4	⤷CHO	1d	-Ph	2a (10)	3da	5	48	62	98
5	⬡CHO	1c	-Ph	2a (10)	3ca	5	48	60	98
6	Ph⤷CHO Ph	1k	-Ph	2a (10)	3ka	5	48	79	99
7	Ph⤷CHO	1l	-Ph	2a (10)	3la	5	48	67 (dr: 2/1)	94
8	TBSO∼CHO	1m	-Ph	2a (10)	3ma	5	96	61	93
9	⤷CHO	1d	furyl	2g (10)	3dg	5	48	66	97
10		1d	OMe-phenyl	2h (10)	3dh	5	48	48	97
11		1d	OMe-phenyl	2i (5)	3di	5	48	36	98
12		1d	naphthyl	2j (5)	3dj	5	48	40	96

[a] Reaction conditions: ligand **13** (5 mol%), Et_2Zn (10 mol%), $Ph_3P=S$ (15 mol%), THF.

(5 °C). Bifunctional Zn catalysis is proposed, with one Zn acting as Lewis acid and another Zn-alkoxide functioning as a Brønsted base to generate a Zn-enolate (Figure 6.4).

Trost also reported that modification of the ligand occasionally led to better results. When ligand **14** was used instead of **13** the direct aldol reaction of acetone (10–15 equiv.) proceeded smoothly with good enantioselectivity (Scheme 6.10) [12]. The results are summarized in Table 6.4. Aldol adducts were obtained in good yield (59–89%) and ee (78–94%). In

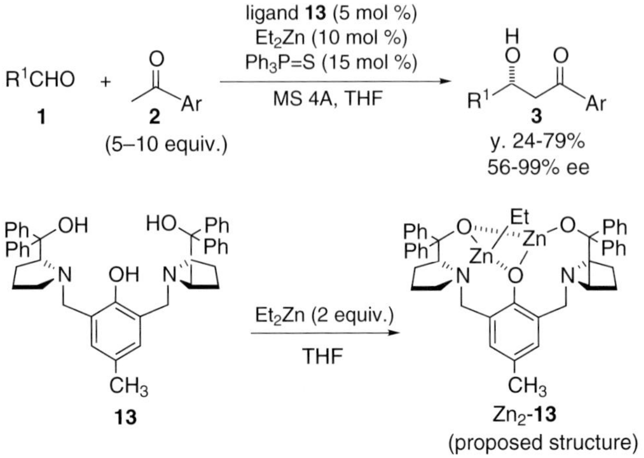

Scheme 6.9
Direct catalytic asymmetric aldol reactions of methyl ketone **2** promoted by dinuclear Zn_2-**13** complex.

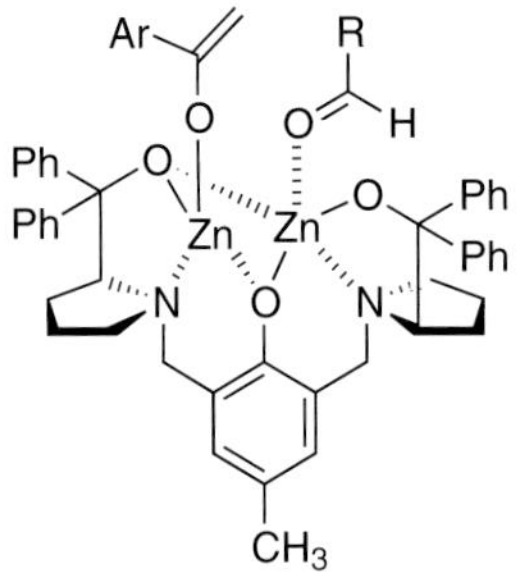

Fig. 6.4
Proposed transition state for the direct aldol reaction of methyl ketone **2**.

general, self-condensation of aldehydes should be suppressed to achieve good yield by using α-unsubstituted aldehydes. It is worthy of note that good yield and high ee were achieved even with α-unsubstituted aldehydes (entries 5–7, yield 59–76%, 82–89% ee).

Noyori prepared a highly active Ca-**15** complex from $Ca[N\{Si(CH_3)_3\}_2]_2$ and **15** (Scheme 6.11) [13]. As shown in Table 6.5, aldol adducts were obtained in good yield (75–88%) in moderate to good ee (66–91% ee) by using as little as 1–3 mol% catalyst loading. The reactivity of the Ca catalyst is higher than those of other catalysts. The enhanced reactivity is ascribed to the high basicity of the calcium alkoxide. Even with this active Ca-**15** catalyst, however, excess ketone was essential to achieve good yield and ee. This problem would be solved in future research.

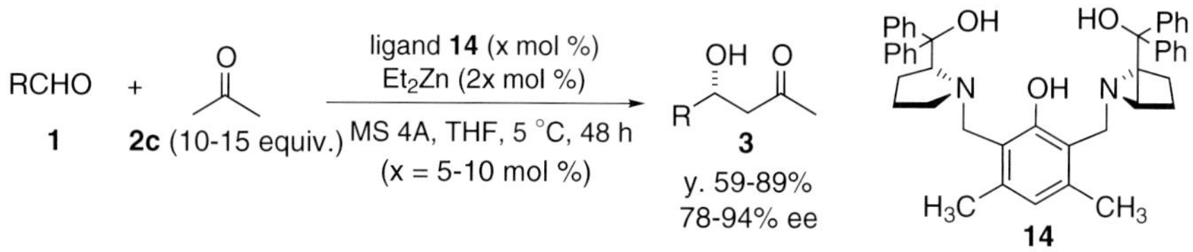

Scheme 6.10
Direct catalytic asymmetric aldol reactions
of acetone (**2c**) promoted by dinuclear
Zn$_2$-**14** complex.

Tab. 6.4
Direct catalytic asymmetric aldol reactions of acetone (**2c**) promoted by dinuclear
Zn$_2$-**14** complex[a].

Entry	Aldehyde		Product	Catalyst (mol%)	Yield (%)	ee (%)
1	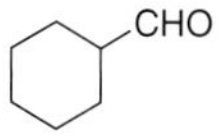 cyclohexyl-CHO	1c	3cc	10	89	92
2	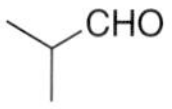 isopropyl-CHO	1d	3dc	10	89	91
3	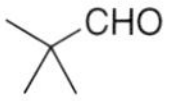 tert-butyl-CHO	1a	3ac	10	72	94
4	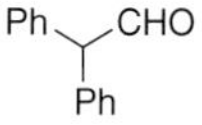 Ph$_2$CH-CHO	1k	3kc	10	84	91
5	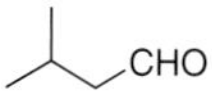 (CH$_3$)$_3$C-CH$_2$-CHO	1j	3jc	10	59	84
6	 Ph-CH$_2$CH$_2$-CHO	1e	3ec	10	76	82
7	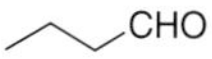 CH$_3$CH$_2$CH$_2$-CHO	1i	3ic	10	69	89
8	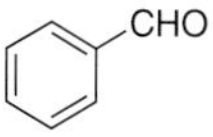 Ph-CHO	1n	3nc	5	78	83
9	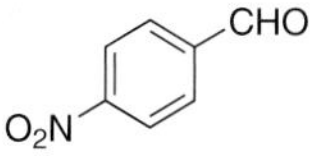 4-O$_2$N-C$_6$H$_4$-CHO	1o	3oc	5	62	78

[a] Reaction conditions: ligand **14** ($\times$ mol%), Et$_2$Zn ($2\times$ mol%), THF, MS
4A, 5 °C, 48 h.

Scheme 6.11
Direct catalytic asymmetric aldol reactions
of acetophenone (2a) promoted by Ca-15
complex.

6.3
Direct Aldol Reactions with Methylene Ketones

The aldol reaction between aldehydes and methylene ketones or propionates
should provide a powerful tool for construction of two continuous chiral
centers and for formation of carbon–carbon bonds. Catalytic asymmetric
syntheses of *syn* and *anti* aldols from latent enolates have already been well
investigated [1]. In contrast, diastereo- and enantioselective synthesis of
aldols, starting from methylene ketones, by means of the *direct* catalytic
asymmetric aldol reaction is still immature. The bulkiness of methylene
ketones was expected to make it more difficult for the catalysts to abstract
an α-hydrogen from the ketones.

Shibasaki reported a strongly basic La-Li-16 complex with an Li alkoxide

Tab. 6.5
Direct catalytic asymmetric aldol reactions of acetophenone (2a) promoted by Ca-15
complex[a].

Entry	Aldehyde		Product	Catalyst (mol%)	Time (h)	Yield (%)	ee (%)
1	CHO (2,2-dimethylpropanal)	1a	3aa	3	22	87	86
2		1a	3aa	1	20	79	82
3	Ph~~CHO	1b	3ba	3	24	75	87
4	BnO~~CHO	1f	3fa	3	24	76	91
5	cyclohexyl-CHO	1c	3ca	3	20	88	66

[a] Reaction conditions: (S,S)-Ca-15 ($\times$ mol%), C_2H_5CN/THF, $-20\ ^{\circ}C$.

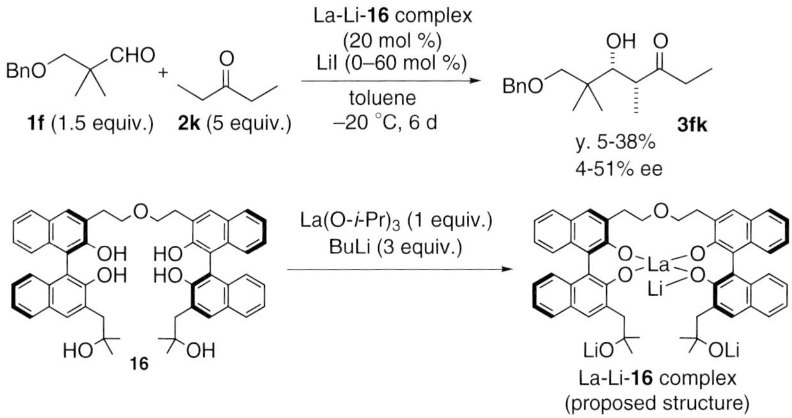

Scheme 6.12
Direct catalytic asymmetric aldol reactions
of 3-pentanone (**2k**) promoted by La-Li-**16**
complex.

moiety. The catalyst promoted the direct aldol reaction of 3-pentanone *anti*-selectively; yield and ee were only modest, however (Scheme 6.12) [14].

A notable advance in the direct aldol reaction of methylene ketone was reported by Mahrwald in 2002 [15]. As shown in Scheme 6.13, a catalytic amount (10 mol%) of Ti-**17**-*rac*-BINOL complex was suitable for promoting the direct asymmetric aldol reaction of 3-pentanone with a variety of aldehydes. Interestingly, combination of chiral mandelic acid (**17**) and *rac*-BINOL afforded a good chiral Ti-catalyst. As summarized in Table 6.6, aldol adducts were obtained *syn*-selectively (*syn/anti* up to 91:9) in moderate to good yield (43–85%) and with good ee (71–93%). Good selectivity was achieved at room temperature. The aldehyde/ketone ratio in this reaction is

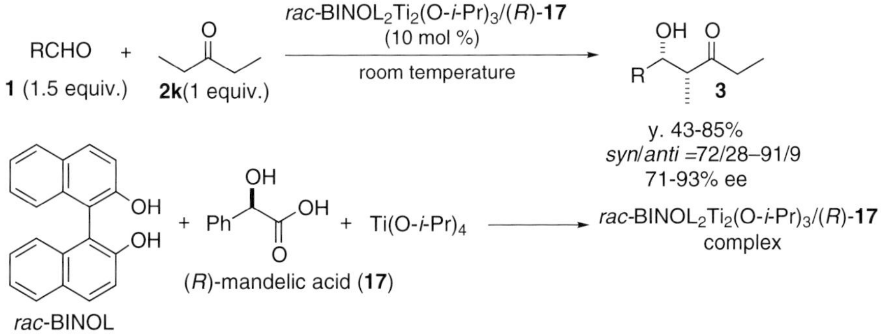

Scheme 6.13
Direct catalytic asymmetric aldol reactions
of 3-pentanone (**2k**) promoted by *rac*-
BINOL$_2$Ti$_2$(O-*i*-Pr)$_3$/(R)-**17** complex.

Tab. 6.6
Direct catalytic asymmetric aldol reactions of 3-pentanone (**2k**) promoted by
rac-BINOL$_2$Ti$_2$(O-i-Pr)$_3$/(R)-**17** complex[a].

Entry	Aldehyde		Product	Catalyst (mol%)	dr (syn/ anti)	Yield (%)	ee (%)
1	PhCHO	**1n**	**3nk**	10	91/9	85	91
2	CHO	**1a**	**3ak**	10	88/12	71	93
3	Ph———CHO	**1p**	**3qk**	10	73/27	68	78
4	CHO	**1d**	**3dk**	10	79/21	43	71
5	CHO	**1q**	**3qk**	10	72/28	78	74

[a] Reaction conditions: rac-BINOL$_2$Ti$_2$(O-i-Pr)$_3$/(R)-**17** (10 mol%), room temperature.

also worthy of note. The aldol reaction proceeded smoothly with 1 equiv. ketone and 1.5 equiv. aldehyde.

6.4
Direct Aldol Reaction with α-Hydroxyketones

α-Hydroxyketones also serve as aldol donors in the direct catalytic asymmetric aldol reaction. In contrast with conventional Lewis acid-catalyzed aldol reactions, protection of the OH group is not necessary. The versatility of the resulting chiral 1,2-diols as building blocks makes this process attractive. The first successful result was reported by List and Barbas with proline catalysis [16]. L-Proline catalyzed a highly chemo-, diastereo- and enantio-selective aldol reaction between hydroxyacetone and aldehydes to provide chiral anti-1,2-diols.

Trost and Shibasaki have made important contributions to metallic catalysis. Trost reported a direct aldol reaction with 2-hydroxyacetophenone and 2-hydroxyacetylfuran using the dinuclear Zn$_2$-**13** catalyst [17] (Scheme 6.14). The aldol reaction between a variety of aldehydes and 1.5 equiv. ketone proceeded smoothly at −35 °C with 2.5–5 mol% of catalyst, in the presence of MS 4A, to afford the products syn-selectively (syn/anti = 3:1 to 100:0) in 62–98% yield and 81–98% ee (Table 6.7). Strikingly, the absolute configuration of the stereocenter derived from the aldehyde is opposite to that obtained with acetophenone (**2a**, Scheme 6.9) and acetone (**2c**, Scheme 6.10) as donors, possibly because of the bidendate coordination of α-hydroxyketone to the catalyst, as depicted in Figure 6.5. Occasionally the

RCHO + [ketone structure with O, Ar, OH] → [product 19 structure] 13

$$RCHO + \underset{\underset{OH}{1}}{\overset{O}{\underset{\|}{C}}}\!\!-Ar$$

ligand **13** (x mol %)
Et$_2$Zn (2x mol %)
MS 4A, THF, −35 °C
(x = 2.5–5 mol %)

18a: Ar = Ph
18b: Ar = 2-furyl
(1.1-1.5 equiv.)

19
y. 65-97%
syn/anti = 4/1–100/0
86-98% ee

13

Scheme 6.14
Direct catalytic asymmetric aldol reaction of
hydroxyketone **18** promoted by Zn$_2$-**13**
complex.

Tab. 6.7
Direct catalytic asymmetric aldol reaction of hydroxyketone **18** promoted by dinuclear Zn$_2$-**13**
complex[a].

Entry	Aldehyde		Ketone (Equiv.)	Product	Catalyst (× mol%)	Yield (%)	dr (syn/ anti)	ee (%) (syn)
1	[cyclohexyl-CHO]	1c	18a (1.5)	19ca	2.5	83	30/1	92
2		1c	18a (1.5)	19ca	5	97	5/1	90
3		1c	18b (1.3)	19cb	5	90	6/1	96
4		1c	18b (1.1)	19cb	5	77	6/1	98
5	[isopropyl-CHO]	1d	18a (1.5)	19da	2.5	89	13/1	93
6		1d	18a (1.1)	19da	5	72	6/1	93
7	Ph—CH(Ph)—CHO	1k	18a (1.5)	19ka	2.5	74	100/0	96
8	[isobutyl-CHO]	1j	18a (1.5)	19ja	2.5	65	35/1	94
9		1j	18a (1.1)	19ja	5	79	4/1	93
10	Ph—CH$_2$CH$_2$—CHO	1e	18a (1.5)	19ea	2.5	78	9/1	91
11	[branched chain ()$_4$ CHO]	1r	18a (1.5)	19ra	5	89	5/1	86
12	[alkenyl ()$_6$ CHO]	1s	18a (1.5)	19sa	5	91	5/1	87

[a] Reaction conditions: ligand **13** (× mol%), Et$_2$Zn (2× mol%), THF, MS 4A,
−35 °C, 24 h.

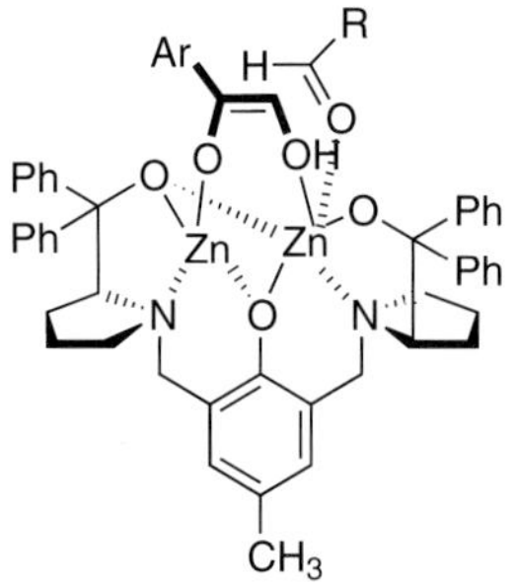

Fig. 6.5
Proposed transition state for the direct aldol reaction of hydroxyketone **18**.

reaction has been performed with a ketone/aldehyde ratio of 1.1:1.0 albeit at the expense of conversion, which comes closest in reaching the ideal atom economical process. The reaction with **18b** resulted in higher ee and, moreover, the furan moiety is suitable for further conversion of the resulting chiral 1,2-diol. Oxidative cleavage of the furan ring was successfully used for asymmetric synthesis of (+)-boronolide, as shown in Scheme 6.15 [18].

Shibasaki developed direct catalytic asymmetric aldol reactions of 2-hydroxyacetophenones, providing either *anti* or *syn* chiral 1,2-diols, by using two types of multifunctional catalyst, (*S*)-LLB–KOH and an Et$_2$Zn/(*S,S*)-linked-BINOL **20** complex [19]. (*S*)-LLB–KOH (5–10 mol%) promoted the direct aldol reaction of 2-hydroxyacetophenones **18** to afford *anti*-1,2-diols in good yields and ee (up to 98% ee), although *anti*-selectivity was occasionally modest. (Scheme 6.16 and Table 6.8) [19, 20]. Enolizable aldehydes were successfully utilized without any self-condensation. The absolute configuration was identical at the α-position of both *anti* and *syn* products, suggesting that the enantioface of the enolates derived from 2-hydroxyacetophenones

18b 1.1 eq   **1h**: 16 mmol scale   **19bh**   **20**

(+)-boronolide   **21**

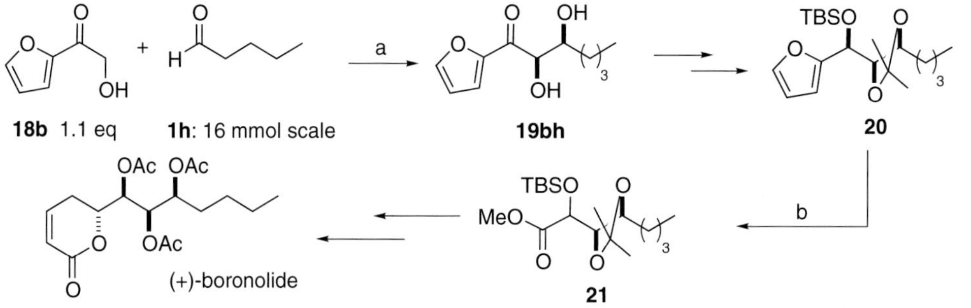

^{a}(a) 5 mol % of Zn catalyst, MS 4A, THF, −35 °C, 12 h, 93% (*syn/anti* = 4.2/1, *syn* = 96% ee);
(b) RuCl$_3$ (cat.), NaIO$_4$, CCl$_4$, CH$_3$CN, H$_2$O; CH$_2$N$_2$, Et$_2$O, 70%.

Scheme 6.15
Stereocontrolled total synthesis of (+)-boronolide.

Scheme 6.16
Direct catalytic asymmetric aldol reaction of
hydroxyketone **18** promoted by (S)-LLB–KOH.

18 was well differentiated. The configuration at the β-position of the major
anti diastereomer was opposite to that of the aldol product from the
methyl ketone (Scheme 6.5) possibly because of bidendate coordination of
α-hydroxyketone **18** to the catalyst. Proposed transition state models are depicted in Figure 6.6. Hydroxyketones would coordinate to La in a bidentate
fashion, resulting in efficient *Si*-face shielding of the *Z* enolate (Figure 6.6,
a and b). In addition, this preferential bidentate coordination of **18** is supposed to suppress the self-condensation of aldehydes.

Shibasaki reported another approach to the direct asymmetric aldol reac-

Tab. 6.8
Direct catalytic asymmetric aldol reaction of hydroxyketone **18** promoted by (S)-LLB–KOH[a].

Entry	Aldehyde		Ketone (R-)		Product	Catalyst (mol%)	Temp (°C)	Time (h)	Yield (%)	dr (anti/ syn)	ee (%) (anti/syn)
1	Ph⌒⌒CHO	1t	H-	18a	19ta	10	−50	24	84	84/16	95/74
2		1t	H-	18a	19ta	5	−50	40	78	78/22	92/70
3		1t	4-MeO-	18c	19tc	10	−40	35	50	81/19	98/79
4		1t	2-Me-	18d	19td	10	−40	35	90	77/23	84/57
5		1t	4-Me-	18e	19te	10	−40	35	90	83/17	97/85
6	⌒⌒⌒CHO	1u	4-Me-	18e	19ue	10	−40	12	96	75/25	96/89
7	⌒⌒CHO	1v	H-	18a	19va	10	−50	28	90	72/28	94/83
8	⋋⌒CHO	1j	H-	18a	19ja	10	−50	24	86	65/35	90/83

[a] Reaction conditions: (S)-LLB ($\times$ mol%), KHMDS ($0.9\times$ mol%), H_2O
($2\times$ mol%), THF.

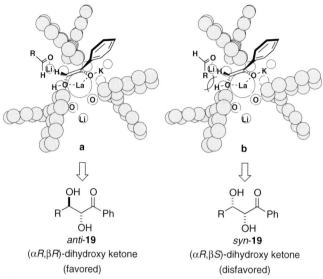

Fig. 6.6

Transition states postulated for formation of *anti* diol and *syn* diol.

tion of 2-hydroxyacetophenones **18**. The Et$_2$Zn/(*S,S*)-linked-BINOL **20** complex promoted the aldol reaction of 2-hydroxyacetophenones **18** to afford *syn*-1,2-diols in good yield (Scheme 6.17) [19, 21, 22]. Reactivity and stereoselectivity depended on the substituent on the aromatic ring of the 2-hydroxyacetophenones **18**. 2-Hydroxy-2′-methoxyacetophenone (**18f**) gave the best result to afford the product in 94% yield and high stereoselectivity (*syn/anti* = 89:11, *syn* = 92% ee, *anti* = 89% ee) with as little as 1 mol% catalyst. The catalyst was applicable to a variety of aldehydes including α-unsubstituted aldehydes (Scheme 6.17). As summarized in Table 6.9, the reaction reached completion within 24 h with 1 mol% catalyst to give *syn*-

R^1 = PhCH$_2$CH$_2$, R^2 = H (**18a**), X = 10: 48 h, y. 81%, *syn/anti* = 67/33, 78% ee(*syn*)
R^1 = PhCH$_2$CH$_2$, R^2 = 2-MeO (**18f**), X = 3: 4 h, y. 94%, *syn/anti* = 90/10, 90% ee(*syn*)
R^1 = PhCH$_2$CH$_2$, R^2 = 2-MeO (**18f**), X = 1: 16 h, y. 94%, *syn/anti* = 87/13, 93% ee(*syn*)

Scheme 6.17

Direct catalytic asymmetric aldol reactions of hydroxyketone **18** promoted by Et$_2$Zn/(*S,S*)-linked-BINOL **20** complex.

Tab. 6.9
Direct catalytic asymmetric aldol reaction of hydroxyketone **18f** promoted by Et_2Zn/
(S,S)-linked-BINOL **20** complex[a].

Entry	Aldehyde		Product	Time (h)	Yield (%)	dr (syn/ anti)	ee (%) (syn/anti)
1	Ph⌒⌒CHO	1e	19ef	20	94	89/11	92/89
2	⌒⌒⌒CHO	1u	19uf	18	88	88/12	95/91
3	(isopropyl)⌒CHO	1j	19jf	18	84	84/16	93/87
4	O=⌒⌒CHO	1w	19wf	12	91	93/7	95/–
5	⌒⌒⌒CHO	1v	19vf	24	94	86/14	87/92
6	BnO⌒⌒CHO	1x	19xf	18	81	86/14	95/90
7	BnO⌒CHO	1y	19yf	16	84	72/28	96/93
8	BOMO⌒⌒CHO	1z	19zf	14	93	84/16	90/84
9	⌒CHO	d	19df	24	83	97/3	98/–
10	⌒⌒CHO	1g	19gf	16	92	96/4	99/–
11	(cyclohexyl)⌒CHO	1c	19cf	18	95	97/3	98/–

[a] Reaction conditions: (S,S)-linked-BINOL **20** (1 mol%), Et_2Zn
(2 mol%), −30 °C, THF.

1,2-diols in excellent yield and stereoselectivity (yield 81–95%, syn/anti = 72:28 to 97:3, syn = 87–99% ee).

Mechanistic investigations by kinetic studies, X-ray crystallography, ^{1}H NMR, and cold-spray ionization mass spectrometry (CSI-MS) analyses shed light on the reaction mechanism and the structure of the active species [22]. X-ray analysis of a crystal obtained from a 2:1 solution of Et_2Zn/(S,S)-linked-BINOL **20** in THF revealed the complex consisted of Zn and ligand **20** in a ratio of 3:2 [trinuclear Zn_3(linked-BINOL)$_2$thf$_3$] with C_2 symmetry (Figure 6.7, **21**). The CSI-MS analysis and kinetic studies revealed that the complex **21** was a precatalyst and that a oligomeric Zn-**20**-**18f** complex would work as the actual active species. The proposed catalytic cycle is shown in Figure 6.8. The product dissociation step is rate-determining.

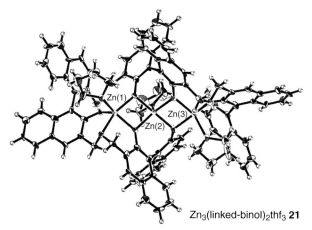

Fig. 6.7
X-ray structure of preformed complex Zn₃ (linked-BINOL)₂thf₃ **21**.

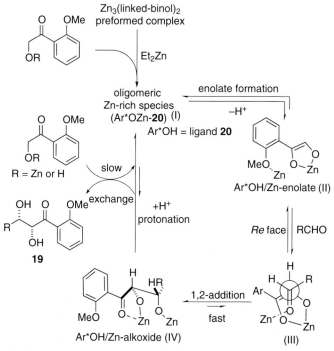

Fig. 6.8
Postulated catalytic cycle for the direct aldol reaction with Et2Zn/(S,S)-linked-BINOL **20** complex.

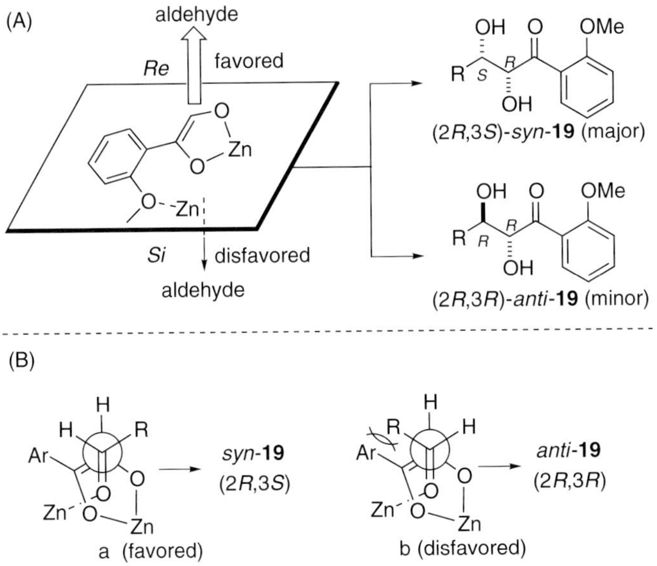

Fig. 6.9
Stereochemical course of direct aldol reaction of hydroxyketone **18f**.

The identical absolute configuration (R) was obtained at the α-position of both the *syn-* and *anti*-aldol products (Figure 6.9), suggesting that the catalyst differentiates the enantioface of the enolate well and aldehydes come from the *Re* face of the zinc enolate (Figure 6.9A). *Syn* selectivity is explained by the transition state shown in Figure 6.9B. The positive effects of the *ortho* MeO group suggested the MeO group coordinated with one of the Zn centers in the oligomeric Zn complex affecting the stereoselection step. The electron-donating MeO group has a beneficial effect on further conversion of the products into esters and amides via regioselective rearrangement as shown in Scheme 6.18 [21, 22].

Mechanistic studies suggested that additional Et_2Zn and MS 3A would accelerate the reaction rate. In the presence of MS 3A the second generation Zn catalyst, prepared from Et_2Zn/linked-BINOL **20** in a ratio of 4:1, promoted the direct aldol reaction of hydrocinnamaldehyde (**1e**) and 1.1 equiv. ketone **18f**, smoothly and with reduced catalyst loading (0.25–0.1 mol%, Scheme 6.19) [22]. The practical utility of the reaction was demonstrated by a large-scale reaction performed on the 200-mmol scale by using 0.25 mol% **1** (0.5 mmol, 307 mg) to afford 53.7 g product **19cf** (yield 96%) in high dr (*syn*/*anti* = 98:2) and ee (94% ee) after 12 h (Scheme 6.19). Considering that the standard catalyst loading for the direct catalytic asymmetric aldol reaction is 2.5 to 20 mol% the exceptionally low catalyst loading in this asymmetric zinc catalysis is remarkable.

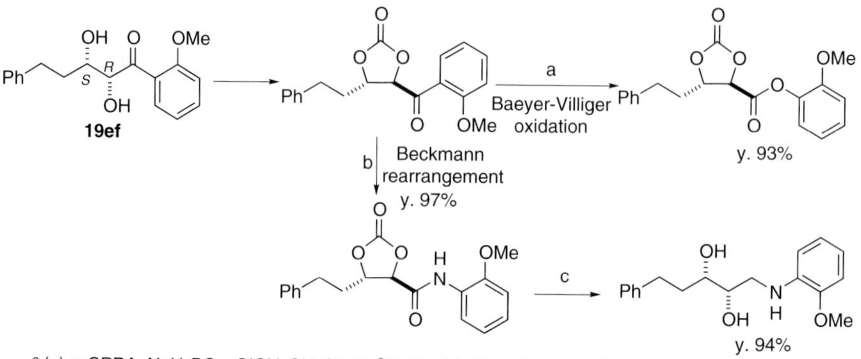

^{a} (a) *m*CPBA, NaH$_2$PO$_4$, ClCH$_2$CH$_2$Cl, 50 °C, 2h; (b) *O*-mesitylenesulfonylhydroxylamime, CH$_2$Cl, rt, 4 h; (c) DIBAL, −78 °C to rt, 2h.

Scheme 6.18
Transformations of aldol adduct via regioselective rearrangement^a.

Et$_2$Zn/linked-BINOL **20**, 4:1, with MS 3A enabled the direct aldol reaction of 2-hydroxy-2′-methoxypropiopheneone (**23**), leading to construction of a chiral tetrasubstituted carbon stereocenter (Scheme 6.20) [22]. Although a higher catalyst loading and 5 equiv. ketone **23** were required, a variety of α-unsubstituted aldehydes afforded the product *syn*-selectively (*syn/anti* = 59:41 to 71:29) in moderate to good yield and ee (yield 72–97%, *syn* = 72–87% ee, *anti* = 86–97% ee) with (*S,S*)-linked-BINOL **20** (Table 6.10). Interestingly, the reaction using (*S,S*)-sulfur-linked-BINOL **22**, a linked-BINOL analog including sulfur in the linker, instead of oxygen, resulted in the op-

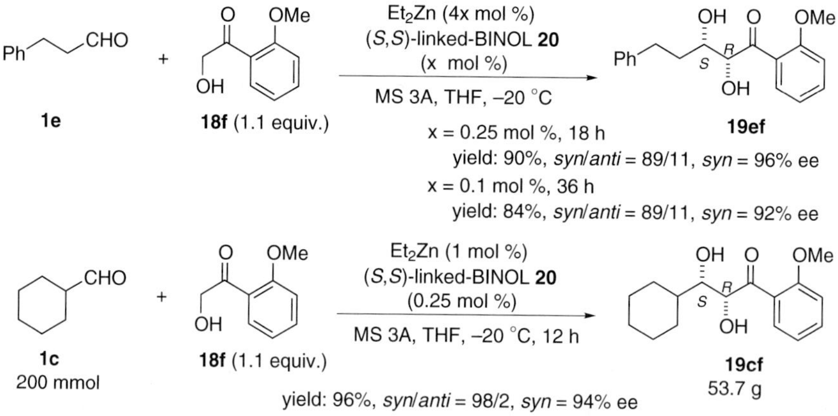

Scheme 6.19
Direct catalytic asymmetric aldol reactions of hydroxyketone **18f** promoted by Et$_2$Zn/(*S,S*)-linked-BINOL **20** = 4:1 complex with MS 3A.

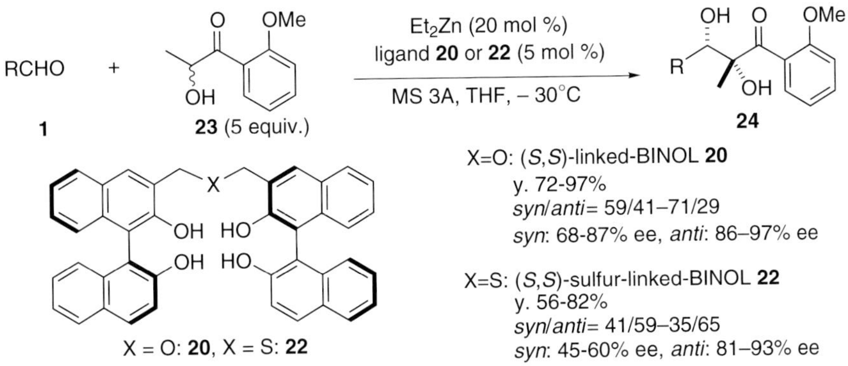

Scheme 6.20
Direct catalytic asymmetric aldol reactions
of 2-hydroxy-2'-methoxypropiophenone (**23**)
promoted by Et$_2$Zn/(S,S)-linked-BINOL
complex.

posite diastereoselectivity (Scheme 6.20). Reactivity with **22** was somewhat lower than that with **20**, and aldol adducts were obtained in moderate to good yield (56–82%) on use of 10 equiv. ketone **23**. Major *anti* isomers were obtained in high ee (81–93% ee), although ee of minor *syn* isomers was rather low (48–60% ee) (Table 6.10).

6.5
Direct Aldol Reaction with Glycine Schiff Bases

Catalytic asymmetric aldol reactions of glycine equivalents with aldehydes afford efficient and direct access to β-hydroxy-α-amino acid derivatives, which serve as useful chiral building blocks, especially in the pharmaceutical industry. A partially successful catalytic asymmetric aldol reaction of a glycine Schiff base was reported by Miller in 1991 [23]. When *N*-benzylcinchoninium chloride **26** was used as chiral phase-transfer catalyst, aldol reaction of Schiff base **25** with aldehydes afforded products *syn*-selectively (14–56% de) in good yield (46–92%) although ee was at most 12% ee (Scheme 6.21). After that work no efficient artificial catalyst was reported for a decade, except for the chemoenzymatic process with glycine-dependent aldolases [2].

In 2002 Shibasaki reported use of the heterobimetallic asymmetric catalyst La$_3$Li$_3$(binaphthoxide)$_3$.LiOH (LLB.LiOH) for *anti*-selective direct aldol reaction of glycine Schiff bases with aldehydes (Scheme 6.22) [24]. Use of 20 mol% of the catalyst, Schiff base **25b**, and aldehydes (3 equiv.) gave the products *anti*-selectively (*anti/syn* = 59:41 to 86:14) in moderate to good yield (71–93%) and with moderate ee (*anti* = 19–76% ee).

Tab. 6.10
Direct catalytic asymmetric aldol reaction with 2-hydroxy-2′-methoxypropiophenone (**23**) promoted by Et$_2$Zn/ (S,S)-linked-BINOL complex[a].

Entry	Aldehyde	Ketone 23 (Equiv.)		Product	Catalyst (× mol%)	Temp. (°C)	Yield (%)	dr (anti/ syn)	ee (%) (anti/syn)
1	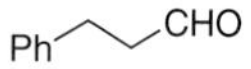 Ph～CHO	1e	5	24e	20 (5)	−30	97	62/38	87/96
2		1e	10	24e	22 (10)	−20	82	35/65	60/92
3	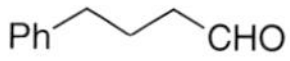 Ph～～CHO	1t	5	24t	20 (5)	−30	72	64/36	78/90
4		1t	10	24t	22 (10)	−20	63	41/59	45/86
5	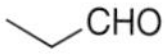 ～CHO	1a′	5	24a′	20 (5)	−30	88	71/29	68/86
6		1a′	10	24a′	22 (10)	−20	56	41/59	48/87
7	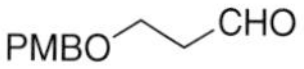 PMBO～CHO	1b′	5	24b′	20 (5)	−30	89	59/41	86/95
8		1b′	10	24b′	22 (10)	−20	73	41/59	58/93
9	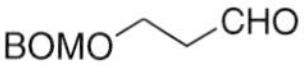 BOMO～CHO	1z	5	24z	20 (5)	−30	92	69/31	87/97
10		1z	10	24z	22 (10)	−20	72	39/61	52/81
11	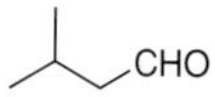 ⅄CHO	1j	5	24j	20 (5)	−30	80	68/32	72/87
12	 BnO～CHO	1y	5	24y	20 (5)	−30	80	65/35	85/92

[a] Reaction conditions: Et$_2$Zn (4× mol%), ligand (× mol%), MS 3A, THF.

Maruoka recently developed an efficient direct aldol reaction of glycine Schiff base **25a** using phase-transfer catalyst (R,R)-**29** [25]. The aldol reaction of Schiff base **25a** with aldehydes was efficiently promoted by 2 mol% of **29** in toluene–aqueous NaOH (1%) at 0 °C to give *anti*-β-hydroxy-α-amino acids in excellent ee (91–98% ee), moderate to good dr (*anti/syn* = 1.2:1 to

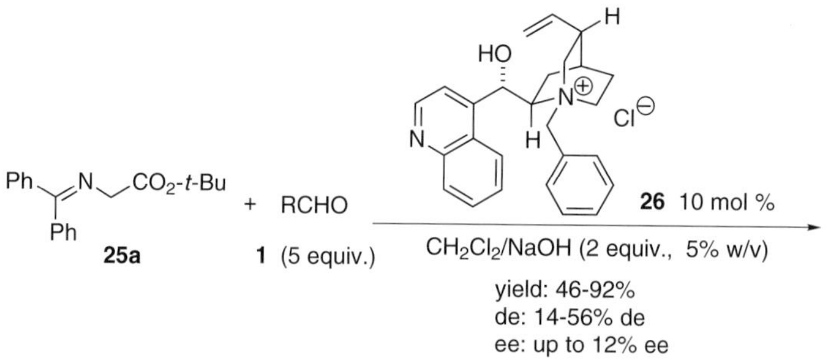

Scheme 6.21
Direct catalytic asymmetric aldol reaction of **25a** promoted by phase-transfer catalyst **26**.

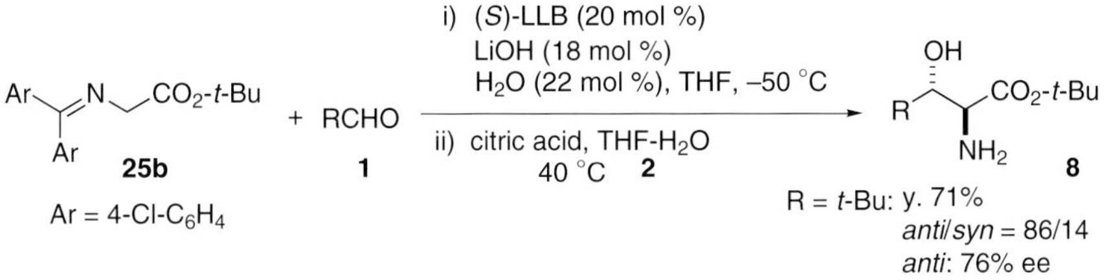

Scheme 6.22
Direct catalytic asymmetric aldol reaction of
25b promoted by heterobimetallic catalyst
(S)-LLB.LiOH.

20:1), and in good yield (58–78%) (Scheme 6.23 and Table 6.11). Low catalyst loading, operationally easy reaction conditions (0 °C, two-phase reaction), and high ee are noteworthy. Use of (R,R)-**29b** as catalyst occasionally significantly enhanced both diastereo- and enantioselectivity in this system.

6.6
Other Examples

Morken reported a catalytic asymmetric reductive aldol reaction in which aldehydes and acrylate **30** were converted into chiral syn-α-methyl-β-hydroxy esters under the catalysis of transition metal (Rh and Ir) complexes (Scheme 6.24) [26]. The method provided a means of catalytic synthesis of active

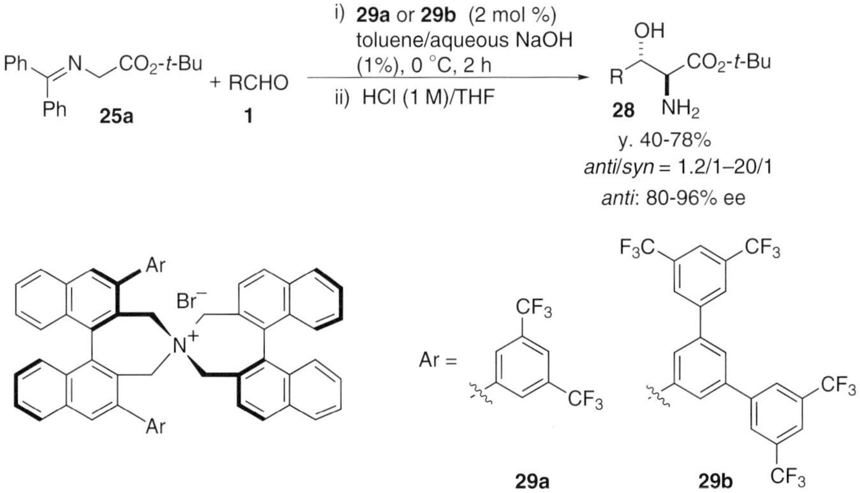

Scheme 6.23
Direct catalytic asymmetric aldol reaction of
25a promoted by chiral phase-transfer
catalyst **29** promoted by chiral phase-
transfer catalyst **29ª**.

Tab. 6.11

Direct catalytic asymmetric aldol reaction of **25a**[a].

Entry	Aldehyde		Product	Catalyst	Yield (%)	dr (anti/ syn)	ee (%) (anti)
1	Ph⌒⌒CHO	1e	28e	29a	76	3.3/1	91
2		1e	28e	29b	71	12/1	96
3	⌒⌒⌒CHO	1t	28t	29b	65	10/1	91
4	TIPSO⌒CHO	1c'	28c'	29b	72	20/1	98
5	⌒⌒CHO	1d'	28d'	29b	62	6.3/1	80
6		1d'	28d'	29a	71	2.4/1	90
7	CH₃CHO	1e'	28e'	29a	58	2.3/1	92
8	⬡-CHO	1c	28c	29a	40	2.8/1	95
9[b]		1c	28c	29a	78	1.2/1	93

[a] Reaction conditions: catalyst **29** (2 mol%), toluene/aqueous NaOH
(1%), 0 °C, 2 h.
[b] Use of dibutyl ether as solvent.

R¹CHO + 30 (1.2 equiv.)

$$\text{R}^1\text{CHO} + \overset{O}{\underset{}{\diagup}}\text{OPh}$$

1   **30** (1.2 equiv.)

[(cod)RhCl]₂ (2.5 mol %)
(R)-BINAP
(6.5 mol %)
Et₂MeSiH (1.2 equiv.)
——————————————→
dichloroethane
rt, 24 h

OH O
R⌒⌒OPh

31
yield: 48-82%
syn/anti = 1.8/1-5.1/1
syn = 45-88% ee, *anti* = 7-99% ee

(R)-BINAP

$$\text{R}^1\text{CHO} + \overset{O}{\underset{}{\diagup}}\text{OR}^2$$

1   **30** (1.2 equiv.)

[(cod)IrCl]₂ (2.5 mol %)
indane-pybox **32**
(7.5 mol %)
Et₂MeSiH (1.2 equiv.)
——————————————→
dichloroethane
rt, 24 h

OH O
R⌒⌒OR'

31
yield: 47-68%
syn/anti = 2.7/1-9.9/1
syn = 82-96% ee

32

Scheme 6.24
Catalytic asymmetric reductive aldol
reactions promoted by Rh–(R)-BINAP
complex and Ir-**32** complex.

Tab. 6.12
Catalytic asymmetric reductive aldol reaction promoted by Ir-**32** complex[a].

Entry	Aldehyde		R'	Yield (%)	dr (syn/anti)	ee (%) (syn)
1[b]	PhCHO	**1n**	Et	68	6.6/1	94
2	BnO⌒CHO	**1y**	Me	49	9.9/1	96
3[c]		**1y**	Me	59	9.5/1	96
4	TBSO⌒CHO	**1f'**	Me	47	8.2/1	96
5	BnO⌒⌒CHO	**1x**	Me	65	2.7/1	82

[a] Reaction conditions: [(cod)lrCl]$_2$ (2.5 mol%), **32** (7.5 mol%),
Et$_2$MeSiH (2 equiv.), rt, 24 h.
[b] [(coe)lrCl]$_2$ was used instead of [(cod)lrCl]$_2$.
[c] Reaction carried out on 35 mmol scale with 1 mol% [(cod)lrCl]$_2$ and
3 mol% of **32**.

enolate species from unmodified substrates. Combination of 2.5 mol%
[(cod)RhCl]$_2$ and 6.5 mol% (R)-BINAP promoted the reductive aldol reaction
between aldehydes **1** and phenyl acrylate (**30**) in the presence of Et$_2$MeSiH
at room temperature to afford a diastereomixture (syn/anti = 1.7:1 to 5.1:1)
of β-hydroxy esters **31** in good to moderate yield (48–82%) and ee (45–88%
ee) [26a]. The stereoselectivity was improved by using the Ir complex derived from [(cod)IrCl]$_2$ and indane-pybox **32**, affording the product syn-
selectively (up to 9.9:1) in high ee (up to 96% ee). The results obtained with
Ir-**32** are summarized in Table 6.12 [26b]. Use of H$_2$ gas as reducing reagent
instead of silanes was recently reported by Krische in an achiral reductive
aldol reaction [27].

Evans reported a direct catalytic diastereoselective aldol reaction of *N*-
acyloxazolidinones **33**. In the presence of a catalytic amount of Mg salt
(10 mol%) stoichiometric amounts of Et$_3$N (2 equiv.) and (CH$_3$)$_3$SiCl (1.5
equiv.) the enolate species was generated in situ and *anti*-aldol adducts
were obtained in excellent stereoselectivity (yield 36–92%, dr 3.5:1 to 32:1;
Scheme 6.25) [28a]. Stoichiometric amounts of silylating regents were essential to achieve efficient catalyst turnover. Because mechanistic study
revealed that the reaction does not involve an enol silyl ether as an intermediate, a Mukaiyama-type reaction pathway was reasonably excluded. By
using *N*-acylthiazolidinethiones **35** instead of **33** the aldol adducts were
obtained with different stereoselectivity [28b]. MgBr$_2$–OEt$_2$ (10 mol%) afforded the best results for **35** (Scheme 6.25). *anti*-Aldol adducts were obtained in excellent selectivity (yield 56–93%, dr 7:1 to 19:1). Catalytic asymmetric variants of these reactions on the basis of the Mg catalysis seem
promising.

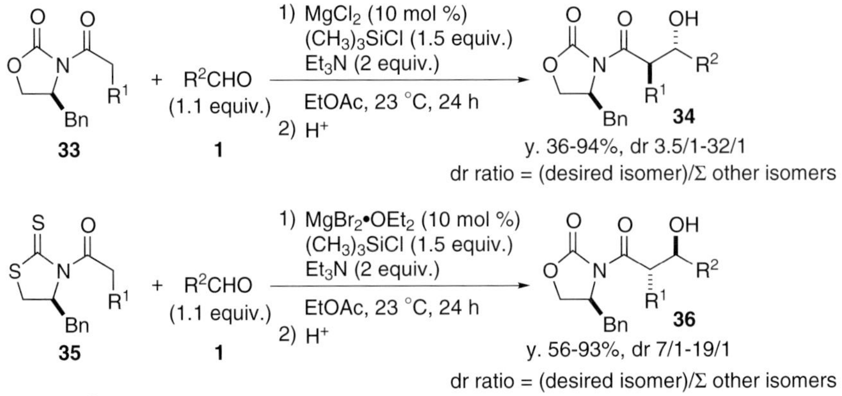

Scheme 6.25
Direct catalytic diastereoselective aldol reactions promoted by Mg salts.

Shair recently reported an achiral direct aldol reaction starting from malonic acid half thioester **37** (Scheme 6.26). The active enolate species were generated under extremely mild conditions with Cu catalysis. Development of a direct catalytic asymmetric aldol reaction on the basis of this strategy is also promising [29].

6.7
Conclusion

Representative examples of direct catalytic asymmetric aldol reactions promoted by metal catalysis have been summarized. The field has grown rapidly during past five years and many researchers have started to investigate this "classical" yet new field using metal catalysts and organocatalysts. Application of these systems to direct Mannich reactions and Michael reactions has also been studied intensively recently. Further investigations will be needed to overcome problems remaining with regard to substrate generality, reaction time, catalyst loading, volumetric productivity, etc. The development of a direct catalytic asymmetric aldol reaction with unmodified esters as a donor is particularly required.

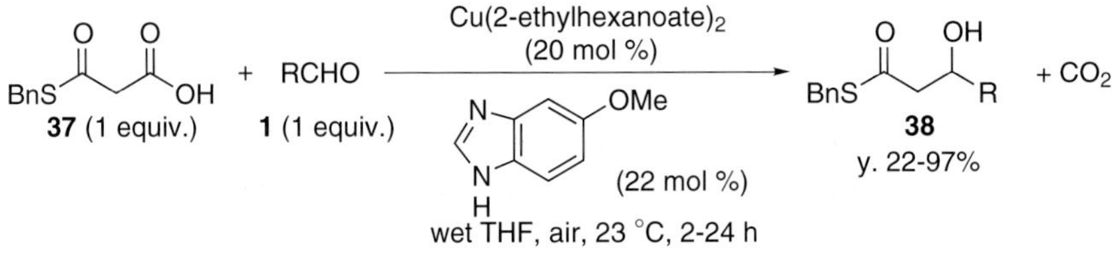

Scheme 6.26
Catalytic thioester aldol reactions prompted by Cu(II) salt.

6.8
Experimental Section

Procedure for the Preparation of (S)-LLB Complex. A solution of La(O-*i*-Pr)$_3$ (20.4 mL, 4.07 mmol, 0.2 M in THF, freshly prepared from La(O-*i*-Pr)$_3$ powder and dry THF) was added to a stirred solution of (*S*)-binaphthol (3.50 g, 12.2 mmol) in THF (39.7 mL) at 0 °C. (La(O-*i*-Pr)$_3$ was purchased from Kojundo Chemical Laboratory, 5-1-28 Chiyoda, Sakado-shi, Saitama 350-0214, Japan; Fax: +81-492-84-1351). The solution was stirred for 30 min at room temperature and the solvent was then evaporated under reduced pressure. The resulting residue was dried for 1 h under reduced pressure (ca. 5 mmHg) and dissolved in THF (60.5 mL). The solution was cooled to 0 °C and *n*-BuLi (7.45 mL, 12.2 mmol, 1.64 M in hexane) was added. The mixture was stirred for 12 h at room temperature to give a 0.06 M solution of (*R*)-LLB which was used to prepare (*S*)-LLB–KOH catalyst.

General Procedure for Direct Catalytic Asymmetric Aldol Reactions of Methyl Ketone 2 Using (S)-LLB–KOH. A solution of water in THF (48.0 μL, 0.048 mmol, 1.0 M) was added to a stirred solution of potassium bis(trimethylsilyl)amide (KHMDS, 43.2 μL, 0.0216 mmol, 0.5 M) in toluene at 0 °C. The solution was stirred for 20 min at 0 °C and then (*S*)-LLB (400 μL, 0.024 mmol, 0.06 M in THF, prepared as described above) was added and the mixture was stirred at 0 °C for 30 min. The resulting pale yellow solution was cooled to −20 °C and acetophenone (**2a**) (175 μL, 1.5 mmol) was added. The solution was stirred for 20 min at this temperature then 2,2-dimethyl-3-phenylpropanal (**1b**) (49.9 μL, 0.3 mmol) was added and the reaction mixture was stirred for 28 h at −20 °C. The mixture was then quenched by addition of 1 M HCl (1 mL) and the aqueous layer was extracted with ether (2 × 10 mL). The combined organic layers were washed with brine and dried over Na$_2$SO$_4$. The solvent was removed under reduced pressure and the residue was purified by flash chromatography (SiO$_2$, ether–hexane 1:12) to give **3ba** (72 mg, 85%, 89% ee).

General Procedure for Direct Catalytic Asymmetric Aldol Reaction of Methyl Ketone 2 Using Dinuclear Zn$_2$-13. The prepare the catalyst a solution of diethyl zinc (1 M in hexane, 0.2 mL, 0.2 mmol) was added to a solution of ligand **13** (64 mg, 0.1 mmol) in THF (1 mL) at room temperature under an argon atmosphere. After stirring for 30 min at the same temperature, with evolution of ethane gas, the resulting solution (ca. 0.09 M) was used as catalyst for the aldol reaction.

To perform the aldol reaction a solution of the catalyst (0.025 mmol) was added, at 0 °C, to a suspension of aldehyde (0.5 mmol), triphenylphosphine sulfide (22.1 mg, 0.075 mmol), powdered 4 Å molecular sieves (100 mg, dried at 150 °C under vacuum overnight), and ketone **2** (2.5 or 5 mmol) in THF (0.8 mL). The mixture was stirred at 5 °C for 2 days then poured on to

1 M HCl and extracted with ether. After normal work-up, the crude product was purified by silica gel column chromatography.

General Procedure for Catalytic Asymmetric Aldol Reaction of Hydroxyketone 18f Promoted by Et₂Zn/(S,S)-linked-BINOL, 4:1, with MS 3A. MS 3A (200 mg) in a test tube was activated before use under reduced pressure (ca. 0.7 kPa) at 160 °C for 3 h. After cooling, a solution of (S,S)-linked-BINOL (1.53 mg, 0.0025 mmol) in THF (0.6 mL) was added under Ar. The mixture was cooled to −20 °C and Et₂Zn (10 μL, 0.01 mmol, 1.0 M in hexanes) was added to the mixture at this temperature. After stirring for 10 min at −20 °C, a solution of **18f** (182.8 mg, 1.1 mmol) in THF (1.1 mL) was added. Aldehyde **1e** (1.0 mmol) was added and the mixture was stirred at −20 °C for 18 h and then quenched by addition of 1 M HCl (2 mL). The mixture was extracted with ethyl acetate and the combined organic extracts were washed with sat. aqueous NaHCO₃ and brine and dried over MgSO₄. Evaporation of the solvent gave a crude mixture of the aldol products. The diastereomeric ratios of the aldol products were determined by ¹H NMR of the crude product. After purification by silica gel flash column chromatography (hexane–acetone 8:1 to 4:1), **19ef** was obtained (269.6 mg, 0.898 mmol, yield 90%, dr *syn/anti* = 89:11, 96% ee (*syn*)).

General Procedure for Catalytic Asymmetric Aldol Reaction of Glycine Schiff Base 25a Promoted by Phase-transfer Catalyst 29. Aqueous NaOH (1%, 2.4 mL) was added at 0 °C, under Ar, to a solution of Schiff base **25a** (88.6 mg, 0.3 mmol) and (R,R)-**29b** (9.9 mg, 2 mol%) in toluene (3 mL). Aldehyde **1e** (79 μL, 0.6 mmol) was then introduced dropwise. The whole mixture was stirred for 2 h at 0 °C, and water and diethyl ether were then added. The ether phase was isolated, washed with brine, dried over Na₂SO₄, and concentrated. The crude product was dissolved in THF (8 mL) and treated with HCl (1 M, 1 mL) at 0 °C for 1 h. After removal of THF in vacuo the aqueous solution was washed three times with diethyl ether and neutralized with NaHCO₃. The mixture was then extracted three times with CH₂Cl₂. The combined extracts were dried over MgSO₄ and concentrated. After purification by silica gel column chromatography (CH₂Cl₂–MeOH 15:1) **28e** was obtained (56.8 mg, 0.214 mmol, yield 71%, dr *anti/syn* = 12:1, 96% ee (*anti*)).

References and Notes

1 Recent review: C. PALOMO, M. OIARBIDE, J. M. GARCÍA, *Chem. Eur. J.* **2002**, *8*, 37.
2 Review: T. D. MACHAJEWSKI, C.-H. WONG, *Angew. Chem. Int. Ed.* **2000**, *39*, 1352.
3 Recent review for the direct catalytic asymmetric aldol

reactions: B. Alcaide, P. Almendros, *Eur. J. Org. Chem.* **2002**, 1595.

4 Review: M. Sawamura, Y. Ito, *Chem. Rev.* **1992**, *92*, 857.

5 Review: B. List, *Tetrahedron* **2002**, *58*, 5573.

6 Y. M. A. Yamada, N. Yoshikawa, H. Sasai, M. Shibasaki, *Angew. Chem. Int. Ed. Engl.* **1997**, *36*, 1871.

7 K. Fujii, K. Maki, M. Kanai, M. Shibasaki, *Org. Lett.* **2003**, *5*, 733.

8 N. Yoshikawa, Y. M. A. Yamada, J. Das, H. Sasai, M. Shibasaki, *J. Am. Chem. Soc.* **1999**, *121*, 4168.

9 (a) D. Sawada, M. Kanai, M. Shibasaki, *J. Am. Chem. Soc.* **2000**, *122*, 10521; (b) D. Sawada, M. Shibasaki, *Angew. Chem. Int. Ed.* **2000**, *39*, 209.

10 Y. M. A. Yamada, M. Shibasaki, *Tetrahedron Lett.* **1998**, *39*, 5561.

11 B. M. Trost, H. Ito, *J. Am. Chem. Soc.* **2000**, *122*, 12003.

12 B. M. Trost, E. R. Silcoff, H. Ito, *Org. Lett.* **2001**, *3*, 2497.

13 T. Suzuki, N. Yamagiwa, Y. Matsuo, S. Sakamoto, K. Yamaguchi, M. Shibasaki, R. Noyori, *Tetrahedron Lett.* **2001**, *42*, 4669.

14 N. Yoshikawa, M. Shibasaki, *Tetrahedron* **2001**, *57*, 2569.

15 R. Mahrwald, B. Ziemer, *Tetrahedron Lett.* **2002**, *43*, 4459.

16 (a) W. Notz, B. List, *J. Am. Chem. Soc.* **2000**, *122*, 7368; (b) K. Sakthivel, W. Notz, T. Bui, C. F. Barbas, III, *J. Am. Chem. Soc.* **2001**, *123*, 5260.

17 B. M. Trost, H. Ito, E. R. Silcoff, *J. Am. Chem. Soc.* **2001**, *123*, 3367.

18 B. M. Trost, V. S. C. Yeh, *Org. Lett.* **2002**, *4*, 3513.

19 N. Yoshikawa, N. Kumagai, S. Matsunaga, G. Moll, T. Ohshima, T. Suzuki, M. Shibasaki, *J. Am. Chem. Soc.* **2001**, *123*, 2466.

20 N. Yoshikawa, T. Suzuki, M. Shibasaki, *J. Org. Chem.* **2002**, *67*, 2556.

21 N. Kumagai, S. Matsunaga, N. Yoshikawa, T. Ohshima, M. Shibasaki, *Org. Lett.* **2001**, *3*, 1539.

22 N. Kumagai, S. Matsunaga, T. Kinoshita, S. Harada, S. Okada, S. Sakamoto, K. Yamaguchi, M. Shibasaki, *J. Am. Chem. Soc.* **2003**, *125*, 2169.

23 C. M. Gasparski, M. J. Miller, *Tetrahedron* **1991**, *47*, 5367.

24 N. Yoshikawa, M. Shibasaki, *Tetrahedron* **2002**, *58*, 8289.

25 T. Ooi, M. Taniguchi, M. Kameda, K. Maruoka, *Angew. Chem. Int. Ed.* **2002**, *41*, 4542.

26 (a) S. J. Taylor, M. O. Duffey, J. P. Morken, *J. Am. Chem. Soc.* **2000**, *122*, 4528. (b) C.-X. Zhao, M. O. Duffey, S. J. Taylor, J. P. Morken, *Org. Lett.* **2001**, *3*, 1829. Achiral reaction: (c) S. J. Taylor, J. P. Morken, *J. Am. Chem. Soc.* **1999**, *121*, 12202.

27 H.-Y. Jang, R. R. Huddleston, M. J. Krische, *J. Am. Chem. Soc.* **2002**, *124*, 15156.

28 (a) D. A. Evans, J. S. Tedrow, J. T. Shaw, C. W. Downey, *J. Am. Chem. Soc.* **2002**, *124*, 392. (b) D. A. Evans, C. W. Downey, J. T. Shaw, J. S. Tedrow, *Org. Lett.* **2002**, *4*, 1127.

29 G. Lalic, A. D. Aloise, M. D. Shair, *J. Am. Chem. Soc.* **2003**, *125*, 2852.

7
Catalytic Enantioselective Aldol Additions with Chiral Lewis Bases

Scott E. Denmark and Shinji Fujimori

7.1
Introduction

7.1.1
Enantioselective Aldol Additions

The aldol addition reaction is one of the most powerful carbon–carbon bond-construction methods in organic synthesis and has achieved the exalted status of a "strategy-level reaction". The generality, versatility, selectivity, and predictability associated with this process have inspired many reviews and authoritative summaries and constitute the theme of this treatise [1].

The primary objective in the evolution of the aldol addition is the striving for exquisite diastereo- and enantioselectivity from readily available enolate precursors. The ideal aldol reaction would provide selective access for all four isomers of the stereochemical dyad that make up the aldol products. This has given way to more ambitious investigation of the triads and tetrads that accrue from double and triple diastereoselection processes [2]. The solutions to these challenges have been imaginative and diverse, and have pioneered the contemporaneous development of asymmetric synthesis as a core discipline. A secondary and more recent objective is the development of "direct aldol additions" that mimic enzymatic processes (aldolases) and obviate the independent activation of the nucleophilic partner.

The number and variety of inspired and elegant solutions for perfecting the aldol addition are expertly described in the accompanying chapters of this volume. This chapter differs somewhat, however, in that it describes a conceptually distinct process that has been designed to address some of the shortcomings inherent in the more classic approaches involving chiral Lewis acid catalysis of aldol addition in its many incarnations. Thus, to assist the reader in understanding the distinctions and to provide the conceptual framework for invention of Lewis-base-catalyzed addition, the introduction will outline briefly the stereocontrolling features of the main families

Modern Aldol Reactions. Vol. 2: Metal Catalysis. Edited by Rainer Mahrwald
Copyright © 2004 WILEY-VCH Verlag GmbH & Co. KGaA, Weinheim
ISBN: 3-527-30714-1

Scheme 7.1
Aldol additions with chirally modified lithium enolates.

of enantioselective aldol additions and thus, the basis for inventing a new process.

7.1.1.1 Background

Early examples of asymmetric aldol addition reactions involved lithium enolates of chiral carbonyl compounds that reacted with aldehydes to give good diastereoselectivity [3]. The chirality of the enolate translated to enantiomerically enriched products when the auxiliaries were destroyed or removed. Thus, using enolates of modified ketones [3a], esters [3b], and sulfonamides [3c], high enantioselectivity and diastereoselectivity can be achieved if enolates are generated in geometrically defined form (Scheme 7.1). Although high selectivity is obtained, these reactions are not practical because they require stoichiometric amounts of covalently bound auxiliaries. In addition, the high reactivity of the lithium enolates did not ensure reaction via closed, organized transition structures, a feature crucial for stereochemical information transfer.

A revolutionary advance in aldol technology was the use of less reactive metalloenolates (boron [4a], titanium [4b–d], and zirconium [4d]) that organize the aldehyde, enolate, and auxiliary in a closed transition structure (Scheme 7.2). Although these reagents are similar to those described above in that an auxiliary is needed in stoichiometric amounts, the use of boron and titanium enolates enable attachment of the modifier by an acyl linkage or directly around the metal of the enolate. Geometrically defined enolates react with aldehydes to give the *syn* or *anti* diastereomers with high enantiomeric excess. This variant is best exemplified by the acyl oxazolidinone boron enolates [1a], the diazaborolidine derived enolates [5], titanium enolates derived from diacetone glucose [6], the diisiopinylcampheyl boron enolates for ketone aldolizations [2c], and proline-derived silanes for *N,O*-ketene acetals [7].

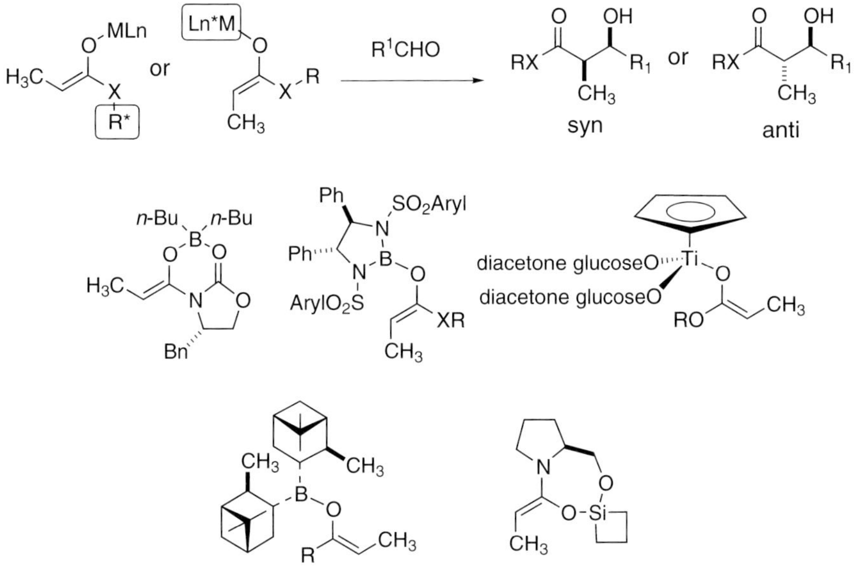

R¹CHO

syn   anti

n-Bu   n-Bu

H₃C

Bn

Ph   SO₂Aryl

Ph

ArylO₂S

CH₃

diacetone glucoseO⋯Ti—O

diacetone glucoseO

RO   CH₃

CH₃   CH₃

B—O

R   CH₃

N

O—Si

CH₃

Scheme 7.2
Chirally modified boron, titanium, and silicon enolates.

The key stereocontrolling features common to these agents are:

- the organizational role of the metal center;
- the close proximity of the electrophile, nucleophile and asymmetric modifier in coordination sphere of the metal assuring high stereochemical information transfer, and
- the high stereochemical influence of enolate geometry on product diastereoselectivity.

The major disadvantage of these variants is the inability to operate catalytically. Indeed, it is the high metal affinity of the aldehyde, enolate, and chiral auxiliary that interferes with the turnover.

Catalytic processes have, over the past decade, dominated the development of enantioselective aldol addition reactions [1j,k]. This category can be subdivided into five main classes:

- chiral-Lewis-acid-catalyzed aldol additions of silicon or tin enol ethers (Mukaiyama aldol addition);
- in-situ generated metalloenolates from silicon or tin enol ethers;
- in-situ generated metalloenolates directly from ketones;
- in-situ generated enolate equivalents (enamines) directly from carbonyl compounds; and
- enzyme- and antibody-catalyzed aldol additions.

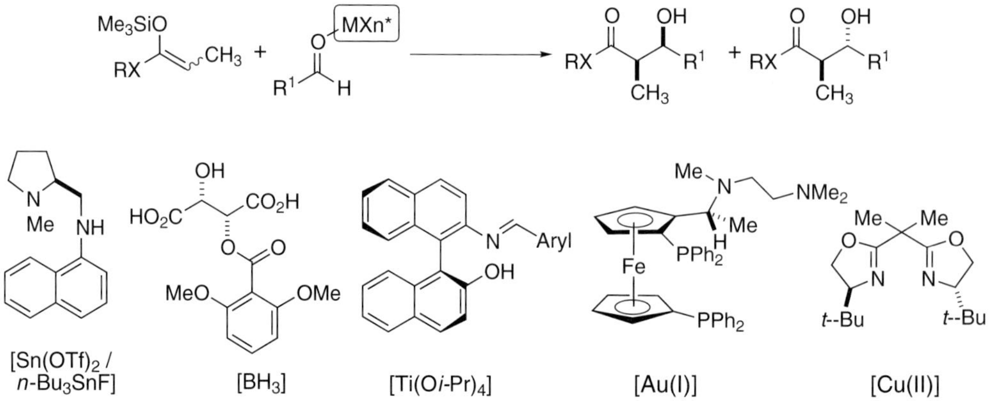

Me₃SiO–C(RX)=CH₃ + R¹CHO —[MXn*]→ RX–C(O)–CH(CH₃)–CH(OH)–R¹ + RX–C(O)–CH(CH₃)–CH(OH)–R¹

[Sn(OTf)₂ / n-Bu₃SnF]   [BH₃]   [Ti(Oi-Pr)₄]   [Au(I)]   [Cu(II)]

Scheme 7.3
Representative metal-based chiral Lewis acids.

The first three only will be discussed in the context of the origins of stereo-induction.

The use of chiral Lewis acids [8] has received by far the most attention and is amply discussed in the many chapters dedicated to various metals in this volume. Some of the more commonly used and selective chiral Lewis acids are shown here, for example diamine complexes of tin(II) triflate [9], borane complexes of a monoester of tartaric acid (CAB catalysts) [10], sulfonamido amino acid borane complexes [11], titanium binaphthol [12] and binaphthylimine complexes [13], ferrocenylphosphine–gold [5d] and BINAP–silver [14] complexes, and copper(II) bisoxazoline and pyridyl(bisoxazoline) complexes [15], (Scheme 7.3).

These variants of the aldol reaction have several key features in common:

- the additions have been demonstrated for aldehydes and enol metal derivatives with sub-stoichiometric loading of the chiral Lewis acid;
- the diastereo- and enantioselectivity are variable although they can be high; and
- these reactions are not responsive to prostereogenic features – when the configuration of the enolsilane nucleophile changes, the diastereoselectivity of the product does not change [16].

In these reactions, the metal center is believed to activate the aldehyde to addition and the enol addition subject primarily to steric approach control, i.e. it is lacking the pre-organization associated with the stoichiometric aldol addition reactions of the boron and titanium enolates.

This problem has been addressed in part by the recently developed class of aldol additions that involve the use of chirally modified metalloids in a catalytic process [17]. In these reactions it is proposed a metal–phosphine complex undergoes transmetalation with TMS enol ethers or tributylstannyl ketones to provide chiral metalloid enolates in situ. The aldol addition then proceeds, with turnover of the metalloid species to another latent enol donor.

In addition, in the third class the metalloenolate is generated in situ from either heterobimetallic (lanthanide/alkali metals) or chiral zinc phenoxide complexes and promotes the addition of unmodified ketones to aldehydes [18]. In these reactions it is postulated that the aldehyde is coordinated to the metal after generation of the metalloenolate. However, because the enolates are generated in situ, the enolate geometry is not known and geometry has not been correlated with product configuration.

Despite the power and clear synthetic applicability of these families, deficiencies are still apparent:

- lack of a catalytic variant of the boron or titanium enolate family; and
- lack of controllable selectivity in the chiral Lewis acid family.

Lewis base-activation provides a mechanism enabling devising of a class of aldol addition that addresses these concerns. This chapter describes, in detail, the formulation, development, and understanding of a Lewis-base-catalyzed aldol reaction process that embodies both the selectivity and versatility of the stoichiometric reactions in combination with the efficiency of the catalytic methods.

7.1.2
Lewis Base Catalysis

The design criteria for Lewis basic catalysis of the aldol addition are outlined in Figure 7.1. This approach differs from Lewis acid catalysis of aldol addition in that it postulates activation of the enoxymetal derivative by preassociation with a chiral Lewis basic (LB) group bearing a non-bonding pair of electrons. This complex must be more reactive than the free enolate for ligand accelerated catalysis to be observed [19]. Next, association of this -ate complex with the Lewis basic carbonyl oxygen of the aldehyde produces a hyper-reactive complex in which the metal has expanded its valence by two. It is expected that this association complex between enolate, aldehyde, and the chiral Lewis basic group reacts through a closed-type transition structure to produce the metal aldolate product. For turnover to be achieved the aldolate must undergo the expulsion of the LB group with formation of the chelated metal aldolate product. *Thus, Lewis base-catalysis involves simultaneous activation of the nucleophile and the electrophile within the coordination sphere of the metal. The reaction must occur in a closed array and be capable of releasing the activating group by chelation or change in the Lewis acidity.*

To realize this process selection of the appropriate enoxymetal and activator moieties is crucial. For the metal, the MX_n subunit must be able expand

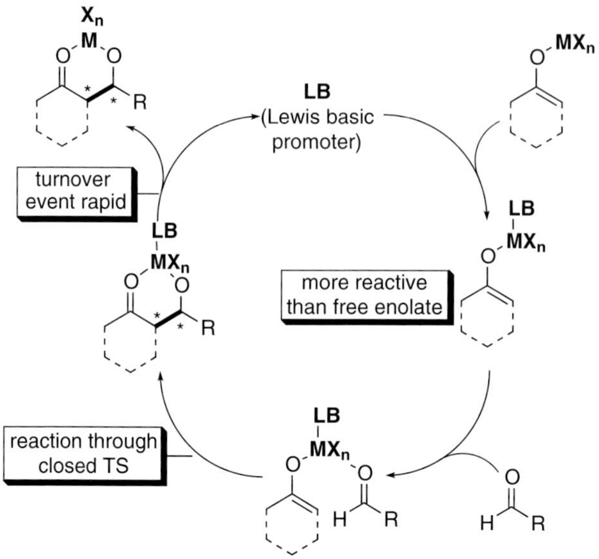

Fig. 7.1
Hypothetical catalytic cycle for Lewis-base-catalyzed aldol addition.

its valence by two and balance the nucleophilicity of the enolate with electrophilicity to coordinate both the Lewis basic aldehyde and the chiral LB group. To impart sufficient Lewis acidity to that metal group and accommodate the valence expansion such that two Lewis basic atoms may associate, the ligands (X) should be small and strongly electron-withdrawing. The criteria necessary for the chiral Lewis basic group LB are that it must be able to activate the addition without cleaving the $O-MX_n$ linkage and provide an effective asymmetric environment. Candidates for the Lewis basic group include species with high donicity properties as reflected in solvent basicity scales [20].

The inspiration to propose the possibility of nucleophilic catalysis of aldol additions and guide selection of the appropriate reaction partners is found in the cognate allylation process by allyl- and 2-butenyltrichlorosilanes. Inspired by the pioneering observations of Sakurai [21] and Kobayashi [22] that allyltrihalosilanes can be induced to add to aldehydes in the presence of nucleophilic activators (fluoride ion or DMF solvent) it was first shown in 1994 that chiral Lewis bases (phosphoramides) are capable of catalyzing the addition of allyltrichlorosilanes [23]. Thus, by analogy, reducing this plan to practice required the invention of a new class of aldol reagent, trichlorosilyl enolates, in conjunction with one of the most Lewis basic neutral functional groups, the phosphoramide group, Figure 7.2. Trichlorosilyl enolates of esters had been reported in the literature [24] and (because of the electron-withdrawing chloride ligands on silicon) were expected to be highly electro-

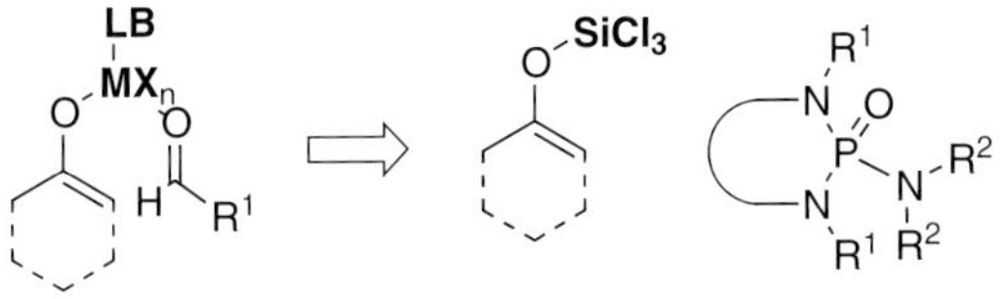

Fig. 7.2
Reaction components required for chiral Lewis-base-catalyzed aldol addition.

philic and thus able to stabilize the hypercoordinate silicon species necessary in such a process. The phosphoramides can be seen as chiral analogs of HMPA the Lewis basicity of which is well documented [20], especially toward silicon-based Lewis acids [25].

7.1.3
Organization of this Chapter

On the basis of the design criteria outlined above, the first, chiral-Lewis-base-catalyzed, enantioselective aldol addition was reported in 1996 [26]. This disclosure, which reported the reaction of the trichlorosilyl enolate of methyl acetate with a variety of aldehydes in the presence of several chiral phosphoramides, was significant not so much for the results obtained (enantioselectivity was modest −20 to 62% ee) but rather as a proof of principle for this conceptually new approach to the aldol addition reaction (Scheme 7.4).

This early success launched a broad-ranging program on the scope, synthetic application, and mechanistic understanding of chiral Lewis base catalysis of the aldol addition. A chronological recounting of the evolution of this program has already appeared [27]. For this chapter, a more comprehensive treatment of the various components of the process is presented, and thus, a more structurally based organization is employed. The main section begins with the preparation of the two new reaction components, namely the enoxytrichlorosilanes of ester, ketones, and aldehydes and the chiral Lewis basic catalysts (phosphoramides and N-oxides).

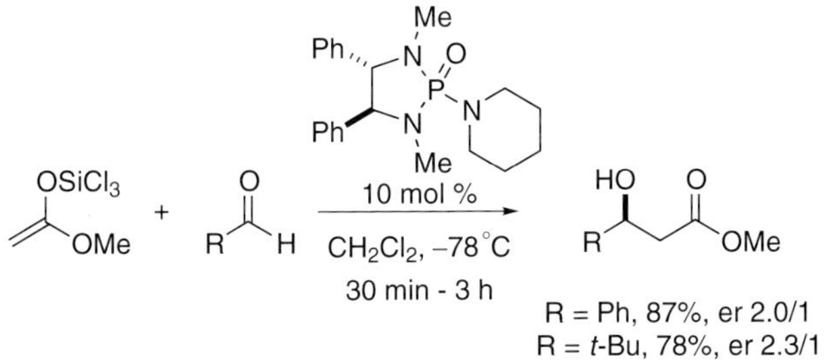

R = Ph, 87%, er 2.0/1
R = t-Bu, 78%, er 2.3/1

Scheme 7.4
The first enantioselective, chiral-Lewis-base-catalyzed aldol addition.

The bulk of the chapter is dedicated to describing the diversity of enolate structural subtypes in order of increasing structural complexity. Beginning with trichlorosilyl enolates of simple achiral methyl, ethyl, and cyclic ketones the survey then addresses chirally modified enolates of ketones and the phenomenon of double $(1,n)$-diastereoinduction. The next sections outline the use of trichlorosilyl enolates derived from aldehydes and esters and the features unique to these structures.

The final preparative section is dedicated to the newest variation on the theme, namely, the use of chiral Lewis bases to activate simple, achiral Lewis acids for enantioselective aldolization.

To facilitate more fundamental understanding of the development of the reaction variants, the chapter ends with an overview of the current mechanistic picture. Although this aspect is still evolving, the basic features are well in hand and enable integrated understanding of the behavior of trichlorosilyl enolates under these conditions.

Representative procedures for all the asymmetric processes described herein are provided at the end of the chapter.

7.2
Preparation of Enoxytrichlorosilanes

Silyl enol ethers (enoxysilanes) derived from carbonyl compounds are among the most important reagents in synthetic organic chemistry, because of their ability to form carbon–carbon bonds when combined with a myriad of carbon electrophiles [28]. The first silyl enol ethers, reported in 1958, were obtained by hydrosilylation of unsaturated carbonyl compounds (Scheme 7.5) [29]. Since then silyl enol ethers have become particularly versatile synthetic intermediates, and a number of reviews on preparation and reactions of these compounds have appeared [30]. The synthetic utility of enoxysilanes was not fully recognized until pioneering work by Mukaiyama on Lewis acid-catalyzed aldol additions of trimethylsilyl enol ethers to different carbonyl compounds (Scheme 7.6) [31].

The physical properties of simple enoxytrialkylsilanes were thoroughly investigated by Baukov and Lutsenko [30d]. Unlike conventional metal enolates, enoxytrialkylsilanes are stable and isolable covalent species. These

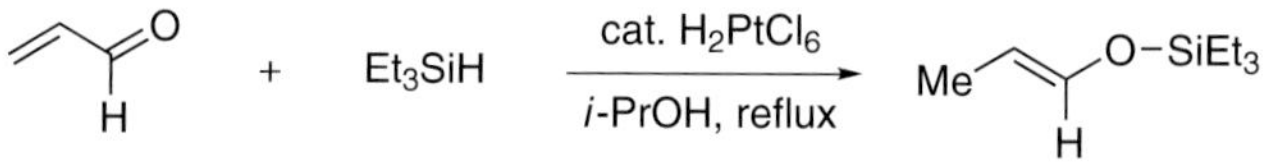

Scheme 7.5
First reported synthesis of a silyl enol ether.

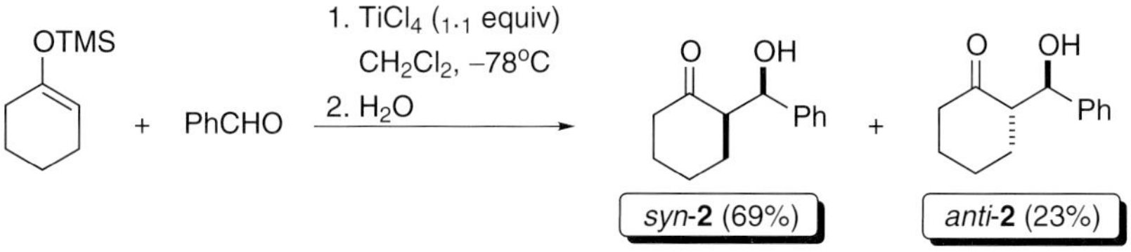

Scheme 7.6
Application of a silyl enol ether in a directed aldol addition.

species can be stored under non-acidic conditions for a long period of time but can also can be readily hydrolyzed to the parent carbonyl compounds under acidic conditions.

Trialkylsilyl enol ethers were originally introduced as precursors for regioisomerically-defined metal enolates. As enol derivatives they have reasonable nucleophilicity, although the most common use of these reagents involved regeneration of the metal enolate under basic conditions followed by reaction with electrophiles [30]. The nucleophilicities of a variety of enoxytrialkylsilanes have recently been correlated with other nucleophiles by Mayr [32]. The established order indicates that the nucleophilicity of enoxytrialkylsilanes is greater than that of allylic trialkylsilanes and less than that of commonly used enamines. Mayr also showed that silyl ketene acetals are much more nucleophilic than silyl enol ethers.

Trialkylsilyl enol ethers are extensively utilized in chiral Lewis acid-catalyzed stereoselective aldol additions [33]. On the other hand, silyl enol ethers with other groups on the silicon have been less widely applied in synthesis [34]. Heteroatom-functionalized silyl enol ethers can be prepared by methods similar to those used to prepare their trialkylsilyl counterparts (Scheme 7.7). Walkup and coworkers reported a convenient procedure for syntheses of a variety of non-alkyl-substituted enoxysilanes such as **4** and **5** [34a]. Hydrosilylation of α,β-unsaturated carbonyl compounds is also a viable method for preparation of such enoxysilanes [35]. Although a variety of silicon-functionalized silyl enol ethers have appeared in the literature, their application in synthetically useful reactions is still limited. A recent exception disclosed by Yamamoto and coworkers is a Lewis-acid-catalyzed enantioselective aldol reaction of an enoxy(trimethoxy)silane (Scheme 7.8) [36].

Because development of an effective Lewis-base-catalyzed aldol addition required access to electrophilic enolates, the preparation, properties and reactivity of enoxytrichlorosilanes became important areas of investigation. Pioneering studies by Baukov et al. ensured the possibility of generating trichlorosilyl ketene acetals, but it was subsequent studies by Denmark et al. that elevated these and related species to the status of useful synthetic reagents [30d, 37].

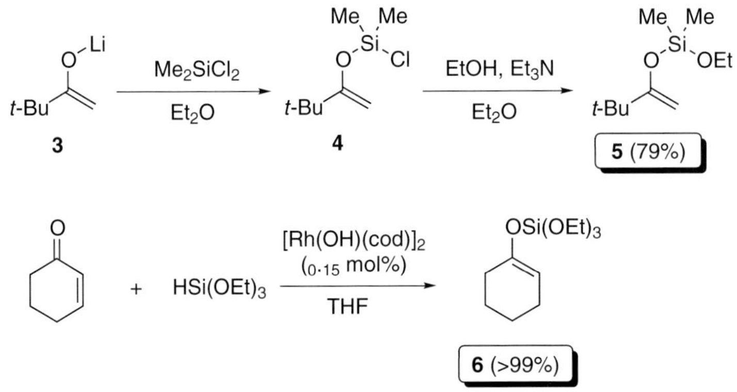

Scheme 7.7
Preparation of silicon-functionalized silyl enol ethers.

7.2.1
General Considerations

Trichlorosilyl enolates (enoxytrichlorosilanes) are typically viscous oils and can be obtained in the pure form by simple distillation. These silyl enolates can be stored under anhydrous conditions at low temperature for an appreciable time without decomposition. Exclusion of moisture is essential when working with trichlorosilyl enolates. Trace amounts of water leads to hydrolysis of the chlorosilane unit, and the resulting HCl is deleterious to the trichlorosilyl enolate. Degradation of the trichlorosilyl enolates can also be initiated by trace impurities such as metal salts and ammonium salts which promote formation of di- and polyenoxysilane species [37].

The thermal stability of trichlorosilyl enolates depends on their structure. For ketone- and aldehyde-derived trichlorosilyl enolates the O-silyl and C-silyl isomerism strongly favors the O-silyl species [37]. Although ketone- and aldehyde-derived trichlorosilyl enolates can be heated to 140 °C, trichlorosilyl enolates can disproportionate into dienoxysilanes and silicon tetrachloride at higher temperatures [37]. On the other hand, trichlorosilyl ketene acetals are not as thermally stable as the other trichlorosilyl enolates

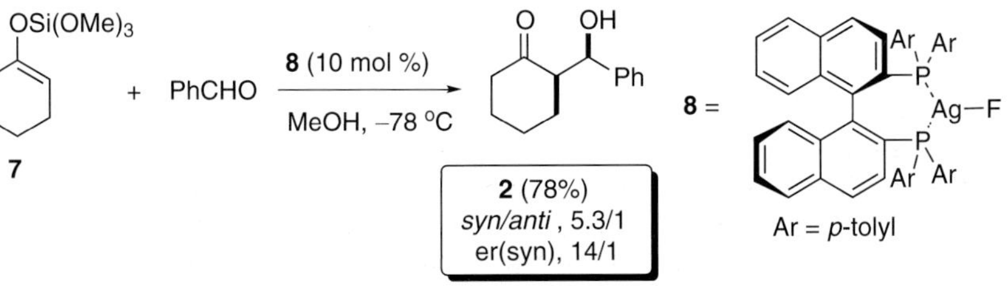

Scheme 7.8
Asymmetric aldol addition of a trialkoxysilyl enol ether.

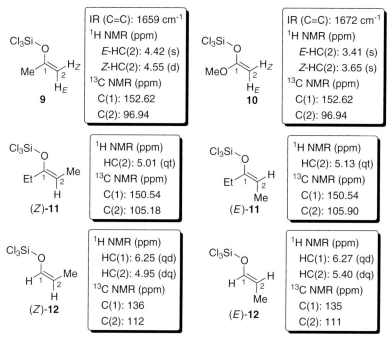

Fig. 7.3
Spectroscopic properties of enoxytrichlorosilanes.

and tend to isomerize to the corresponding carbon-bound, α-trichlorosilyl esters upon heating [37]. Distillation of these reagents should therefore be performed under vacuum at a temperature as low as possible.

The spectroscopic properties of several enoxytrichlorosilanes are summarized in Figure 7.3 [37, 38]. The ^{1}H NMR chemical shifts of the vinylic protons in enoxytrichlorosilanes are usually higher than those of the corresponding trimethylsilyl enol ethers. The vinylic protons for (E)- and (Z)-**11** derived from ethyl ketones are sufficiently different that the E/Z ratios for these enolates are readily obtained by ^{1}H NMR analysis. The vinylic proton for the E enolate is typically found at lower field than for the corresponding Z enolate. In the aldehyde-derived enoxytrichlorosilane the former aldehydic proton appears above 6 ppm. The IR stretching frequencies for the enol double bonds appear between 1630 and 1660 cm^{-1} for ketone-derived enoxytrichlorosilanes and at 1677 cm^{-1} for the acetate-derived trichlorosilyl ketene acetal.

One of the major differences between trialkylsilyl enol ethers and trichlorosilyl enolates is their reactivity toward aldehydes. Trialkylsilyl enol ethers usually do not react with aldehydes in the absence of nucleophilic or electrophilic activators [39]. On the other hand, trichlorosilyl enolates undergo aldol additions spontaneously with aldehydes at or below ambient

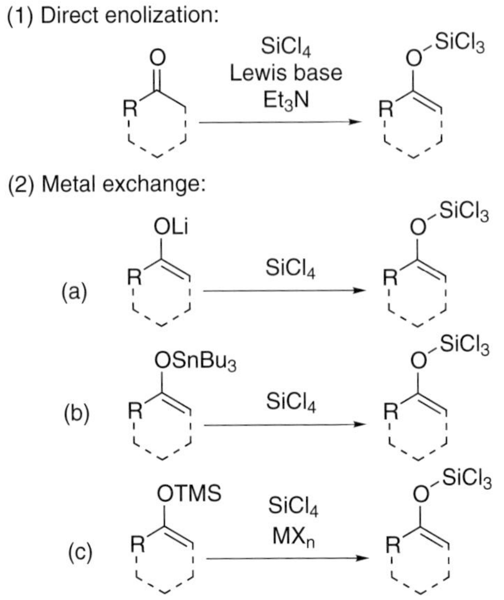

(1) Direct enolization:

(2) Metal exchange:

(a)

(b)

(c)

Fig. 7.4
General methods for preparation of enoxytrichlorosilanes.

temperature to afford aldol adducts in good yields. More importantly, tri-chlorosilyl enolates are susceptible to ligand-accelerated catalysis in the presence of Lewis bases [26]. The development, scope, utility, and mechanism of this process will be covered in subsequent sections.

In this section, the preparation and properties of trichlorosilyl enolates, classified by enolate structure, are described. Preparations of enoxytrichlorosilanes can be generalized to several categories: direct enolization of parent carbonyl compounds, trapping of corresponding metal (lithium) enolates, and metathesis of tin(IV) or trialkylsilyl enol ether with silicon tetrachloride (Figure 7.4). The optimum method for a given different class depends on the structure of the enolate. Synthetically viable methods only are discussed herein; other approaches are found in earlier review articles [30d].

7.2.2
Preparation of Ketone-derived Trichlorosilyl Enolates

One of the earliest reports on the synthesis of a trichlorosilyl enolate described the reduction of an α-chloroketone using trichlorosilane and a tertiary amine. Benkeser employed a combination of trichlorosilane and tri-*n*-butylamine for reduction of polyhalogenated organic compounds [40]. For example, α-chloroketone **13** is smoothly converted to trichlorosilyl enolate **14** in good yield by use of this procedure (Scheme 7.9). Under similar con-

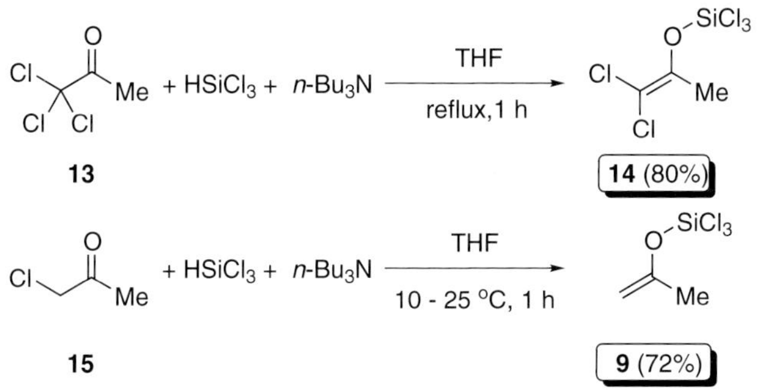

Scheme 7.9
Reductive silylation of α-chloroketones.

ditions the monochloro ketone **15** provides the corresponding trichlorosilyl enolate **9** in good yield.

Surprisingly, the reaction cannot be effected by triethylamine or diisopropylethylamine in place of tri(*n*-butyl)amine. The use of pentane as solvent was found to be superior to use of tetrahydrofuran because it enabled easier removal of solvent from these volatile trichlorosilyl enolates. These enolates are purified first by vacuum-transfer of the reaction mixture to separate them from ammonium salt and then by redistillation to remove solvent. Despite the operational simplicity and high yields obtained by use of this procedure, the scope of the reaction is somewhat limited by the availability of the corresponding α-chloro ketone and the volatility of the resulting enolate, which is necessary for vacuum-transfer. For the acetone-derived enolate **9**, however, this is the method of choice.

Another useful method of preparation of trichlorosilyl enolates involves metathesis of the corresponding enol stannane with silicon tetrachloride (Scheme 7.10) [37, 41]. Enol stannanes can be prepared by treatment of enol acetates with tributylmethoxystannane at elevated temperature [42]. Reaction of enol stannanes with silicon tetrachloride at low temperature provides trichlorosilyl enolates in modest to good yield. Excess silicon tetrachloride is recommended to prevent formation of polyenoxysilanes. These two steps can be performed without purification of the intermediate enol stannane, and this makes the method more practical.

Trichlorosilyl enolates are efficiently generated from methyl and cyclic ketones by this method. For the propiophenone-derived enolate **18**, excellent geometric selectivity for the *Z* isomer is observed.

Although this method is general for the preparation of ketone-derived trichlorosilyl enolates, the use of a stoichiometric amount of the tin reagent and the limited availability of structurally homogeneous enol acetates make this procedure less practical. Distillation of the trichlorosilyl enolate in the presence of tin residues and excess chlorosilane during purification is sometimes difficult.

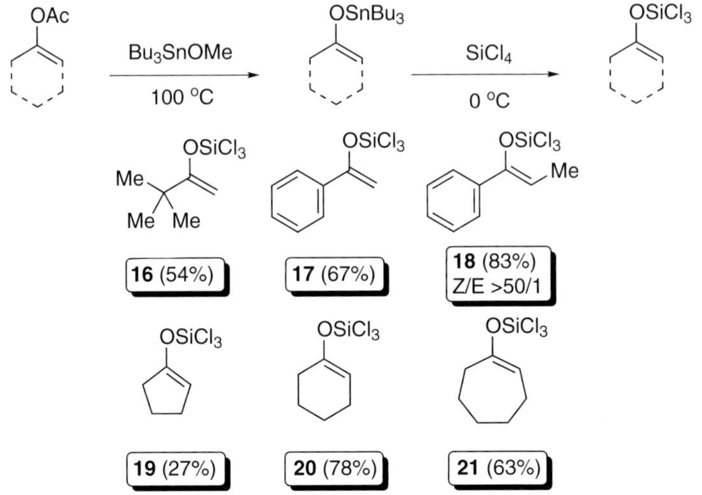

Scheme 7.10
Preparation of enoxytrichlorosilanes from enoxystannanes.

To avoid the use of tin reagents several other methods have been developed for preparation of trichlorosilyl enolates. Among these the most general method for ketone-derived trichlorosilyl enolates is the metal-catalyzed trans-silylation of trimethylsilyl enol ethers. It is known that mercury(II) and tin(IV) salts react with trimethylsilyl enol ethers to generate α-mercurio- and α-stannyl-ketones [43, 44]. Also, as shown previously, tin enolates can be readily converted into trichlorosilyl enolates by the action of silicon tetrachloride. From these observations, a metal-catalyzed process for conversion from trimethylsilyl enol ethers to trichlorosilyl enolates could be devised (Figure 7.5) [37].

A survey of different metal salts revealed that soft Lewis acids such as Hg(OAc)$_2$ and Pd(OAc)$_2$ are effective catalysts of this transformation [37]. Optimization studies indicated that the stoichiometry of the reagents and the reaction concentration are critical to the rate of trans-silylation and to control the amount of bisenoxysilane species formed. A bis(enoxy)-dichlorosilane is a common impurity associated with many aspects of enoxytrichlorosilane chemistry. The formation of a bis(enoxy)dichlorosilane can be explained by disproportionation of a monoenoxytrichlorosilane. At

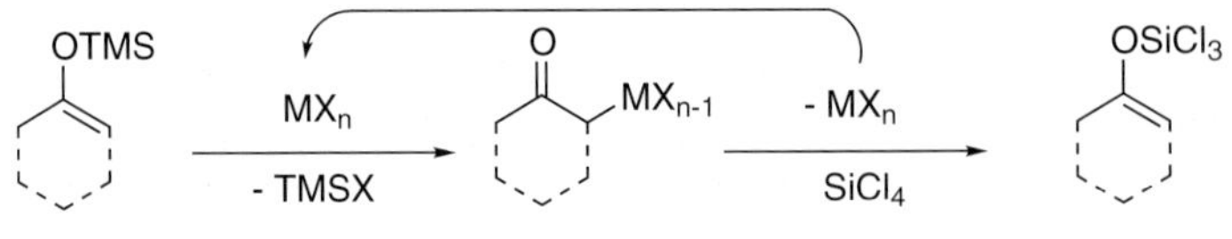

Fig. 7.5
Metal-catalyzed transsilylation.

the end of the reaction the crude reaction mixture usually contains 10–15% bis(enoxy)dichlorosilane species. The amount of bis(enoxy)silane depends on the metal catalyst used, and mercury(II) acetate is the most selective for production of enoxytrichlorosilanes. Although a slight excess of silicon tetrachloride can reduce the amount of bis(enoxy)dichlorosilane formed, use of a large excess (more than 3 equiv.) leads a significant rate deceleration owing to catalyst deactivation.

Optimum conditions are use of 2–3 equiv. of silicon tetrachloride and a concentration below 1.0 M in dichloromethane [37]. The loading of the metal catalyst can be as low as 0.25 mol%, but usually 1–5 mol% of the metal salt can be employed. Several trichlorosilyl enolates have been prepared by this method (Scheme 7.11). This transformation is general for a variety of enolate structures and it is synthetically appealing, because the precursor trimethylsilyl enol ether can be readily prepared in regiochemically pure form. This transformation is extremely facile, especially for preparation of methyl ketone-derived trichlorosilyl enolates, and enables complete conversion in less than 2 h with 1 mol% Hg(OAc)$_2$. The cyclic ketone-derived enol ethers require longer reaction times ranging from 18 to 24 h [37].

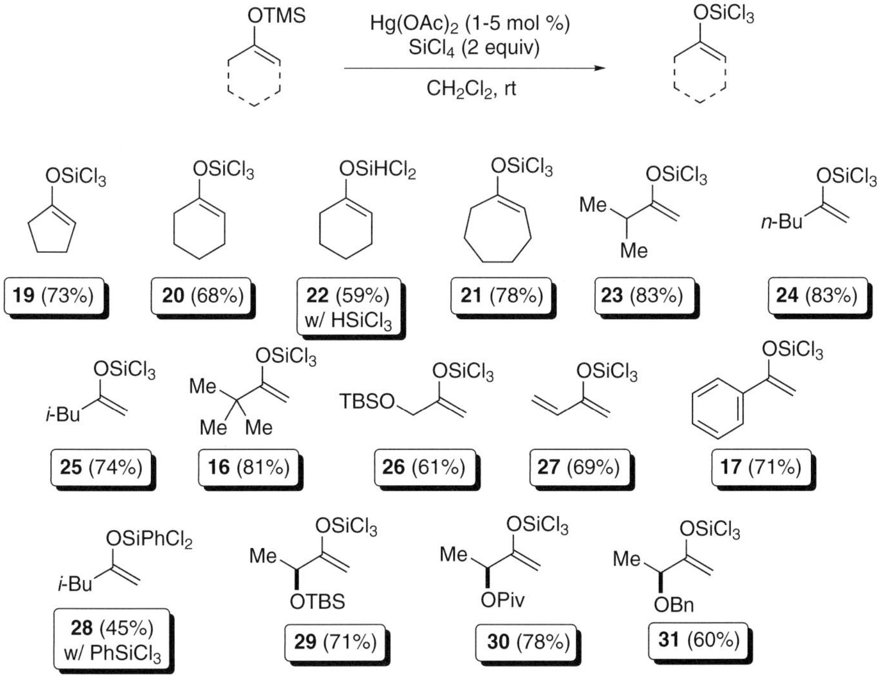

Scheme 7.11
Preparation of enoxytrichlorosilanes by Hg(II)-catalyzed metathesis.

Tab. 7.1

Metal-catalyzed transsilylation of 3-pentanone-derived trimethylsilyl ether **32**.

$$\text{Me}\diagdown\overset{\text{OTMS}}{\underset{\underset{\mathbf{32}}{\text{Me}}}{\diagup}}\diagup\text{Me} \quad\xrightarrow[\text{CH}_2\text{Cl}_2,\text{ rt}]{\substack{\text{SiCl}_4\text{ (2 equiv)}\\ \text{MX}_2\text{ (5 mol \%)}}}\quad \text{Me}\diagdown\overset{\text{OSiCl}_3}{\diagup}\diagup\text{Me}$$

$$\mathbf{11}$$

Entry	MX$_2$	32, E/Z	Time, h	Yield, %[a]	11, E/Z[b]
1	Hg(OAc)$_2$	3/1	5	72	1/2
2	Hg(OAc)$_2$	1/4	5	60	1/2
3	Pd(OAc)$_2$	3/1	5	76	1/6
4	Pd(OAc)$_2$	1/4	5	69	1/6
5	Pd(TFA)$_2$	3/1	15	70	1/7
6	Pd(TFA)$_2$	1/4	15	43	1/6

[a] Yield of distilled material. [b] Determined by ^{1}H NMR analysis.

Common functional and protecting groups can be tolerated under the reaction conditions and the resulting trichlorosilyl enolates are sufficiently pure for use in the phosphoramide-catalyzed aldol addition (*vide infra*). The use of other chlorosilanes enables the preparation of different classes of chlorosilyl enolate. For example, when trichlorosilane is used in place of silicon tetrachloride, dichlorohydridosilyl enolate **22** can be obtained.

A major drawback to the metal-catalyzed trans-silylation is lack of control over the geometry of the resulting trichlorosilyl enolate. Starting from either *E*- or *Z*-enriched trimethylsilyl enol ether **32**, the *E/Z* ratio of the enoxytrichlorosilane **11** is always 1:2 when Hg(OAc)$_2$ is used as the catalyst (Table 7.1) [45]. Use of Pd(II) salts results in slightly higher *Z* selectivity, but again the *E/Z* ratio of the enoxytrichlorosilane does not mirror the *E/Z* ratio of the trimethylsilyl enol ether.

The *E/Z* ratio also depends on the structure of enolates (Table 7.2). *Z*-Trichlorosilyl enolates are always selectively formed. The general trend of *Z/E* selectivity is related to the size of the R group. For larger R groups, higher *Z* selectivity is observed. These trends are also observed in the trans-silylation catalyzed by Pd(OAc)$_2$ and Pd(TFA)$_2$.

These observations can be rationalized by the following mechanism (Figure 7.6). The overall process consists of electrophilic attack of the metal salt to afford α-metalloketone **34**. Coordination of silicon tetrachloride to the carbonyl group of **34** and loss of metal salt gives the enoxytrichlorosilane. The initial formation of **34** is presumably reversible, and this event can account for the randomization of enolate geometry. The *E/Z* ratio of the resulting trichlorosilyl enolate is determined by the relative rate of breakdown of the two limiting conformers **i** and **ii**. The avoidance of steric interaction between Me and R in **i** makes this conformer more favorable, leading to the preferred formation of the *Z* enoxytrichlorosilane. This explanation is consistent with the trend observed in the relationship between steric

Tab. 7.2

Hg(II)-catalyzed transsilylation of a variety of trimethylsilyl enol ethers.

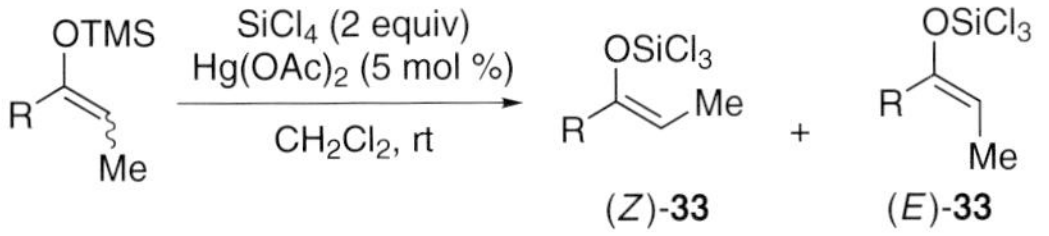

Entry	R	Time, h	Yield, %[a]	33, E/Z[b]
1	H	16	50	1/8
2	Me	16.5	58	1/2
3	Et	5	72	1/2
4	i-Pr	18	65	1/8
5	t-Bu	24	55	1/>20
6[c]	Ph	18	66	1/99

[a] Yield of distilled material. [b] Determined by ^{1}H NMR analysis.
[c] 10 mol% of Hg(OAc)$_2$ was used.

demand of R and the E/Z ratio. In fact, the presence of bulky R groups enables highly selective preparation of Z enolates under these conditions. E-Configured trichlorosilyl enolates cannot, however, be obtained selectively by this method.

A method has been developed that avoids the use of a metal catalyst to prepare geometrically defined trichlorosilyl enolates. It involves generation of a lithium enolate, by treatment of an isomerically enriched trimethylsilyl enol ether with methyllithium, and subsequent capture of the configurationally defined lithium enolate with silicon tetrachloride [46]. Both E and Z trichlorosilyl enolates can be prepared by means of this method, without loss of geometrical purity (Scheme 7.12). Addition of methyllithium to a trimethylsilyl enol ether leads to smooth conversion to the lithium enolate [46]. The E/Z ratio of the trichlorosilyl enolates mirrors the E/Z ratio of the starting trimethylsilyl enol ether.

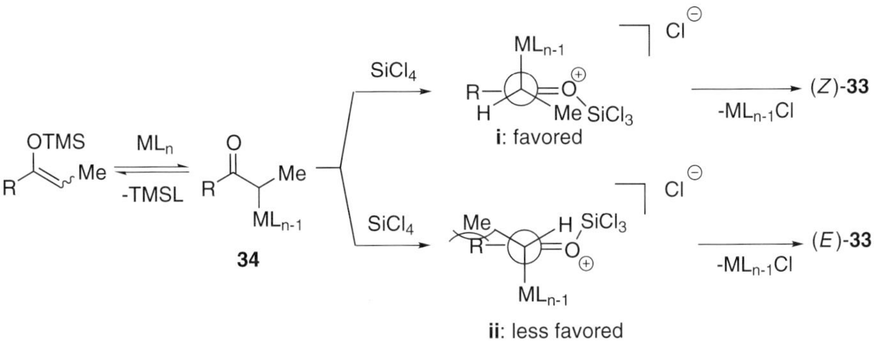

Fig. 7.6
Proposed mechanism for metal-catalyzed transsilylation.

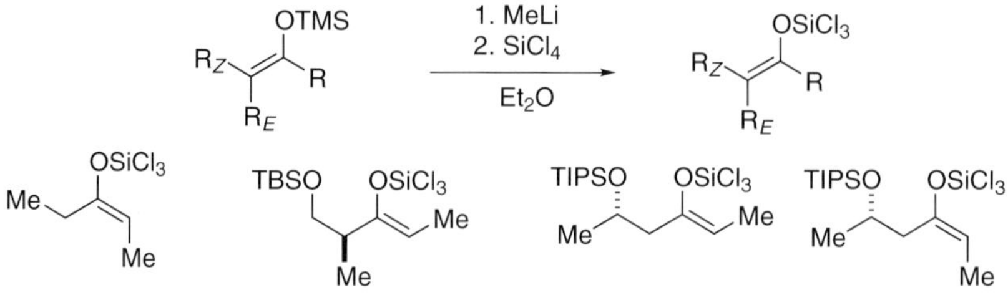

Scheme 7.12
Preparation of geometrically defined enoxytrichlorosilanes.

Numerous methods are used to prepare geometrically defined trimethylsilyl enol ethers. For example, Z trimethylsilyl enol ethers can be prepared by using dibutylboron triflate and subsequent treatment of the boron enolate with trimethylsilyl chloride [47]. The E isomers are typically prepared by use of lithium tetramethylpiperidide as described by Collum [48]. The geometrically defined trimethylsilyl enol ethers are converted into the corresponding trichlorosilyl enolates by the above-mentioned procedure. Unfortunately, the resulting enolate is often contaminated with the bis-(enoxy)dichlorosilane thus reducing the overall yield of the process. Nonetheless, the ability to prepare geometrically defined enoxytrichlorosilanes makes the metal-exchange method synthetically attractive.

7.2.3
Preparation of Aldehyde-derived Trichlorosilyl Enolates

In this section, three methods used to generate aldehyde-derived enoxytrichlorosilanes are described [49]. The first is the metathetical route from the corresponding trimethylsilyl enol ether using a catalytic amount of Pd(OAc)$_2$ and excess silicon tetrachloride (Scheme 7.13). In this method, the geometry of the resulting trichlorosilyl enolate is not dependent on the trimethylsilyl enol ether for the same reason as discussed above (Figure 7.6). Thus, only unsubstituted or symmetrically substituted enolates are suitable.

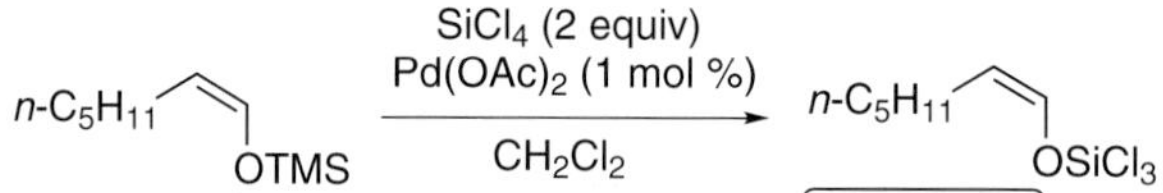

Scheme 7.13
Preparation of an aldehyde-derived enoxytrichlorosilane.

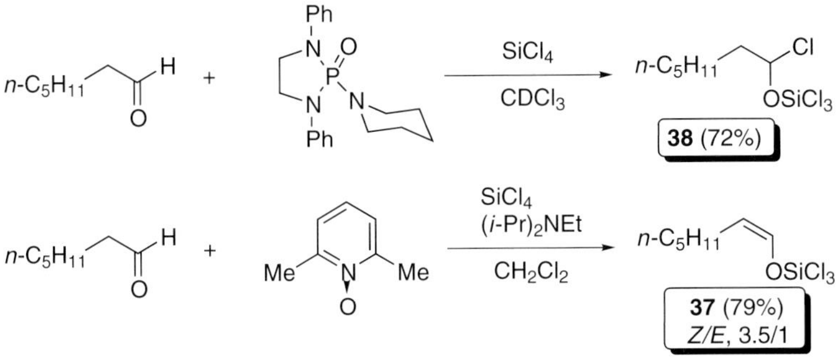

Scheme 7.14
Direct silylation of aldehydes using $SiCl_4$ and a Lewis base.

The second procedure is direct silylation from an aldehyde with phosphoramides or N-oxides and a base (Scheme 7.14). In the presence of silicon tetrachloride and a catalytic amount of a Lewis base aldehydes are rapidly transformed into α-chloro trichlorosilyl ethers. The formation of such intermediates has recently been documented and observed by means of ^1H NMR spectroscopic analysis [50]. Addition of an amine base promotes elimination of HCl to yield the trichlorosilyl enolate. Before use the resulting trichlorosilyl enolate must be distilled from the ammonium salt generated by the reaction. Although this procedure provides the trichlorosilyl enolate directly from a given aldehyde, the enol geometry cannot be controlled, thus limiting the utility of this process.

Generation of stereodefined trichlorosilyl enolates of aldehydes can also be accomplished by the direct O-to-O trans-silylation via lithium enolates (Scheme 7.15). The geometrically-defined trimethylsilyl enol ethers of heptanal react with methyllithium to yield the configurationally stable lithium enolates. After trapping with a large excess of silicon tetrachloride the geometrically enriched trichlorosilyl enolates of aldehydes are prepared in good yield.

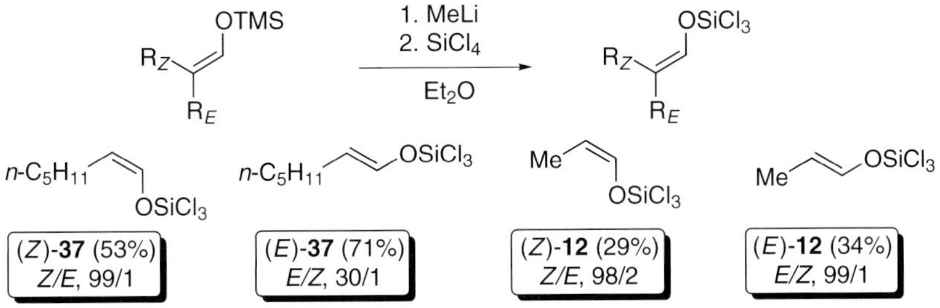

Scheme 7.15
Preparation of geometrically defined enoxytrichlorosilanes.

Scheme 7.16
Preparation of acetate-derived trichlorosilyl ketene acetals.

39

RR'SiCl₂ → 0 °C

40 (48%)   10 (65%)   41 (57%)

42 (23%)   43 (18%)   44 (19%)

7.2.4
Preparation of Trichlorosilyl Ketene Acetals

The first reported enoxytrichlorosilanes were derived from esters [38]. Those ketene acetals are prepared by reaction of a chlorosilane and an α-stannyl ester (Scheme 7.16). This method is still the most general preparation of acetate-derived trichlorosilyl ketene acetals. In the presence of excess silicon tetrachloride the stannyl ester **39** is smoothly converted to the ketene acetal **10**. The ketene acetal can be distilled at ambient temperature under reduced pressure. These species cannot be heated because isomerization to a C-trichlorosilyl ester occurs at higher temperatures [51]. This isomerization is also a problem when these ketene acetals are stored for a long time, because even at room temperature isomerization occurs in a month. Unlike the tri-chlorosilyl enolates derived from ketones and aldehydes, which exist ex-clusively as the O-silyl isomers, trichlorosilyl ketene acetals can isomerize to the thermodynamically more stable C-silyl isomer [38]. The C-silyl esters are not reactive in phosphoramide-catalyzed aldol reactions and these spe-cies do not revert to the corresponding trichlorosilyl ketene acetal under common reaction conditions.

The use of different chlorosilanes enables preparation of structurally diverse chlorosilyl ketene acetals (Scheme 7.16). Although this procedure is relatively simple, the method suffers from low yields because of the difficulty of separating the trichlorosilyl ketene acetal from tributylchlo-rostannane and from the C-trichlorosilylacetate. The purity of the chlorosilyl ketene acetal is critical because tin residues from the reaction promote oligo-merization of the ketene acetal.

In the reaction with tributylstannylpropanoates under similar conditions the major products obtained are, unfortunately, the C-trichlorosilyl propanoate derivatives. Thus, so far only acetate-derived trichlorosilyl ketene acetals have been prepared by this method.

In summary, practical and efficient methods are now available for preparation of enoxytrichlorosilanes. The most general method is the transition metal-catalyzed trans-silylation of trimethylsilyl enol ethers with silicon tetrachloride. Geometrically defined enoxytrichlorosilanes are best prepared by silylation of lithium enolates. The configuration of the lithium enolate precursor is preserved in this process. The metathesis of methyl tributyl-stannylacetate with silicon tetrachloride is the most efficient route for preparation of trichlorosilyl ketene acetals.

7.3
Preparation of Chiral Lewis Bases

The structure of the Lewis base greatly affects its catalytic activity and selectivity in the aldol addition. Moreover, different types of trichlorosilyl nucleophile require different types of chiral Lewis base catalyst. To examine a wide range of structures, general methods are needed for synthesis of phosphoramides and N-oxides [52].

The four most commonly used Lewis-base catalysts are shown in Chart 7.1. The phosphoramide **45** is the most general and selective catalyst for aldol addition of ketone-derived trichlorosilyl enol ethers to aldehydes [53]. In

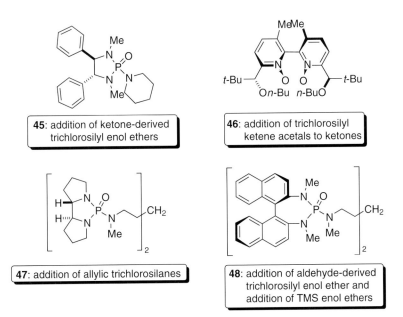

45: addition of ketone-derived trichlorosilyl enol ethers

46: addition of trichlorosilyl ketene acetals to ketones

47: addition of allylic trichlorosilanes

48: addition of aldehyde-derived trichlorosilyl enol ether and addition of TMS enol ethers

Chart 7.1
Commonly used chiral Lewis bases for aldol additions.

recent studies dimeric phosphoramide catalysts such as **47** and **48** have been shown to be highly selective in additions of allyltrichlorosilane and allyltributylstannane (with silicon tetrachloride) to aldehydes [54]. These catalysts are also effective in the addition of aldehyde-derived trichlorosilyl enolates and in additions of trialkylsilyl ketene acetals and enol ethers to aldehydes [49, 50]. For addition of trichlorosilyl ketene acetal to ketones the bis-*N*-oxide **46** has proven to be the most selective catalyst [55]. Syntheses of these Lewis base catalysts are briefly described in the following sections.

7.3.1
Preparation of Chiral Phosphoramides

The basic strategy for synthesis of chiral cyclic phosphoramides is to couple a chiral 1,2-, 1,3-, or 1,4-diamine to either a phosphorus(V) or phosphorus(III) reagent. There are three general routes (Scheme 7.17). Method A is the most straightforward strategy for preparation of chiral cyclic phosphoramides. A chiral diamine is combined with an aminophosphoric dichloride in the presence of triethylamine [56]. The reaction is typically conducted in a halogenated solvent under reflux. This method works well for preparation of sterically less bulky phosphoramides and for coupling aliphatic diamines. For sterically demanding coupling partners, elevated temperatures and longer reaction times are required. For example, phosphoramide **45** is obtained in good yield from (R,R)-N,N'-dimethyl-1,2-diphenylethylenediamine [57] by method A (Scheme 7.18).

For less reactive diamines a more electrophilic phosphorus(III) reagent is needed to enhance the reaction rate (Methods B and C, Scheme 7.17) [58]. In these methods the diamine is first lithiated by use of *n*-BuLi at low temperature. The lithiated diamine is combined with the mono-

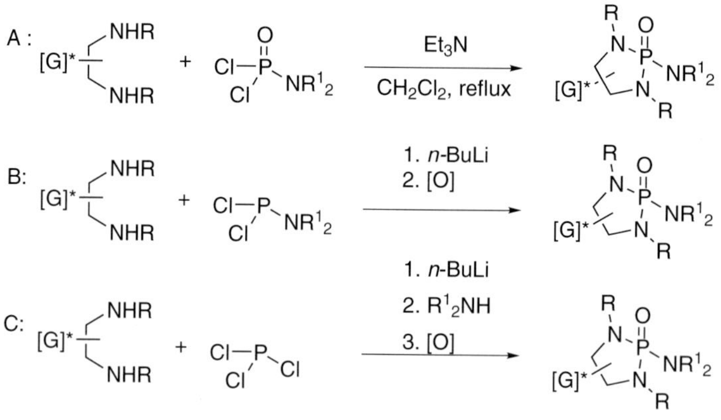

Scheme 7.17
General preparations of chiral phosphoramides.

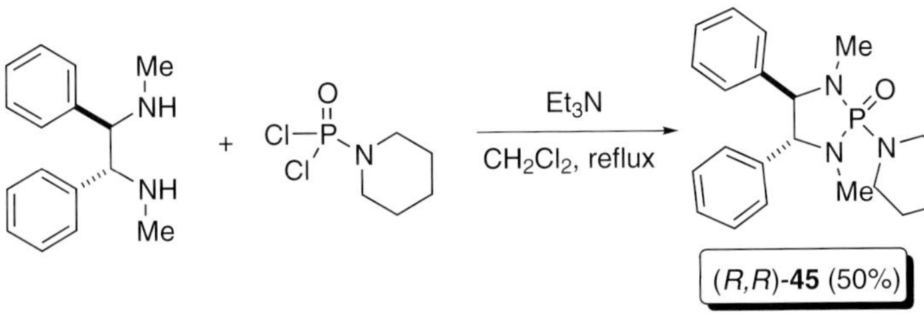

Scheme 7.18
Preparation of monophosphoramide **45**.

aminophosphorus(III) dichloride reagent (Method B, Scheme 7.17) or with phosphorus(III) chloride followed by treatment with an amine (Method C, Scheme 7.17). The resulting phosphorus(III) triamine species is oxidized with *m*-CPBA to give the desired phosphoramide. For example, the bisphosphoramide **48** is prepared in good yield from N,N'-dimethyl-1,1'-binaphthyl-2,2'-diamine by a three-step sequence [59] (Scheme 7.19). In this example, the N,N'-dimethylpentanediamine (**49**) is used as the linker [60].

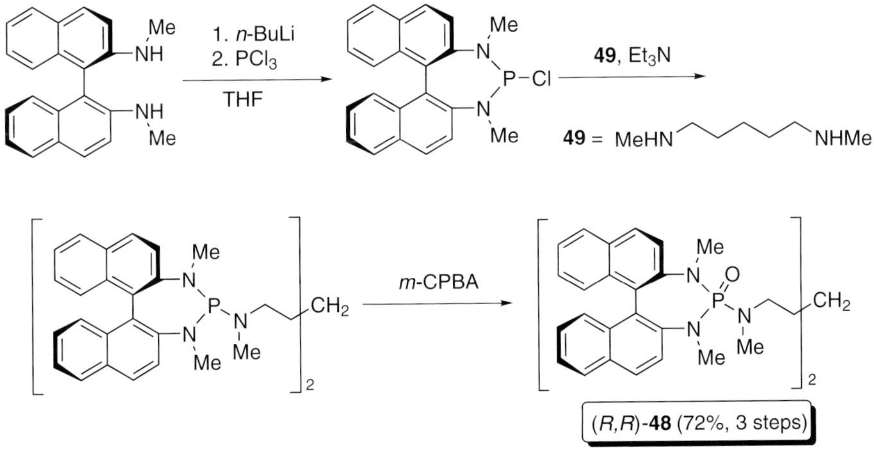

Scheme 7.19
Preparation of bisphosphoramide **48**.

The bisphosphoramide **47** can be prepared from 2,2'-bispyrrolidine (Scheme 7.20) [61]. In the presence of triethylamine, enantiomerically pure bispyrrolidine reacts with phosphorus oxychloride to provide diaminophosphoryl chloride **50**. The lithiated linker **49** is combined with (*R,R*)-**50** to afford the bisphosphoramide **47** in excellent yield.

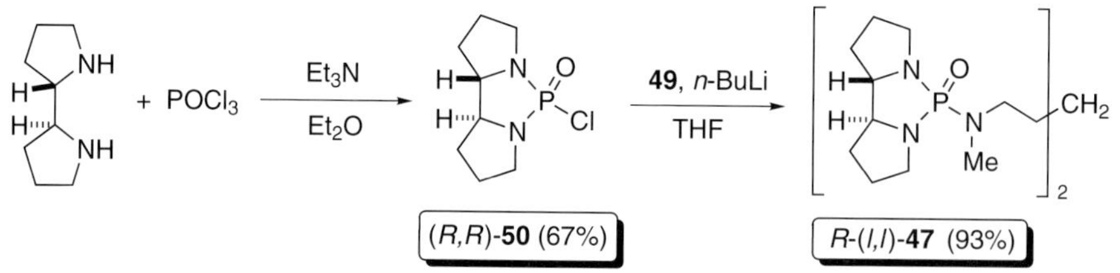

Scheme 7.20
Preparation of bisphosphoramide R-(*l*,*l*)-**47**.

7.3.2
Synthesis of Chiral bis-N-Oxides

N-Oxides are readily obtained by oxidation of tertiary amines [62]. Accordingly, chiral N-oxides are usually prepared by oxidation of chiral tertiary amines. Several axially chiral bis-N-oxides have been synthesized; these are known to promote addition reactions of chlorosilane species [62].

The chiral bis-N-oxide **46** contains both central and axial elements of chirality (Scheme 7.21). These two features are essential for the stereoselectivity observed in promoted aldol reactions and are also helpful in enantio- and diastereoselective synthesis of the catalysts. Introduction of the stereogenic center is achieved by reduction of the *tert*-butyl ketone by (Ipc)₂BCl [63]. The N-oxide obtained after etherification and oxidation undergoes diastereoselective oxidative dimerization to afford (*P*)-**46** [55].

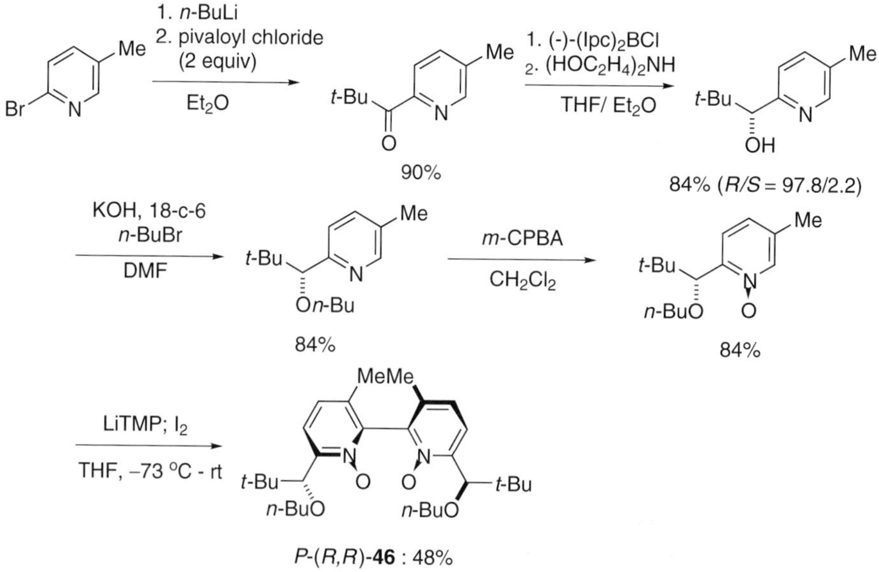

Scheme 7.21
Preparation of chiral bis-N-oxide P-(R,R)-**46**.

7.4
Enantioselective Aldol Addition of Achiral Enoxytrichlorosilanes

Trichlorosilyl enolates (enoxytrichlorosilanes) derived from ketones undergo additions to aldehydes spontaneously at or below ambient temperature without external activation [53]. The intrinsic reactivity of these reagents contrasts with that of trialkylsilyl enol ethers in the aldol addition, for which a promoter is usually required [64]. The reactivity of trichlorosilyl enolates is not because of the inherent nucleophilicity of the enolate but rather the high electrophilicity of the silicon atom [27]. The silicon atom of a trichlorosilyl enolate is highly electropositive, because of the effect of chlorine ligands. Lewis basic functions, including aldehydes, can bind to the Lewis acidic silicon and form a hypercoordinate complex. On binding of an aldehyde to the Lewis-acidic silicon atom the aldehyde is electrophilically activated (Figure 7.7, **iii**). The enolate moiety is concurrently activated by increased polarization of the enolate Si–O bond. This dual activation results in the high reactivity of trichlorosilyl enolates. A similar rationale is also proposed for the aldol reaction of boron enolates [2c] and strained-ring alkyl silyl enolates [65].

Aldol additions of trichlorosilyl enolates are catalyzed by Lewis bases, most notably phosphoramides (Figure 7.8). It is believed that binding of a phosphoramide to a trichlorosilyl enolate leads to ionization of a chloride, forming a cationic silicon–phosphoramide complex [66]. The binding of an aldehyde to the silicon complex leads to aldolization through a closed transition structure. There are two catalyzed pathways – one involves the intermediacy of a pentacoordinate, cationic silicon complex in which only one phosphoramide is bound to silicon and the other involves a hexacoordinate, cationic silicon complex in which two phosphoramide molecules are bound to the silicon [66]. In the former pathway aldolization occurs through a boat-like transition structure, whereas in the latter pathway, the transition structure is chair-like (Figure 7.8, **iv** and **v**). This mechanistic duality in the catalyzed process is analyzed in more detail in Section 7.9.

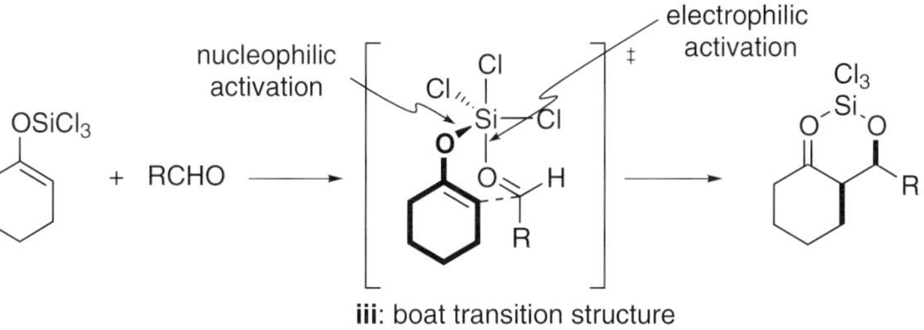

iii: boat transition structure

Fig. 7.7
Hypothetical assembly for uncatalyzed aldol addition of a trichlorosilyl enolate.

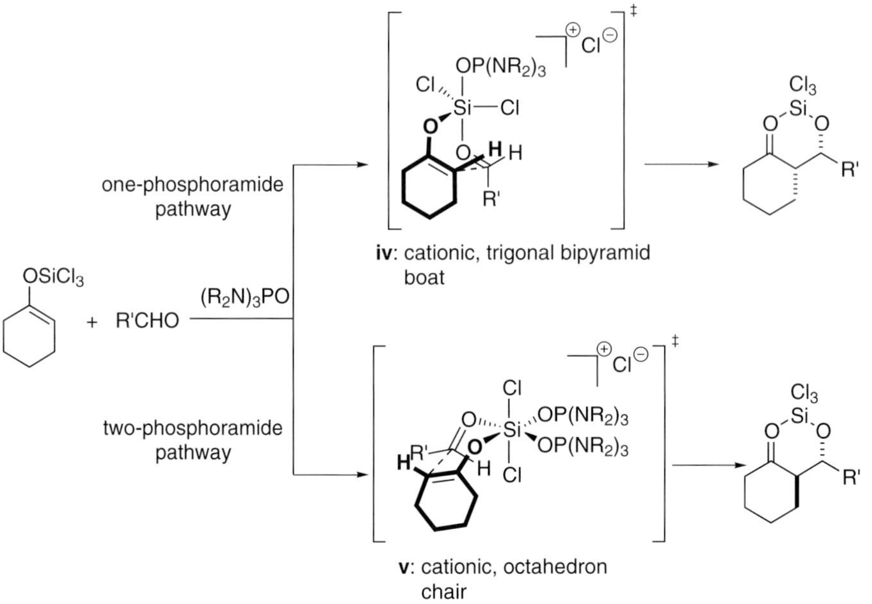

Fig. 7.8
Divergent pathways for catalyzed aldol addition of a trichlorosilyl enolate.

The appeal of aldol additions of trichlorosilyl enolates is the selective and predictable diastereocontrol that probably arises from a closed transition structure. For substituted enolates the diastereomeric ratio of aldol products can be directly correlated with the enolate geometry as predicted by the Zimmerman–Traxler model [67]. Thus, the dominant reaction pathway in the catalyzed reactions of trichlorosilyl enolates involves a chair-like transition structure organized around the silicon. By employing chiral phosphoramides, enantioselection can be controlled. Thus highly stereocontrolled aldol addition can be envisaged in Lewis-base-catalyzed aldol addition of trichlorosilyl enolates.

In this section aldol additions of achiral trichlorosilyl enolates derived from ketones are described. The inherent reactivity of these species and their potential use in asymmetric, catalytic processes will be discussed.

7.4.1
Aldol Additions of Achiral Methyl Ketone-derived Enolates

Trichlorosilyl enolates derived from methyl ketones are reactive toward aldehydes *in the absence of Lewis base catalysts* at ambient temperature (Scheme 7.22) [68]. Trichlorosilyl enolates bearing a broad range of non-participating substituents react with benzaldehyde to give excellent yields of the aldol

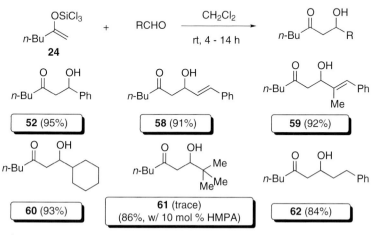

Scheme 7.22
Uncatalyzed aldol addition of methyl ketone-derived trichlorosilyl enolates.

products in several hours. The steric and electronic properties of the enolates do not have a large influence on the rate of aldol addition.

Under similar conditions the trichlorosilyl enolate **24** undergoes aldol
addition to a wide range of aldehydes at room temperature with excellent
yields (Scheme 7.23). Aromatic and conjugated aldehydes are typically more
reactive than aliphatic aldehydes, presumably because of their smaller size
and higher Lewis basicity [69]. The structure of the aldehyde significantly
affects the rate of aldol addition, however. Reactions with bulky aldehydes

Scheme 7.23
Uncatalyzed aldol additions of **62** to a variety of aldehydes.

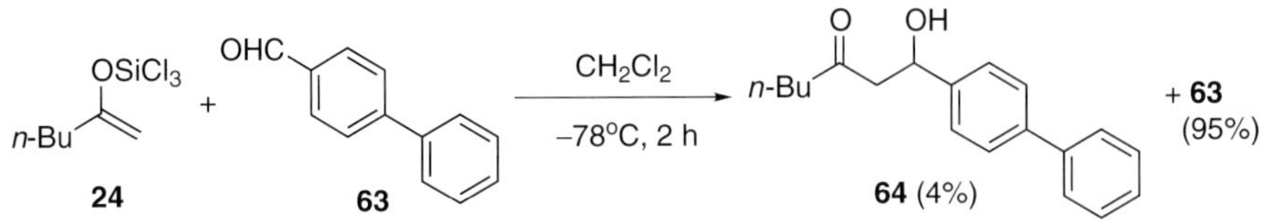

Scheme 7.24
Uncatalyzed aldol addition of **24** at −78 °C.

are slower, as is evidenced by the reaction of pivalaldehyde. Addition of 10 mol% HMPA leads to a dramatic increase in the rate of addition, enabling isolation of **61** in good yield.

The rate of reaction of trichlorosilyl enolates with aldehydes is greatly attenuated at low temperature. For example, in the addition of **24** to 4-biphenylcarboxaldehyde, only 4% of **64** was isolated and 95% of unreacted aldehyde was recovered after 2 h (Scheme 7.24). This behavior is crucial for optimization of the asymmetric process, because suppression of the achiral background reaction is important for achieving high enantioselectivity in the catalyzed reaction.

In the presence of several structurally diverse chiral phosphoramides the aldol addition proceeds smoothly at low temperature (Scheme 7.25). Although all of the phosphoramides are effective in promoting the aldol addition of **24** to benzaldehyde, the stilbene-1,2-diamine-derived phosphoramide **45** is the most active and selective catalyst. The structure of the chiral diamine backbone clearly has a large effect on the enantioselectivity of the process. Also, substitution of a diisopropylamino group for the piperdinyl group in **63** leads to dramatic drop in enantioselectivity.

Correct choice of solvent is a critical aspect of obtaining high enantioselectivity in the aldol addition of trichlorosilyl enolates [68]. Dichloromethane is the most suitable solvent for the reaction, in terms of both reactivity and selectivity (Table 7.3). Other halogenated solvents such as trichloroethylene or the more polar propionitrile are good solvents for this reaction and pro-

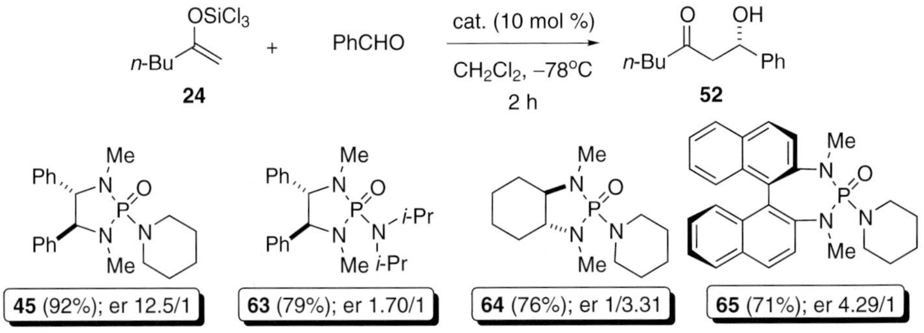

Scheme 7.25
Catalyzed aldol additions of **24** to benzaldehyde.

Tab. 7.3
Effect of solvent on aldol addition of **24** to benzaldehyde.

Entry	Solvent	ε^a	er	Yield, %
1	CH$_2$Cl$_2$	9.08	12.5/1	92
2	trichloroethylene	3.4	4.03/1	88
3	Et$_2$O	4.34	2.25/1	37
4	toluene	2.38	1.85/1	48
5	THF	7.52	3.37/1	59
6[b]	EtCN	27.7	1/8.71	88

[a] Ref. 20a. [b] Performed with 10 mol% (*R,R*)-**45**.

vide good yield of the aldol product. Coordinating solvents such as ether and THF give low yields and attenuated selectivities. In ethereal solvents coordination of the Lewis basic oxygen with the silicon species might compete with binding of **45** or benzaldehyde, thus accounting for the lower yields and attenuated selectivity observed. In a non-polar medium such as toluene, the aldol product is formed in significantly lower yield with only marginal enantioselectivity. The lack of reactivity observed in the non-polar solvent might reflect difficulties in generating the kinetically relevant, ionized silyl cation.

Catalyst loading also has a large effect on selectivity. Although high catalyst loadings (more than 10 mol%) do not improve enantioselectivity significantly, reducing the catalyst loading to less than 3 mol% leads to a slower rate of reaction and attenuated selectivity. At low catalyst loadings the one-phosphoramide pathway that involves a poorly selective boat-like transition structure becomes competitive with the highly selective, two-phosphoramide pathway that involves a chair-like transition structure (c.f. Scheme 7.6). Thus, a catalyst loading of 5 mol% or more is necessary for optimum enantioselectivity.

The scope of the enolate structure in this aldol reaction is significant (Scheme 7.26). Catalyzed reactions with benzaldehyde proceed in excellent yield with modest to good selectivity. The enolate structure does not have a significant effect on the rate of the reaction but greatly affects the enantioselectivity of the process. Enolates with larger substituents on the non-participating side, for example phenyl and *tert*-butyl, result in significantly lower enantioselectivity than those bearing smaller substituents such as methyl and *n*-butyl. In these reactions the lower enantioselectivity can be attributed to a lack of facial selectivity at the aldehyde.

For this aldolization to be synthetically useful it is important for highly functionalized enolates to tolerate the reaction conditions. For example, ad-

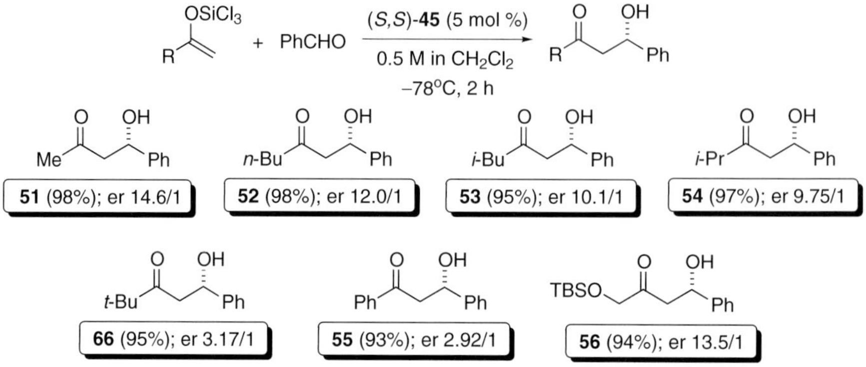

Scheme 7.26
Addition of different trichlorosilyl enolates to benzaldehyde.

dition of enolate bearing a TBSO group results in an excellent yield of **56** with high enantioselectivity.

The scope of the reaction with regard to the aldehyde component is also broad. Good to high enantioselectivity can be achieved in the addition of **24** to a variety of aldehydes using **45** (Scheme 7.27). Under these reaction conditions uniformly excellent yields and good enantioselectivity are obtained

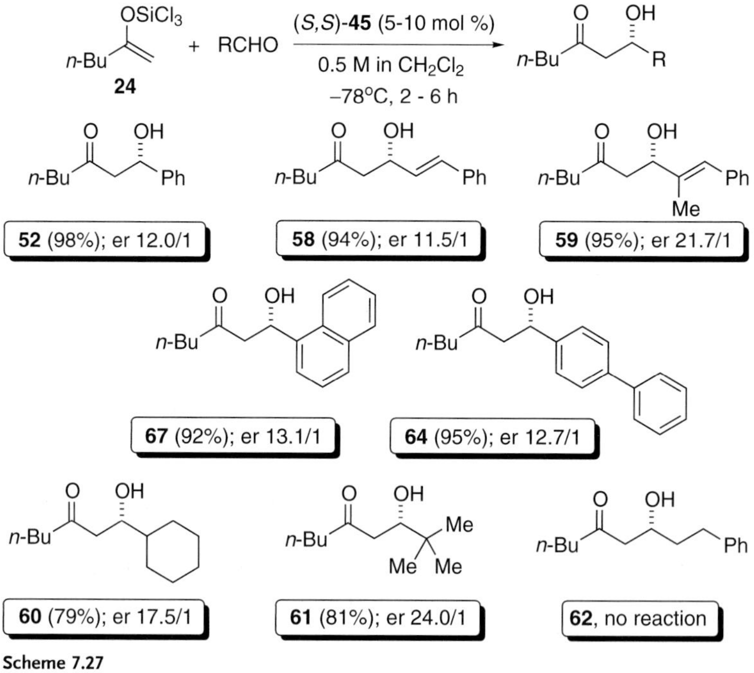

Scheme 7.27
Catalyzed addition of **24** to a variety of aldehydes.

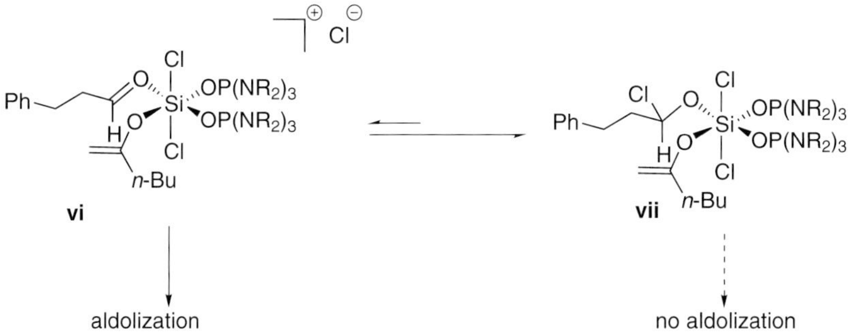

Fig. 7.9
Reaction of aliphatic aldehydes under the action of phosphoramide catalysis.

from aromatic and unsaturated aldehydes. Aliphatic aldehydes, on the other hand, react significantly more slowly than their unsaturated counterparts. When hydrocinnamaldehyde is used as acceptor no aldol product is isolated under standard conditions.

Aldehyde structure clearly has a significant effect on enantioselectivity. Interestingly, sterically congested aldehydes, especially those with α-branching, result in higher enantioselectivity than unbranched substrates. Although the origin of this effect is unclear, sterically bulky aldehydes presumably increase the energy difference between the competing chair- and boat-like transition structures.

Several explanations have been proposed for the relatively low reactivity of aliphatic aldehydes. This behavior was originally rationalized by considering the low Lewis basicity of these aldehydes compared with related unsaturated aldehydes [69]. Another consideration was the competitive enolization of these substrates in the presence of phosphoramide catalysts [68b]. It has, however, now been shown that aliphatic aldehydes rapidly form α-chloro trichlorosilyl ethers under the reaction conditions and these species are unreactive towards nucleophiles (Figure 7.9) [50]. In the mechanism of phosphoramide-catalyzed aldol additions the cationic hexacoordinate silicon complex has a chloride counterion. When the aldehyde carbonyl is activated in this complex chloride is a potential nucleophile that competes with the enolate. Aldolization of aliphatic aldehydes is, apparently, slow compared with addition of chloride, resulting in formation of the α-chloro silyl ether. The role of ionized chloride in suppressing the aldolization is in agreement with the observation that aliphatic aldehydes are reactive under conditions without phosphoramide. Because these uncatalyzed reactions do not involve ionization, the formation of α-chlorosilyl ether is no longer competitive.

Efforts to construct a stereochemical model of absolute stereoselection in these aldol additions have so far been unsuccessful. The absolute configuration of aldol adduct **52** has been determined by formation of the corresponding bromobenzoate then single-crystal X-ray analysis and the absolute configurations of other aldol products have been assigned by analogy. In aldol additions catalyzed by phosphoramide (S,S)-**45**, the nucleophile attacks the *Si* face of the complexed aldehyde, usually forming the *S* configuration at the hydroxy-bearing carbon. The solution and solid-state structures of several Sn(IV)–phosphoramide complexes have been examined to obtain better understanding of the stereochemical environment crafted by the chiral phosphoramide [70]. Despite this effort no transition structure has yet been proposed that can rationalize the absolute configuration observed in these catalyzed aldol additions. One major concern in development of a model is understanding the mode of ligand binding around silicon. Among the challenges facing the formulation of reasonable transition structures are:

- multiple configurational possibilities with two phosphoramides around silicon;
- the conformational flexibility of the phosphoramides; and
- deducing the reactive conformations of the ternary complex from the vast number of potential configurations and conformations.

Computational solution of these problems is currently untenable. Because of the number of heavy atoms and rotatable bonds present in the complex, prediction of the reactive conformation in the ternary complex is difficult at the current level of understanding.

In general, performing aldol addition of a trichlorosilyl enolate involves two discrete steps – generation and isolation of the trichlorosilyl enolate then aldolization. In an effort to streamline the process, a method for in situ formation of the reactive enolate has been developed (Scheme 7.28). As described in Section 7.2, trichlorosilyl enolates can be prepared from the corresponding trimethylsilyl enol ethers by use of silicon tetrachloride and a catalytic amount of transition metal salt. In this procedure the trichlorosilyl enolates generated in situ from trimethylsilyl enol ethers by Hg(OAc)$_2$-catalyzed metathesis can, after removal of Me$_3$SiCl and excess SiCl$_4$, be used directly in the subsequent aldol addition. The yield and enantioselectivity of the aldol reaction are not affected by the presence of the mercury salt. This procedure obviates purification and handling of the moisture-sensitive trichlorosilyl enolate and enables use of shelf-stable trimethylsilyl

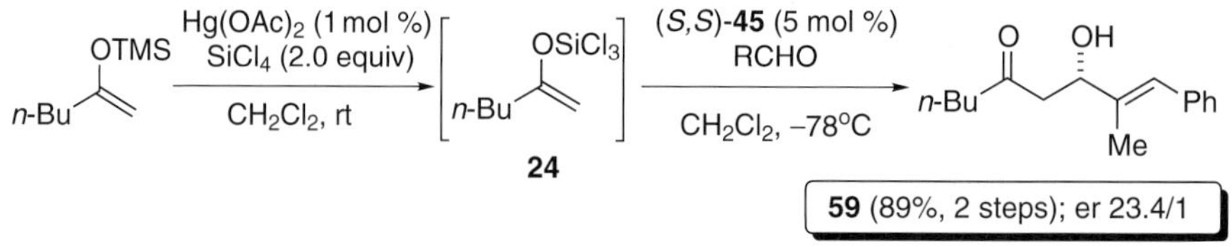

Scheme 7.28
Aldol addition of **24** generated in situ.

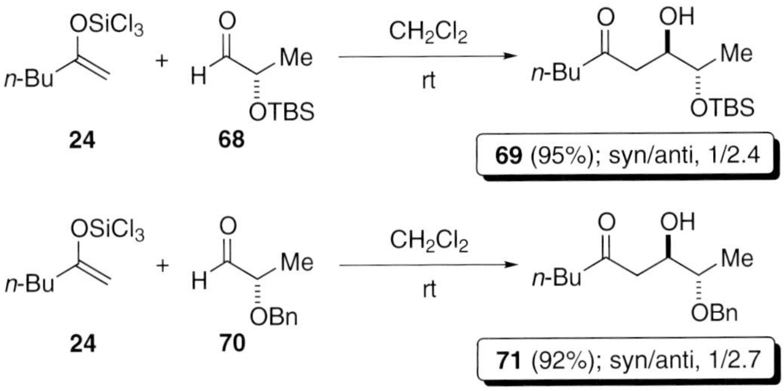

Scheme 7.29
Uncatalyzed aldol additions to chiral aldehydes.

enol ethers. This in situ generation of trichlorosilyl enolates and their use has enhanced the synthetic utility of the phosphoramide-catalyzed aldol reaction.

The possibility of substrate-controlled aldol additions of trichlorosilyl enolates has been investigated using lactate-derived chiral aldehydes [68b]. Uncatalyzed reactions of **24** with chiral aldehydes **68** and **70** proceed in high yield at room temperature (Scheme 7.29). The compatibility of common protecting groups on the aldehyde with trichlorosilyl enolates has been demonstrated in these examples. The internal diastereoselection (Section 7.5, Figure 7.13) exerted by the chiral aldehyde slightly favors the *anti* isomer, although the diastereomeric ratio obtained in this reaction is not synthetically useful [71].

The intrinsic selectivity can be rationalized by use of the Felkin–Ahn model, if boat-like transition structures are considered (Figure 7.10) [72]. The two possible approaches of the nucleophile can be envisaged with the oxygen *anti* to the incoming nucleophile, as suggested in the Heathcock model [73]. These two conformers **viii** and **ix** lead to the two diastereomeric transition structures. Addition of the nucleophile would occur on the sterically less hindered face (H rather than Me), which leads to the observed major *anti* diastereomer.

The nature of the group on the α-oxygen seems to have little effect on the selectivity of the aldolization. This indicates that stereoelectronic factors control the orientation of the oxygen atom in the stereochemistry-determining step [74]. This observation is also consistent with Heathcock's analysis of non-chelation-controlled additions to simple α-oxygenated aldehydes [73].

The aldol reaction of an achiral enolate with a chiral aldehyde in the presence of a chiral phosphoramide is an interesting opportunity for double diastereoselection [75]. Diastereoselectivity in aldol additions to chiral aldehydes can be significantly enhanced when the sense of internal and external

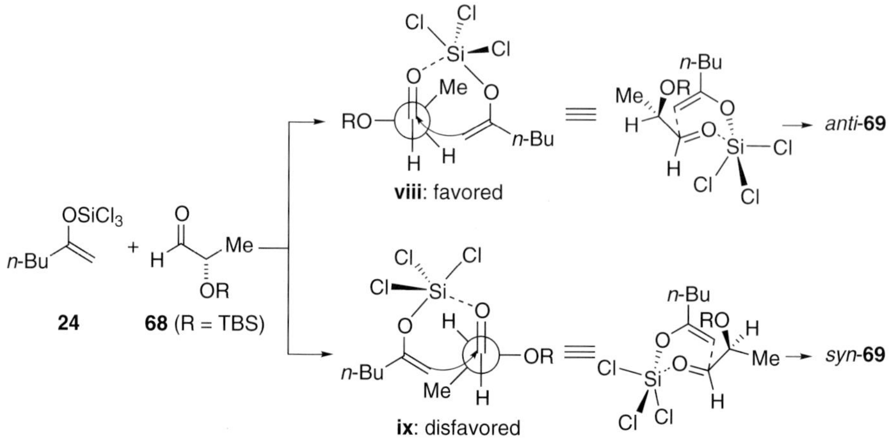

Fig. 7.10
Diastereoselection for addition of **24** to **68**.

diastereoselection are matched (Scheme 7.30). When (*S,S*)-**45** is used in the addition of **24** to **68**, the aldol product obtained is the *syn* diastereomer, albeit with low selectivity. Use of (*R,R*)-**45**, on the other hand, provides *anti*-**69** with good diastereoselectivity.

In catalyzed additions to the chiral aldehyde **68** the configuration of the catalyst dominates the stereochemical course of the addition. The inherent selectivity in the catalyzed pathway can be examined by employing an achiral catalyst. Modest *anti* selectivity is observed when **72** is used. The *anti* selectivity observed with phosphoramide (*R,R*)-**45** is therefore a result of matching internal (from the chiral aldehyde) and external stereoinduction (from chiral catalyst) [71].

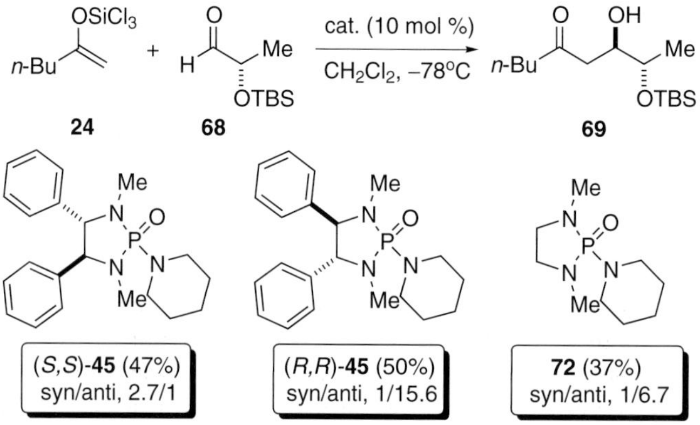

Scheme 7.30
Catalyzed aldol additions to chiral aldehydes.

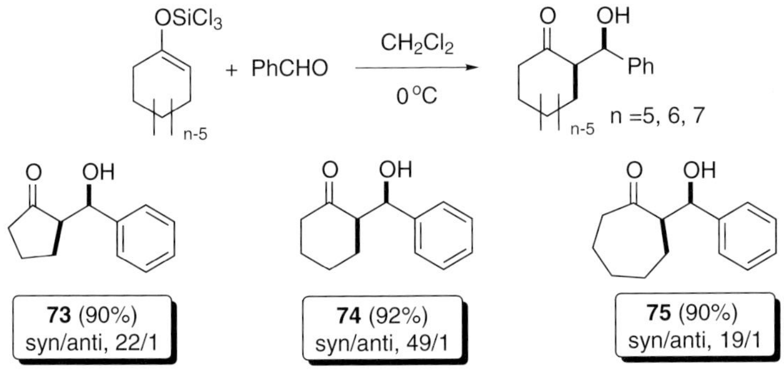

Scheme 7.31
Uncatalyzed aldol additions of cyclic trichlorosilyl enolates.

7.4.2
Aldol Additions of Cyclic Trichlorosilyl Enolates

In the absence of a Lewis base catalyst, cycloalkanone-derived trichlorosilyl enolates undergo aldol additions to benzaldehyde even at 0 °C (Scheme 7.31) [53, 76, 77]. Excellent yields of *syn* aldol products can be obtained under these conditions. The *syn* selectivity derived from the *E*-configured enolate suggests that the reaction proceeds through a closed, boat-like transition structure.

The generality of uncatalyzed aldol additions with **20** has been demonstrated in reactions with a variety of aldehydes (Scheme 7.32). Additions with **20** provide high yields of aldol products with modest to high *syn* selectivity [76]. The rate of the reaction can be correlated with aldehyde structure. The reactions are significantly slower for bulky aldehydes than for smaller ones. Aliphatic aldehydes are less reactive, presumably because of their attenuated Lewis basicity. The aldehyde structure also affects the diastereoselectivity of the aldol reaction. The steric bulk around the aldehyde carbonyl group has a deleterious effect on diastereoselectivity, as is illustrated by comparison of cinnamaldehyde and α-methylcinnamaldehyde. Aliphatic aldehydes are also poorly selective, as is exemplified by aldol addition to cyclohexane carboxaldehyde. In such reactions the energy difference between the boat- and chair-like transition structures is assumed to be small (vide supra). From simple analysis of the transition structure model for the uncatalyzed process it is clear that the steric bulk of the aldehyde increases steric congestion in the favored boat transition structure (Figure 7.8, **iv**).

The reactivity of **20** has been examined at low temperature to assess the rate of the background reaction in the context of the asymmetric, catalytic process. Aldol addition of **20** to benzaldehyde at −78 °C results in 19% conversion after 2 h, indicating that **20** is more reactive than the methyl ketone-derived enolates (c.f. Scheme 7.24). Although the background reac-

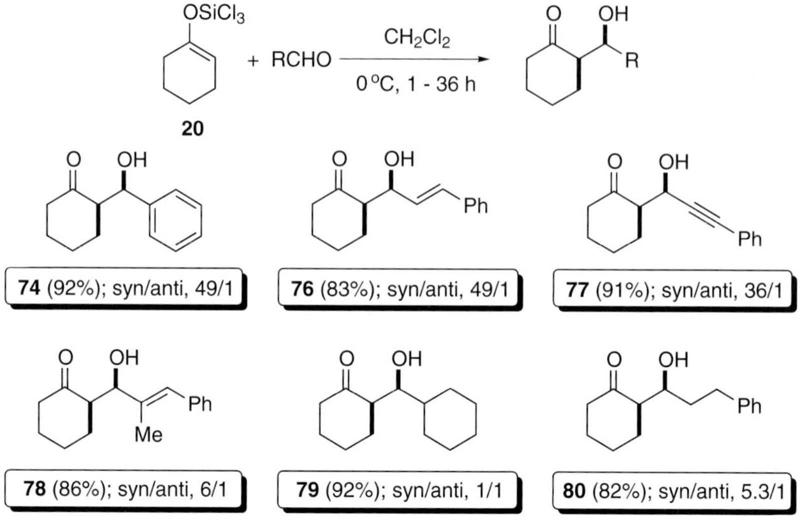

OSiCl₃

$$20 \quad + \quad RCHO \xrightarrow[\text{0 °C, 1 - 36 h}]{\text{CH}_2\text{Cl}_2}$$

74 (92%); syn/anti, 49/1

76 (83%); syn/anti, 49/1

77 (91%); syn/anti, 36/1

78 (86%); syn/anti, 6/1

79 (92%); syn/anti, 1/1

80 (82%); syn/anti, 5.3/1

Scheme 7.32
Uncatalyzed addition of **20** to different aldehydes.

tion is rather significant in the aldol addition of **20** at low temperature, in the presence of a phosphoramide the catalyzed pathway becomes dominant.

An extensive survey of catalysts for the addition of **20** to benzaldehyde showed that several chiral phosphoramides catalyzed the reaction efficiently (Scheme 7.33). The stereoselectivity was highly dependent on catalyst structure. Among the catalysts shown below, the stilbene-1,2-diamine-derived catalyst **45** is the most enantioselective. For other catalysts diastereo- and enantioselectivity were lower. The N,N'-diphenylphosphoramide **81** results in remarkably high *syn* diastereoselectivity, albeit with modest enantioselectivity. As shown in the reactions using **82**, the phosphorus stereogenic center has no significant effect on stereoselectivity.

The strong dependence of the diastereomeric ratio on catalyst structure again implies competition between chair- and boat-like transition structures. As discussed in Section 7.9, the primary pathway in the catalyzed aldol addition of trichlorosilyl enolates using **45** involves a chair-like transition structure organized around a cationic, hexacoordinate silicon atom bound by two phosphoramide molecules. Thus for an E-configured enolate the corresponding *anti* aldol product is expected. From this diastereoselectivity it is clear that in the reaction catalyzed by **45** the two-phosphoramide pathway is favored, because the *anti* product is obtained from the E enolate. The reaction with **81**, on the other hand, proceeds predominantly through a boat-like transition structure involving one phosphoramide, enabling formation of the *syn* aldol product from the E enolate. For other catalysts the energy difference between the two transition structures is very small and so poor diastereoselectivity is obtained.

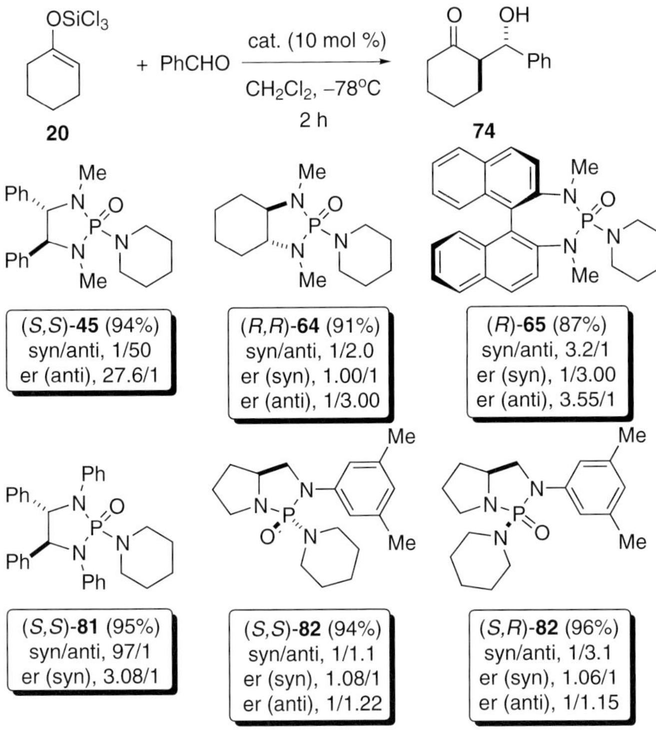

Scheme 7.33
Catalysts for the addition of **20** to benzaldehyde.

Analysis of the absolute configurations of the products enables construction of a crude transition state model that explains the overall arrangement of the components. The absolute configuration of the aldol product **74** from the reaction using (S,S)-**45** has been established to be (2R,1'S) by single-crystal X-ray analysis of the corresponding 4-bromobenzoate. This observation, in conjunction with the results obtained for acyclic Z enolates, offers important insights into the arrangement of the aldehyde relative to the enolate in the transition structure, and into factors that determine the absolute configuration of the aldol adduct (Figure 7.11). The configuration of the chiral phosphoramide determines the face of enolate that undergoes aldolization. On the other hand, the chair or boat transition structure determines the face of aldehyde to be attacked. For example, when (S,S)-**45** is used as catalyst, the *Si* face of the enolate is blocked (Figure 7.11). Placement of aldehyde in a chair-like transition structure will then correctly predict the absolute configuration of the major diastereomer.

The phosphoramide **45** is also an effective catalyst for the additions of other cyclic ketone-derived enolates (Scheme 7.34) [77]. Both cyclopentanone- and cycloheptanone-derived enolates provide *anti* products with good

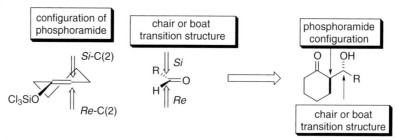

Fig. 7.11
Factors leading to the observed configuration.

enantioselectivity. It is interesting to note that diastereoselectivity for these reactions is sensitive to the rate of mixing; slow addition of aldehyde is therefore necessary to obtain reproducible and high diastereoselectivity.

The synthetic utility of the aldol addition of **20** has been expanded by examining a wide range of aldehydes of different structure (Scheme 7.35). Enantioselectivity is usually good to excellent for the *anti* diastereomers. For addition of **20** to different aldehydes excellent *anti* selectivity is always obtained except for use of phenylpropargyl aldehyde.

In this particular system there is no obvious relationship between stereoselectivity and aldehyde structure. Steric bulk around the aldehyde carbonyl seems to enhance diastereoselectivity and enantioselectivity in additions to benzaldehyde and to 1-naphthaldehyde. The lower diastereoselectivity observed in the addition to phenylpropargyl aldehyde can be attributed to the lack of facial differentiation for the aldehyde, because the substituents (H and acetylenic groups) are similar in size.

Addition of **20** to aliphatic aldehydes does not, unfortunately, furnish the corresponding aldol products under catalysis by chiral phosphoramides. This may be caused by competitive enolization of the aldehyde by the basic

OSiCl₃

+ PhCHO →(S,S)-**45** (10 mol %) / CH₂Cl₂, –78°C → product, n = 5, 6, 7

74 (95%)	**73** (98%)	**75** (91%)
syn/anti, 1/61	syn/anti, 1/22	syn/anti, 1/17
er (anti), 27.6/1	er (anti), 7.13/1	er (anti), 9.87/1

Scheme 7.34
Catalyzed aldol additions of a variety of cyclic enolates to benzaldehyde.

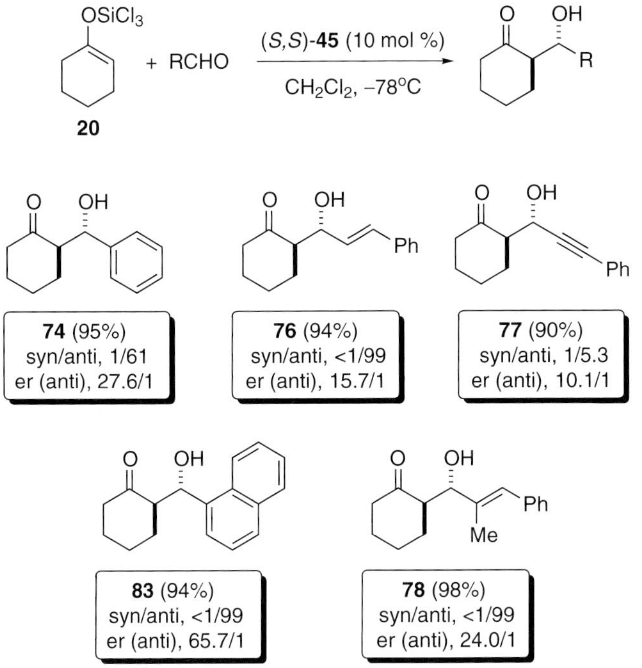
OSiCl3
+ RCHO
(S,S)-45 (10 mol %)
CH2Cl2, –78°C
O OH
R
20
74 (95%)
syn/anti, 1/61
er (anti), 27.6/1
76 (94%)
syn/anti, <1/99
er (anti), 15.7/1
77 (90%)
syn/anti, 1/5.3
er (anti), 10.1/1
Ph
Me
83 (94%)
syn/anti, <1/99
er (anti), 65.7/1
78 (98%)
syn/anti, <1/99
er (anti), 24.0/1

Scheme 7.35
Addition of **20** to a variety of aldehydes catalyzed by **45**.

enolate/phosphoramide complex and/or formation of the α-chlorosilyl ether as discussed for addition of methyl ketone-derived enolates.

7.4.3
Addition of Acyclic Ethyl Ketone-derived Enolates

The control of enolate geometry is a major synthetic challenge in the addition of acyclic, substituted trichlorosilyl enolates, because the diastereomeric composition of the aldol products can reflect the *E*/*Z* ratio in the enolates [67]. Fortunately, the preparation of geometrically-defined trichlorosilyl enolates has been described (Section 7.2) [45]. The strong preference for the chair-like transition structure has been demonstrated for aldol additions of cyclic ketone-derived enolates (*E*-configured) under suitable phosphoramide catalysis. A high degree of diastereocontrol can be achieved for cyclic enolate additions. Use of the *E* enolate correlates with formation of the *anti* aldol product.

The initial study of an acyclic ketone-derived enolate was performed using (*Z*)-**18** derived from propiophenone. At 0 °C, aldol addition of **18** to a variety of aldehydes proceeds at an appreciable rate and provides aldol products in good yield (Scheme 7.36). All of the reactions provide *anti* aldol products as the major diastereomers, albeit with low diastereoselectivity. Correlation

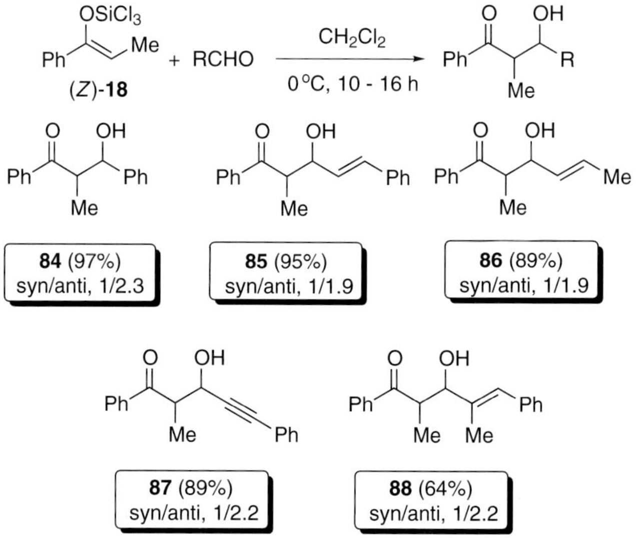

Scheme 7.36
Uncatalyzed addition of (Z)-**18** to a variety of aldehydes.

of the Z enolate with the *anti* aldol product suggests involvement of a boat transition structure. The modest selectivity of uncatalyzed addition of the Z enolate contrasts with uncatalyzed addition of cycloalkanone-derived enolates, which selectively afford *syn* products. The low diastereoselectivity observed is because of the unfavorable interaction between the methyl substituent and an apical chloride in the preferred boat transition structure.

Acyclic ethyl ketone-derived enolates are relatively unreactive toward aldehydes compared with the other types of enolate discussed above. The observed low reactivity of **18** might be because of steric interaction between the α-methyl substituent of the enolate.

Aldol addition of **18** to benzaldehyde is much slower at low temperature. The uncatalyzed aldol addition gives only 1% of isolated aldol product after 6 h at −78 °C [76] whereas in the presence of a catalytic amount of a phosphoramide aldol reactions of (Z)-**18** are significantly accelerated (Scheme 7.37). Although all phosphoramides give good yields of the benzaldehyde-derived aldol products, stereoselectivity is highly variable and depends on catalyst structure. Once again, the stilbene-1,2-diamine-derived phosphoramide **45** is highly selective, providing good *syn* selectivity combined with excellent enantioselectivity. The observed *syn* selectivity suggests that with **45**, the predominant reaction pathway is through a chair-like transition structure.

The scope of this reaction has been explored using phosphoramide **45**. Increasing the catalyst loading from 10 to 15 mol% not only improves yields of aldol products but also leads to higher diastereoselectivity. Addition of **18**

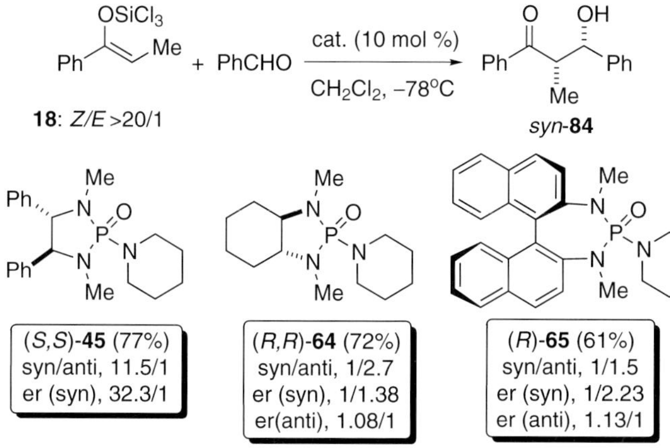

Scheme 7.37
Catalyzed addition of (*Z*)-**18** to benzaldehyde.

to a variety of aldehydes is efficiently catalyzed by **45**, with moderate to good *syn* relative diastereoselectivity and good to excellent enantioselectivity for the *syn* diastereomers (Scheme 7.38).

Good *syn* selectivity has been observed for most conjugated aldehydes. Significantly attenuated diastereoselectivity is observed for sterically de-

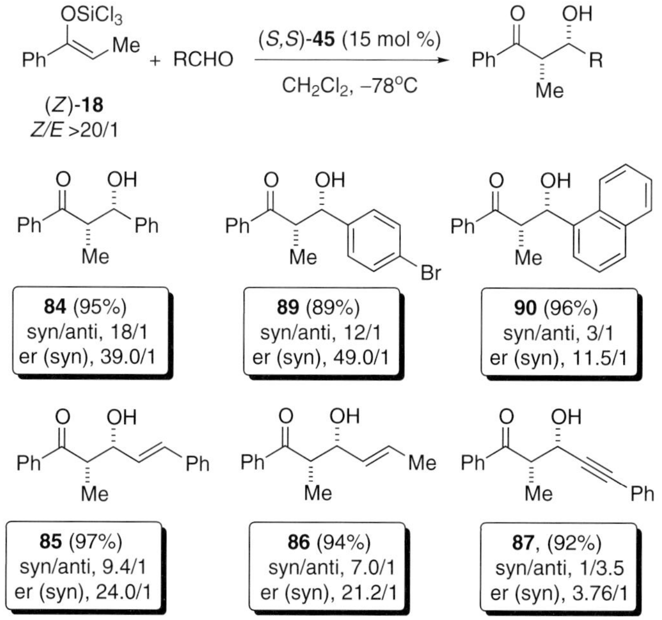

Scheme 7.38
Catalyzed addition of (*Z*)-**18** to various aldehydes.

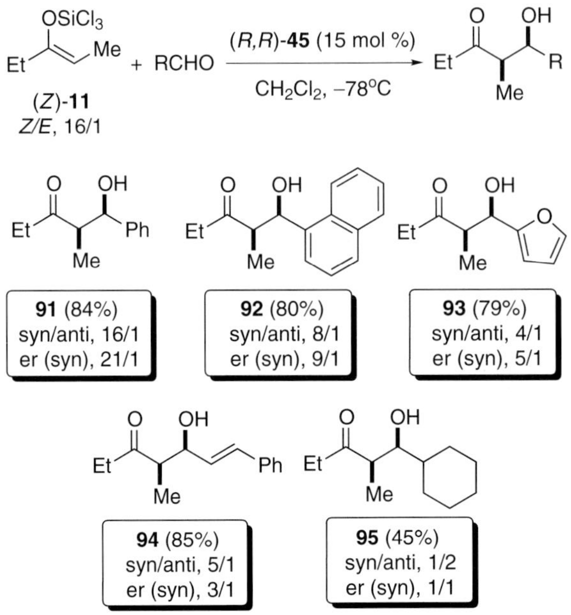

Scheme 7.39
Catalyzed aldol additions of (Z)-**11** to different aldehydes.

manding aldehydes. For example, the diastereoselectivity in the addition to 1-naphthaldehyde is only modestly *syn* selective. In contrast, addition to phenylpropargyl aldehyde is surprisingly *anti* selective.

The sense of stereoinduction enforced by the chiral catalyst (S,S)-**45** is consistent with the model described previously (Figure 7.11). The absolute configuration of the aldol product **84** has been unambiguously assigned as (2S,3S) by X-ray analysis of the corresponding 4-bromobenzoate ester. The phosphoramide (S,S)-**45** blocks the *Si* face of the enolate and the resulting chair-like assembly of the enolate and the aldehyde leads to the stereoselectivity observed.

The reactions of other ethyl ketone enolates have also been investigated. When catalyzed by **45** aldol additions of the 3-pentanone-derived trichlorosilyl enolate (Z)-**11**, Z/E ratio 16:1, give the *syn* aldol products selectively (Scheme 7.39) [45].

For addition to benzaldehyde the diastereoselectivity reflects the Z/E ratio of the starting enolate, indicating that the chair-like transition structure is strongly favored. The attenuated diastereoselectivity in additions to other aldehydes suggests that competitive boat transition structures become operative. Along with decreases in diastereoselectivity, enantioselectivity for other aldehydes is reduced significantly.

Unlike the perfect correlation between starting enolate geometry and aldol product configuration observed in the addition of Z enolates, diastereo-

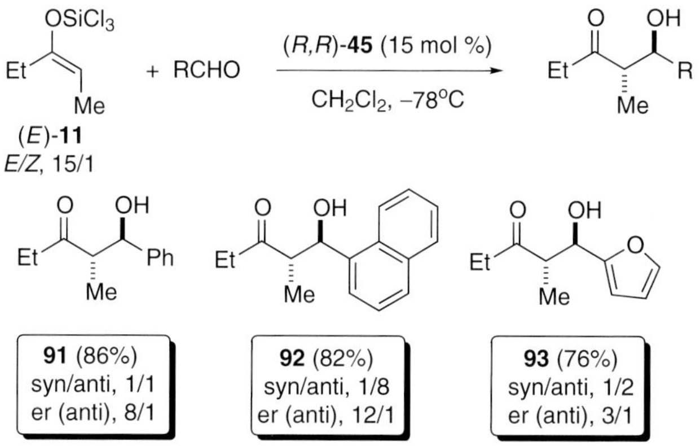

Scheme 7.40
Catalyzed additions of (*E*)-**11** to different aldehydes.

selection is poor for addition of (*E*)-**11** to benzaldehyde, although the enantioselectivity observed in the *anti* pathway is good (Scheme 7.40) [45]. In contrast, addition to 1-naphthaldehyde provides the *anti* aldol product with good diastereo- and enantioselectivity.

The difference between results from the addition of *E* and *Z* enolates is striking. As discussed in the previous section, cycloalkanone-derived trichlorosilyl enolates (*E* enolates) furnish exclusively *anti* aldol products under the action of catalysis by **45**. *Z* Enolates, on the other hand, produce predominantly *syn* aldol products.

Acyclic *E*-trichlorosilyl enolates do not undergo selective aldol additions, however, furnishing aldol products with unpredictable diastereomeric ratios. This lack of diastereoselection is presumably because of competition between the chair and boat transition structures (Figure 7.12). In the structure **xi** the least sterically demanding group would be placed *anti*-periplanar to the enolate C–O bond to minimize the $A^{1,3}$ allylic interaction. The non-participating group will then be in close proximity to the silicon center bearing two, large phosphoramide molecules. This steric congestion might ultimately cause the conformational change from the chair to the boat transition structure.

x (*Z*-enolate,chair)   **xi** (*E*-enolate, chair)   **xii** (*E*-enolate, boat)

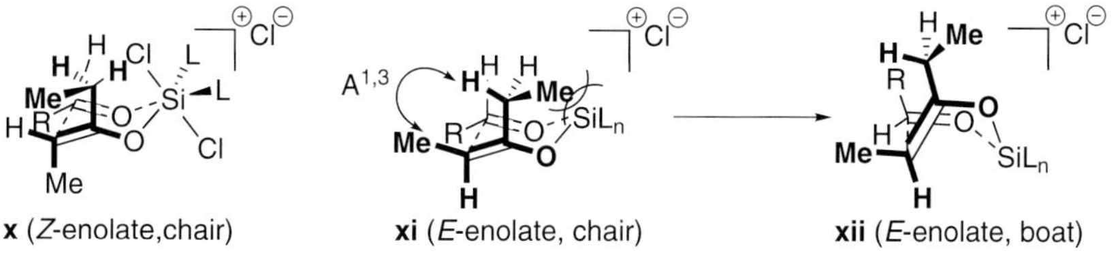

Fig. 7.12
Transition structures for aldol additions of (*Z*)-**11** and (*E*)-**11**.

For these reasons, reactions of cycloalkanone-derived enolates are an exception in *E* enolate additions. Because of the absence of steric congestion on the non-participating side, the reaction can proceed predominantly through the chair transition structure, resulting in high *anti* diastereoselectivity. In the acyclic *E* enolate, unfavorable steric interactions in the chair transition structure enable competition from the boat transition structure. Therefore, only modest *anti* selectivity can be achieved with these enolates.

7.5
Diastereoselective Additions of Chiral Enoxytrichlorosilanes

The use of enoxytrichlorosilanes bearing a stereogenic center is an important extension of Lewis-base-catalyzed aldol additions in organic synthesis. If the reactants contain stereogenic centers the resulting aldol products are diastereomeric. Thus, in theory, stereoinduction can arise from both the resident stereogenic center and the external chiral catalyst.

In additions of achiral trichlorosilyl enolates the stereochemical course of the reaction is governed by two factors (Figure 7.13) [71]. The first is the *relative diastereoselection* which reflects the relative topicity (*like* or *unlike*) [78] of the combination of the two reacting faces (enolate and carbonyl group). In highly organized aldol additions this is often interpreted in terms of chair/boat selectivity in the transition structure. The relative diastereoselec-

(a) relative diastereoselection:

(b) external enantioselection:

(c) internal diastereoselection:

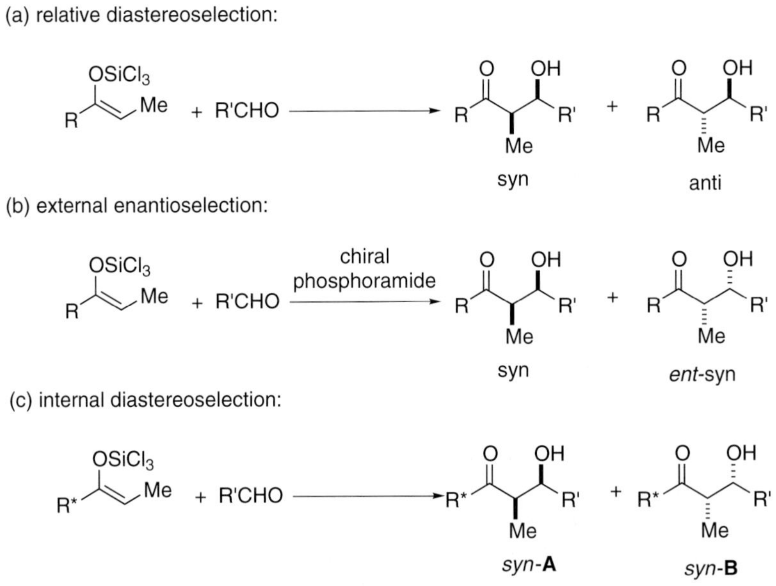

Fig. 7.13
Three types of stereoselection process.

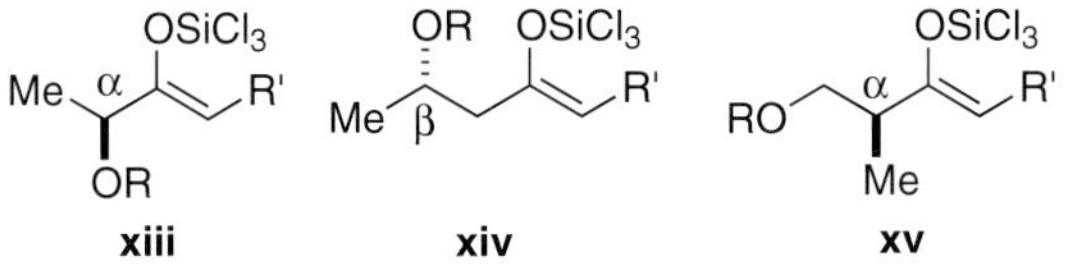

Chart 7.2
Three classes of chiral trichlorosilyl enolate.

tion pertains to α-substituted enolates only, because the term refers to the relative configuration of the two substituents (*like* or *unlike*) at the newly created stereogenic centers. The other is the absolute stereoselection (*external stereoselection*) determined by the chiral Lewis base catalyst. This term is used to describe the enantiofacial outcome at the newly created stereogenic centers.

In additions of chiral enolates there is yet another stereoselection process that is controlled by the resident stereogenic center. The diastereoselection resulting from the effect of the stereogenic centers in either of the reactants is referred as *internal diastereoselection*. When a chiral catalyst is used in conjunction with a chiral enolate there is a possibility of double diastereo-differentiation [2c]. In a matched case in which the sense of external stereoinduction coincides with the internal stereoinduction, the diastereo-selectivity of the reaction can be considerably enhanced [79].

Three classes of chiral trichlorosilyl enolate have been studied to investigate the effect of the resident stereogenic center in the context of Lewis-base-catalyzed aldol addition (Chart 7.2). Two of these enolates (**xiii** and **xiv**) bear heteroatom-based stereogenic centers at the α- and β-carbon atoms on the non-participating side and one (**xv**) bears a carbon-based stereogenic center on the α-carbon of the non-participating side.

7.5.1
Aldol Addition of Lactate-derived Enoxytrichlorosilanes

7.5.1.1 Methyl Ketone-derived Enolates
Aldol additions of lactate-derived trichlorosilyl enolates in the absence of a Lewis base catalyst proceed at room temperature (Scheme 7.41) [80]. The aldol products derived from benzaldehyde are obtained in good yields as mixtures of diastereomers. Although aldol additions of these enolates are poorly selective, the *anti* diastereomers are always favored. Interestingly, the size of the protecting group has some effect on diastereoselectivity; smaller protecting groups result in better selectivity.

The observed *anti* selectivity can be explained by the model depicted in Figure 7.14. Initial coordination of the aldehyde with the Lewis acidic silicon center results in formation of a trigonal bipyramidal species (c.f. Figure 7.7, Section 7.4). Typically, uncatalyzed aldol additions of trichlorosilyl enolates proceed via boat-like transition structures [65, 72]. The oxygen sub-

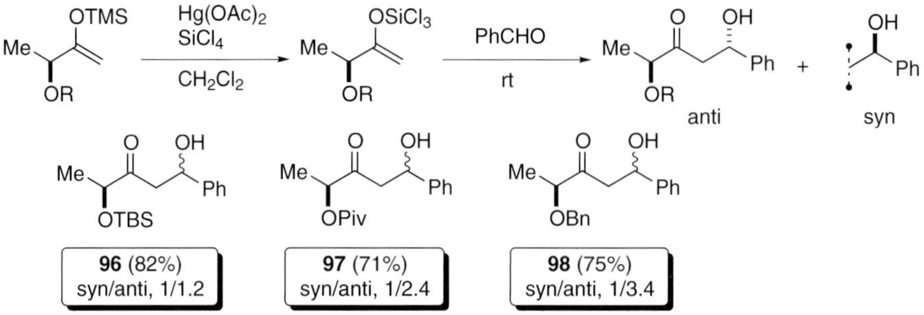

96 (82%)
syn/anti, 1/1.2

97 (71%)
syn/anti, 1/2.4

98 (75%)
syn/anti, 1/3.4

Scheme 7.41
Uncatalyzed aldol additions of lactate-derived enolates.

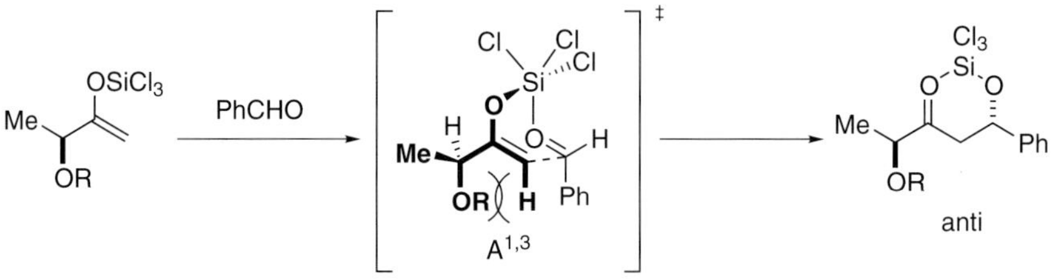

Fig. 7.14
Stereochemical model for uncatalyzed addition of lactate-derived enolates.

stituent on the enolate is expected to be antiperiplanar to the enol oxygen to minimize the net dipole [73]. In this conformation, the *anti* diastereomer is preferentially formed by approach of the aldehyde to the less hindered face of the enolate. This model is consistent with the observation that the bulkier protecting groups result in lower diastereoselectivity, because the $A^{1,3}$-type interaction becomes substantial if the favored conformer contains a large protecting group.

Uncatalyzed additions of **29** to other aldehydes have been examined (Scheme 7.42). Unfortunately, the substrate-induced diastereoselection is

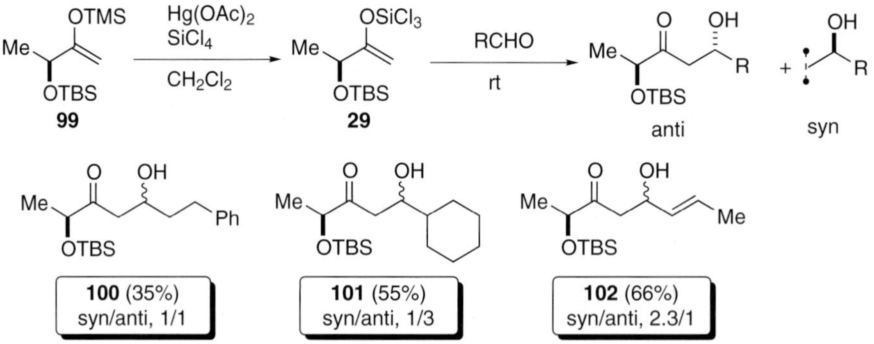

100 (35%)
syn/anti, 1/1

101 (55%)
syn/anti, 1/3

102 (66%)
syn/anti, 2.3/1

Scheme 7.42
Uncatalyzed aldol addition of **29** to a variety of aldehydes.

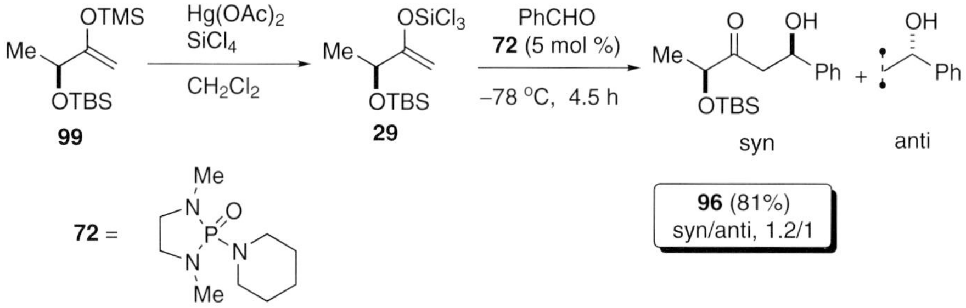

Scheme 7.43
Addition of **29** to benzaldehyde catalyzed by **72**.

only modest and the selectivity cannot easily be rationalized because *syn/anti* selectivity depends on aldehyde structure. Also, for aliphatic aldehydes, significantly reduced yields are obtained.

In the presence of a catalytic amount of a phosphoramide the rate increases significantly. The intrinsic selectivity induced by the stereogenic center has been examined for use of achiral catalyst **72** (Scheme 7.43). Although the product obtained is slightly enriched in the *syn* diastereomer, the *syn/anti* ratio is almost negligible, indicating that the resident stereogenic center has little effect on diastereoselectivity.

Double stereodifferentiation using a chiral catalyst provides a dramatic matched/mismatched effect in the aldol addition of **29**, **30** and **31**. (Scheme 7.44). When (*S*,*S*)-**45** is used marginal improvement of *syn* selectivity is ob-

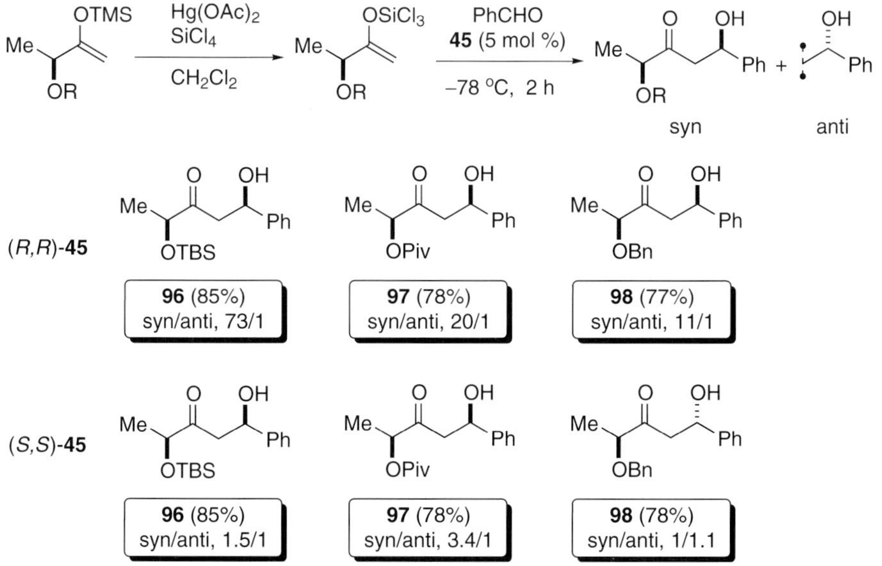

Scheme 7.44
Catalyzed addition of lactate-derived trichlorosilyl enolates to benzaldehyde.

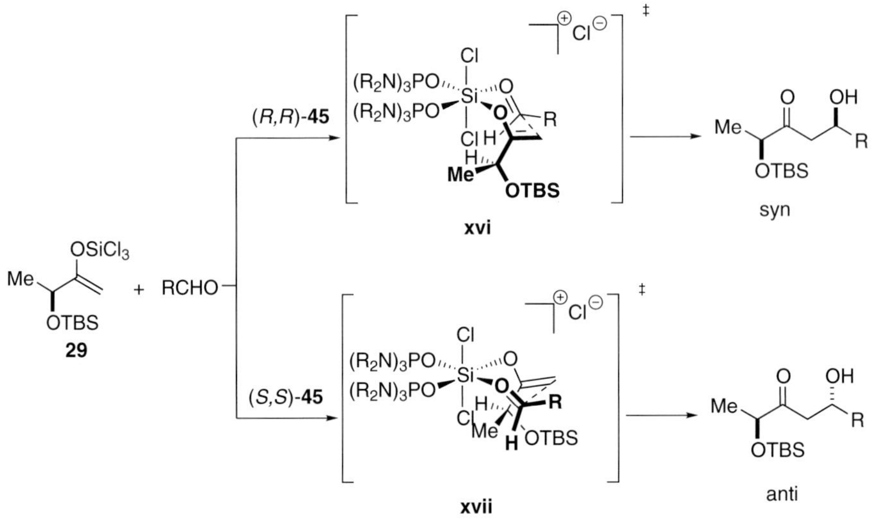

Fig. 7.15
Stereochemical course of aldol addition of **29**.

served. The use of (R,R)-**45**, on the other hand, results in the 1,4-*syn* aldol product with excellent diastereoselectivity. The different protecting groups affect the extent of diastereoselection and diastereoselectivity decreases in the order OTBS > OPiv > OBn. This order suggests that a pathway involving chelation of the cationic silicon might become possible with more coordinating oxygen functions (vide infra) [81].

The use of (R,R)-**45** represents the matched case in which the sense of external stereoinduction is the same as that of internal stereoinduction, whereas in the use of (S,S)-**45** the sense of stereoinduction is opposite, leading to attenuated selectivity. The predominant *syn* selectivity is rationalized by the model depicted in Figure 7.15. The preferred conformation of the resident stereogenic center again places the oxygen substituent in the plane of the enol double bond as explained above. The enolate faces are discriminated not only by the chiral phosphoramide but also by the substituents on the resident stereogenic center. In the chair-like transition structure **xvi** attack of the enolate on the *Re* face of the aldehyde leads to the *syn* diastereomer observed. In this model (R,R)-**45** blocks the more sterically hindered face (*syn* to the methyl group) of the enolate, thus matching the internal and external stereoinduction. On the other hand (S,S)-**45** prevents approach of the aldehyde from sterically less hindered face (*anti* to the methyl group) of the enolate (**xvii**, Figure 7.15). In this case, the internal and external stereoinductions oppose each other, resulting in significantly attenuated diastereoselectivity.

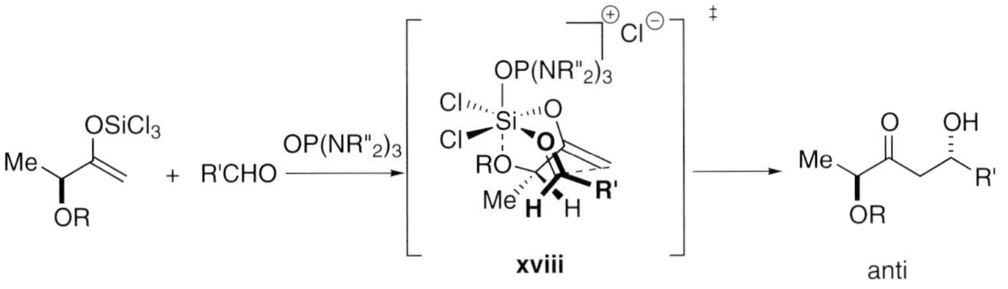

Fig. 7.16
Competitive, chelated transition structure for
catalyzed aldol addition of lactate-derived
trichlorosilyl enolates.

The attenuated diastereoselectivity when protecting groups other than
OTBS are used might indicate the intervention of a competitive chelated
transition structure (Figure 7.16). In the transition structure **xviii**, the oxygen on the non-participating side is coordinated to the Lewis acidic silicon.
As the coordinating capacity of oxygen increases, the transition structure
xviii might be favorable, and attenuated diastereoselectivity is observed.

Catalyzed additions of **29** to olefinic aldehydes have also been demonstrated. For example, addition of **29** to crotonaldehyde catalyzed by **45** gives
the corresponding aldol product in good yield (Scheme 7.45). Although the
diastereoselectivity obtained in these reactions is lower, the stereochemical
trend remains the same. The (R,R)-**45** catalyst is the matched case providing
syn-**102** with good diastereoselectivity.

7.5.1.2 Ethyl Ketone-derived Enolates

The stereochemical course of addition of the corresponding ethyl ketone-
derived enolates incorporates all three forms of stereoselection [82]. Aldol
addition of (Z)-**103** under the action of phosphoramide catalysis provides
the *syn*,*syn* (relative, internal) aldol product with high selectivity (Scheme
7.46). A survey of chiral and achiral phosphoramides shows a remarkable

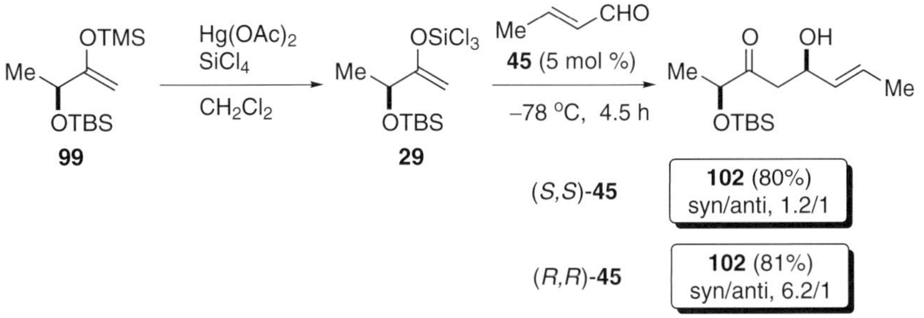

Scheme 7.45
Catalyzed addition of **29** to crotonaldehyde.

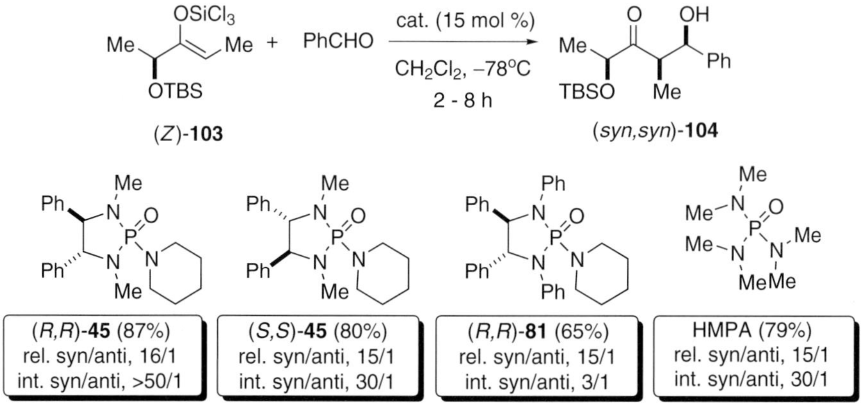

Scheme 7.46
Aldol additions of (Z)-**103** to benzaldehyde catalyzed by different phosphoramides.

trend. All the phosphoramides yield the *syn* relative aldol product with perfect Z/E to *syn/anti* correlation, indicating that a chair-like transition structure is maintained. These results are intriguing considering that bulky phosphoramides such as (R,R)-**81** favor the boat-like transition state in the addition of the cyclohexanone-derived trichlorosilyl enolate [76].

Internal stereoselectivity varies for different catalyst structures. For the stilbene-1,2-diamine derived catalyst, (R,R)-**45**, excellent internal *syn* selectivity is obtained. Use of the enantiomeric catalyst (S,S)-**45** does not reverse the sense of internal diastereoselection and the aldol product is again obtained with good internal *syn* selectivity. These observations indicate the overwhelming influence of the resident stereogenic center and the stereochemical course of the aldol addition is determined solely by this factor. To support this explanation, additions using a variety of achiral phosphoramides including HMPA demonstrate that the internal *syn* aldol product is preferentially formed under catalyzed conditions, irrespective of catalyst configuration. Bulky phosphoramides such as (R,R)-**81** result in attenuated internal diastereoselectivity compared with that resulting from other phosphoramides.

The observed stereochemical outcome can be explained by the non-chelation model that places the OTBS substituent in the enolate in plane with the enolate double bond to minimize the dipole moment (Figure 7.17) [73, 80, 82]. The two enolate faces are differentiated by the size of the groups on the stereogenic center (H compared with Me), and the aldehyde approaches from the less hindered *Si* face of the *Z* enolate. The chair-like arrangement of the aldehyde in the transition structure leads to the formation of the observed *syn,syn* diastereomer. The low selectivity observed when bulky phosphoramides are used can be rationalized by the intervention of another transition structure, **xx** [82]. These phosphoramides are known to

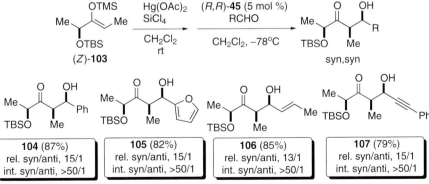

Fig. 7.17
Proposed transition structures for aldol addition of (Z)-**103**.

favor boat-like transition structures via a mechanism that involves only one phosphoramide in the stereodetermining step [66]. In this pentacoordinate species, it is possible that the silyloxy group could coordinate the Lewis acidic silicon to form an octahedral, cationic silicon intermediate. This internal coordination might favor the chair-like arrangement over the usual boat-like transition structure for these phosphoramides. Although the coordinating capacity of the TBS ether is modest at best [81], the proximity of the silyloxy group to the cationic silicon is believed to enhance the possibility of this type of chelation [83]. In this model, the aldehyde now approaches from the *Re* face of the enolate leading to the *syn,anti* diastereomer.

The aldol additions of (Z)-**103** to different aldehydes illustrate the generality of this process (Scheme 7.47). The in-situ generation of trichlorosilyl enolate (Z)-**103** from the corresponding TMS enol ether further demonstrates not only the synthetic utility of this reagent but also the improved

Scheme 7.47
Aldol addition of (Z)-**103** to different aldehydes.

yield and selectivity of the overall process. The addition can be catalyzed by either (R,R)-**45** or HMPA, and *syn,syn*-aldol products can always be obtained selectively. The relative and internal diastereoselectivities are all perfect when (R,R)-**45** is used as the catalyst. The diastereoselectivity and yield are slightly attenuated under the action of catalysis by HMPA.

7.5.2
Aldol Addition of β-Hydroxy-α-Methyl Ketone-derived Enoxytrichlorosilanes

7.5.2.1 Methyl Ketone-derived Enolates
The effects of α-methyl and β-hydroxy groups on the stereochemical course of aldol additions with trichlorosilyl enolates have been investigated. This type of enolate structure is synthetically important because the resulting aldol product resembles the highly oxygenated structural motif for a variety of polypropionate natural products. Not surprisingly, diastereoselective aldol additions of this type of enolate have already been demonstrated for lithium, boron, and tin enolate aldol additions [1d, 79, 84].

The effect of the α-methyl stereogenic center has been determined in aldol additions of the methyl ketone-derived trichlorosilyl enolate (S)-**108** (Scheme 7.48) [85]. The addition of **108** using (R,R)-**45** as catalyst provides the *syn* aldol product selectively. The use of (S,S)-**45** enables formation of *anti*-**109**, albeit with attenuated selectivity. The intrinsic internal selectivity arising from an α-methyl stereogenic center is determined by examining the diastereoselectivity of the aldol addition using the achiral phosphoramide **72**. The internal selectivity is low but slightly favors the 1,4-*syn* diastereomer.

The inherent selectivity is rationalized by means of a transition structure model in which transition structure **xxi** (Figure 7.18) involves octahedral, cationic silicon in a *chair-like* arrangement of groups. To avoid steric interaction between the phosphoramide-bound silicon and the non-participating substituent on the enolate the least sterically demanding substituent (hydrogen) is placed in plane with the enolate C–O bond. This model predicts

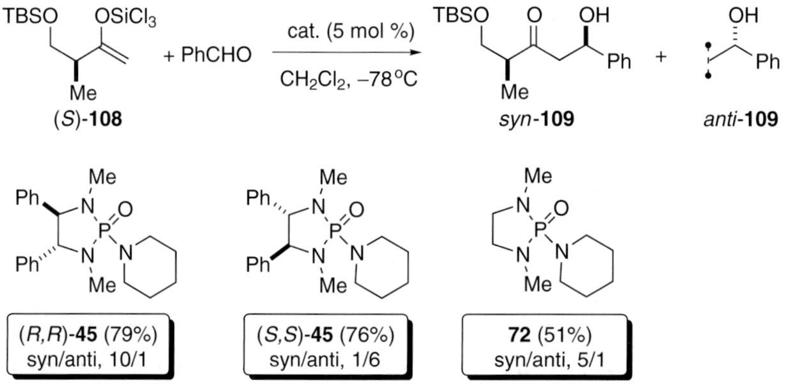

Scheme 7.48
Aldol addition of **108** to benzaldehyde catalyzed by phosphoramides.

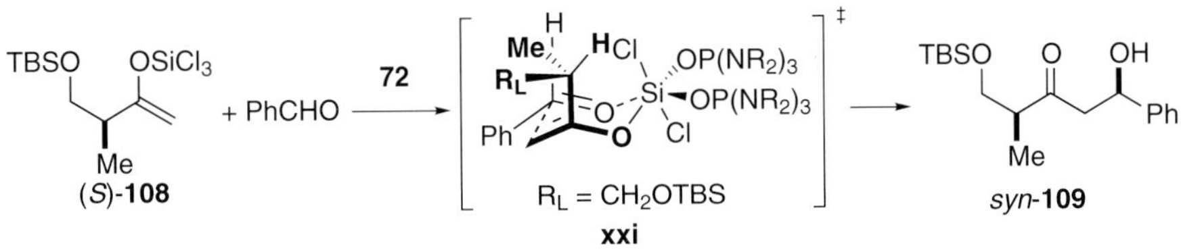

Fig. 7.18
Stereochemical course of aldol addition of (S)-**108** to benzaldehyde.

approach of benzaldehyde from the less sterically demanding methyl group side, leading to the *syn* diastereomer.

The generality of this aldol addition has been investigated with a wide variety of aldehydes (Scheme 7.49). The trichlorosilyl enolate **108** generated in situ (from TMS enol ether **110**) reacts with aromatic, conjugated, and

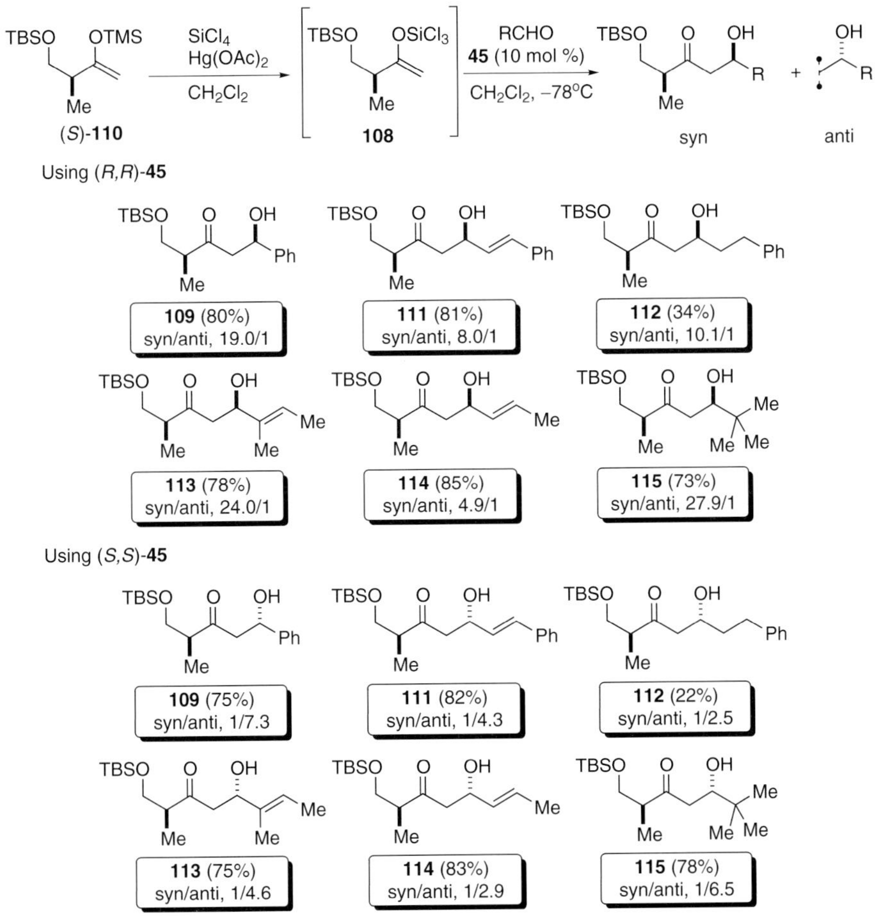

Scheme 7.49
Aldol addition of **108** to different aldehydes catalyzed by **45**.

Tab. 7.4

Catalyzed aldol additions of **35** and **116** to benzaldehyde.

Entry	Enolate	Z/E	Catalyst	Yield	Relative dr (syn/anti)	Internal dr[a] (syn/anti)
1	(Z)-**35**	50/1	(R,R)-**45**	72	9/1	10/1
2	(Z)-**35**	50/1	(S,S)-**45**	82	12/1	1/7
3	(E)-**35**	1/50	(R,R)-**45**	72	1/4	6/1
4	(E)-**35**	1/50	(S,S)-**45**	72	1/2	2/1
5	(Z)-**116**	50/1	(R,R)-**45**	84	53/1	24/1
6	(Z)-**116**	50/1	(S,S)-**45**	82	32/1	1/8
7	(Z)-**116**	50/1	**72**	81	27/1	5/1

[a] Ratio of major relative diastereomer.

aliphatic aldehydes. Additions to aromatic aldehydes result in high yields and good diastereoselectivity. Additions to the olefinic aldehydes always result in good yields, but selectivity is quite variable. Interestingly, steric bulk around the carbonyl group has a beneficial effect on diastereoselectivity. Unhindered aliphatic aldehydes are significantly less reactive, resulting n only modest yields of the aldol products.

7.5.2.2 Ethyl Ketone-derived Enolates

When the corresponding ethyl ketone enolate reacts with aldehydes, an additional stereogenic center is formed (Table 7.4) [86]. The reactions of both (Z)-**35** and (E)-**35** have been examined, enabling the effect of enolate geometry on diastereoselectivity to be probed. Several interesting trends can be noted from aldol additions of **35** and **116** to benzaldehyde. Additions of (Z)-**35** are generally *syn* (relative) selective, indicating that a chair-like transition structure is involved in the dominant pathway (Table 7.4, entries 1 and 2). The diastereoselectivity observed for (E)-**35** is only marginal, however, and the *anti* (relative) diastereomer is preferred (Table 7.4, entries 3 and 4). In both reactions the E/Z ratio of the enolate does not translate strictly into the relative *syn/anti* ratio. This observation can be accounted for by the presence of competitive boat-like transition structures that lead to the minor diastereomers. Fortunately, the relative diastereoselectivity can be significantly improved by changing the protecting group from TBS to the TIPS (Table 7.4, entries 5–7).

The intrinsic selectivity has been determined using the achiral catalyst **72**, and small preference for the internal *syn* diastereomer was observed (Table 7.4, entry 7). The internal diastereoselectivity is also largely determined by catalyst configuration in the addition of *Z* enolates. Reactions with (*R*,*R*)-**45** corresponds to matched cases wherein the sense of internal and external diastereoselection is the same. Thus, higher internal selectivity is obtained with (*R*,*R*)-**45** than with (*S*,*S*)-**45**. In additions of (*E*)-**35**, internal selectivity is modest, and there is no dependence on catalyst configuration.

Additions of (*Z*)-**116** to a variety of aldehydes furnish the *syn* (relative) diastereomers in good yields with good to excellent selectivity (Scheme 7.50). The structure of the aldehyde makes an important contribution to the relative diastereoselectivity. Bulky aldehydes such as 1-naphthaldehyde and tiglic aldehyde result in significantly lower diastereoselectivity. The internal selectivity also depends on the aldehyde structure. Selectivity is significantly higher for aromatic aldehydes than for olefinic aldehydes. The internal diastereoselection is always determined by catalyst configuration. When (*R*,*R*)-**45** is used high *syn* (internal) selectivity can be achieved, and *anti* (internal) diastereomers can be obtained by use of (*S*,*S*)-**45**, albeit with attenuated selectivity.

The dramatic difference between the behavior of acyclic *Z* and *E* enolates in these aldol additions has already been discussed above (Figure 7.12, Section 7.4). Here again, addition of (*Z*)-**35** results in good *syn* relative selectivity and addition of (*E*)-**35** is only slightly *anti* relative selective. In the addition of (*Z*)-**35** the diastereoselectivity observed can be better rationalized by the chair-like transition structure **xxii** (Figure 7.19). The transition structure **xxii** is consistent with the small internal diastereoselection exerted by the stereogenic center on the enolate. The conformation of the stereogenic center in **xxii** minimizes steric interaction between the substituents on the non-participating side of the enolate and the bulky ligands on the hypercoordinate silicon. This transition state model leads to the observed (*syn*,*syn*)-**117**.

The poor selectivity observed for addition of (*E*)-**35** is explained by the competitive transition structure models chair-**xxiii** and boat-**xxiv**. In chair-**xxiii** A^{1,3} strain between the equatorial methyl group and the non-participating substituent of the enolate is minimized [2c]; the disposition of the α-methyl and CH$_2$OTIPS groups toward the bulky silicon center can, however, cause severe steric congestion. This interaction can be significant enough to make the boat-**xxiv** transition structure more favorable. The boat-**xxiv** is easily accessed simply by placing the silicon group in the least crowded quadrant. The *anti* coordination of silicon to the aldehyde places the phenyl group of benzaldehyde in the pseudo-axial position, leading to the *syn* (relative) diastereomer.

7.5.3
Addition of Enoxytrichlorosilanes with a β-Stereogenic Center

Thus far the effect of an α-stereogenic center on the stereochemical course of aldol additions of trichlorosilyl enolates has been described. Diastereo-

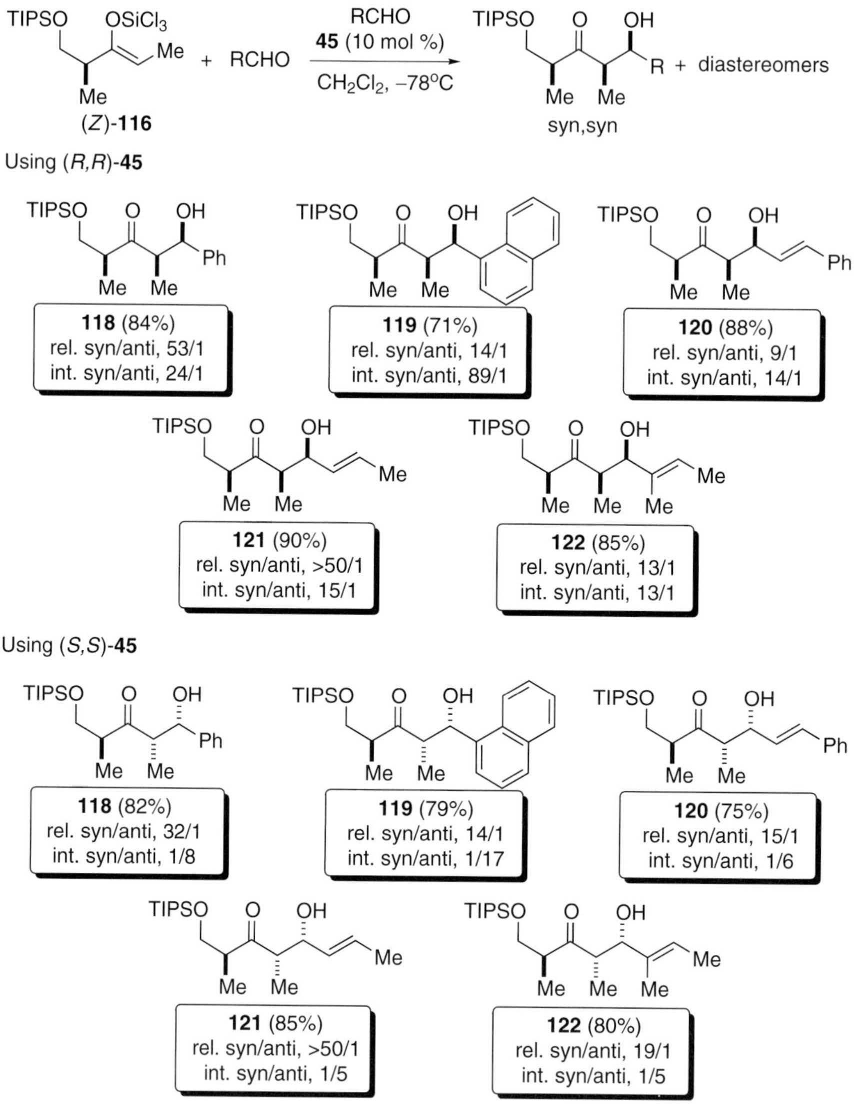
TIPSO OSiCl3 Me + RCHO
RCHO
45 (10 mol %)
CH2Cl2, −78°C
TIPSO O OH R + diastereomers
Me Me
syn,syn
Me
(Z)-116
Using (R,R)-45
118 (84%)
rel. syn/anti, 53/1
int. syn/anti, 24/1
119 (71%)
rel. syn/anti, 14/1
int. syn/anti, 89/1
120 (88%)
rel. syn/anti, 9/1
int. syn/anti, 14/1
121 (90%)
rel. syn/anti, >50/1
int. syn/anti, 15/1
122 (85%)
rel. syn/anti, 13/1
int. syn/anti, 13/1
Using (S,S)-45
118 (82%)
rel. syn/anti, 32/1
int. syn/anti, 1/8
119 (79%)
rel. syn/anti, 14/1
int. syn/anti, 1/17
120 (75%)
rel. syn/anti, 15/1
int. syn/anti, 1/6
121 (85%)
rel. syn/anti, >50/1
int. syn/anti, 1/5
122 (80%)
rel. syn/anti, 19/1
int. syn/anti, 1/5

Scheme 7.50
Aldol additions of (*Z*)-**116** to different aldehydes catalyzed by **45**.

selectivity clearly depends on the nature of the α-substituent. The α-oxygen substituent of lactate-derived enolates has a strong effect on the diastereoselectivity of the catalyzed aldol addition whereas the α-hydroxymethyl stereogenic center of hydroxybutyrate-derived enolates plays a minor role only in the diastereoselection, and catalyst configuration primarily determines the stereochemical course of the aldol addition.

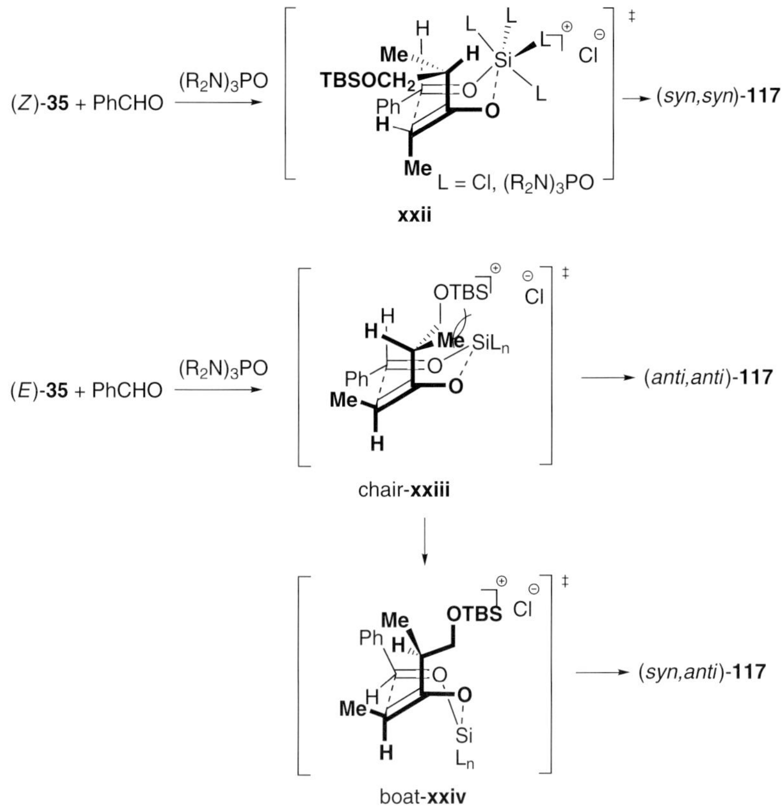

Fig. 7.19
Proposed transition structures for addition of (Z)- and (E)-**35**.

The effect of a remote stereogenic center on diastereoselection in aldol additions is also worth investigation. In the aldol addition of boron enolates it has been demonstrated that a β-oxygen stereogenic center can strongly influence the stereochemical course of the reaction [87]. This class of enolate is also important because these aldol products have a 1,3,5-oxygenated carbon chain, a common motif in a variety of natural products [2c].

The aldol reactions of **123** under phosphoramide catalysis are summarized in Table 7.5 [88]. The intrinsic selectivity determined using **72** is almost negligible, indicating that the β-stereogenic center does not exert significant stereoinduction during addition of this enolate. Interestingly, use of chiral phosphoramides affords only marginal improvement in diastereoselectivity.

Additions of ethyl ketone-derived enolates (Z)- and (E)-**36** are also catalyzed by **45** (Table 7.6). Good relative *syn* diastereoselectivity is observed for addition of (Z)-**125**. As previously observed for addition of (Z)-**35**, changing from the TBS protecting group to TIPS has a beneficial effect on the dia-

Tab. 7.5
Catalyzed aldol addition of **123** to benzaldehyde.

Entry	Catalyst	Yield, %	syn/anti
1	(R,R)-45	72	1/2.5
2	(S,S)-45	75	1.3/1
3	72	55	1/1.4

stereoselectivity and additions of (Z)-**36** result in significantly higher relative diastereoselectivity. Addition of (E)-**36** is again unselective and, surprisingly, *syn* (relative) selective.

The intrinsic internal diastereoselection is again almost negligible (Table 7.6, entry 6). Thus, internal diastereoselectivity is primarily controlled by catalyst configuration. The match/mismatch effect in these aldol additions is not significant. This observation is in contrast to the strong 1,5-*anti* stereoinduction observed for boron enolate aldol additions [87]. The stereochemical model in this reaction should be analogous to that proposed for achiral enolate additions (Figure 7.11, Section 7.4).

Aldol additions of (Z)-**36** to a variety of aldehydes provide *syn* (relative) diastereomers selectively (Scheme 7.51). Excellent *syn* (relative) selectivity is obtained in the reaction with cinnamaldehyde. The internal selectivity is controlled by catalyst configuration, enabling selective preparation of both *syn* (relative) diastereomers.

Tab. 7.6
Catalyzed aldol additions of (Z)-**125**, (Z)-**36**, and (E)-**36** to benzaldehyde.

Entry	Enolate	Z/E	Catalyst	Yield, %	Relative dr (syn/anti)	Internal dr[a] (syn/anti)
1	(Z)-125	12/1	(R,R)-45	59	6/1	14/1
2	(Z)-125	12/1	(S,S)-45	60	12/1	1/14
3	(Z)-36	16/1	(R,R)-45	84	30/1	16/1
4	(Z)-36	16/1	(S,S)-45	86	26/1	1/10
5	(E)-36	1/15	(S,S)-45	80	3/1	1/1
6	(Z)-36	30/1	72	83	29/1	1.4/1

[a] Ratio of major relative diastereomer.

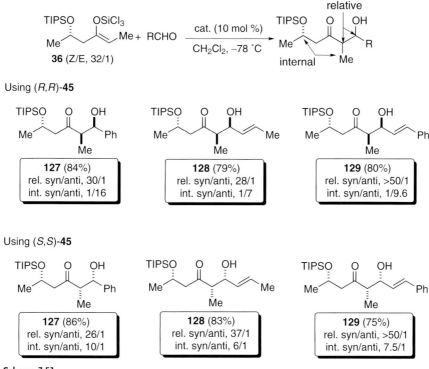

Scheme 7.51
Catalyzed aldol addition of (Z)-**36** to different aldehydes.

These aldol additions using three different classes of chiral trichlorosilyl enolates are interesting examples of double stereodifferentiating aldol additions. In the matched cases, high diastereoselectivity is obtained with the appropriate chiral phosphoramide catalyst. In aldol additions of lactate-derived enolates strong internal stereoinduction dominates the stereochemical course of the reaction. For the other two types of enolate, diastereoselection is primarily determined by catalyst configuration (external diastereoselection), enabling access to two diastereomers.

The effect of the α and β stereogenic centers described above would be very important in the construction of a stereodyad or triad in a predictable manner. The compatibility of common protecting groups with trichlorosilyl reagents is clearly established, and the in-situ generation of trichlorosilyl enolate from the corresponding TMS enol ether further enhances the synthetic utility of this process. In the addition of the substituted enolates, the *syn* (relative) diastereomers can be obtained with high selectivity starting with Z enolates, although, because E enolates do not undergo selective aldol addition, the corresponding *anti* (relative) diastereomers cannot be accessed by these methods.

7.6
Aldol Additions of Aldehyde-derived Enoxytrichlorosilanes

In the previous section the aldol addition of ketone-derived enolates was discussed and illustrated examples were used to document the synthetic utility of these reactions. This section deals with aldol additions of trichlorosilyl enolates derived from aldehydes. This type of aldol addition would be a particularly useful and practical approach to the construction of poly-propionate-derived natural products.

The stereoselective aldol addition of an aldehyde-derived enolate and an aldehyde remains a challenging topic [3a]. The difficulties associated with this process arise from complications inherent in the self-aldol reaction of aldehydes:

- polyaldolization resulting from multiple additions to the aldol products;
- Tischenko-type processes among the products; and
- oligomerization of the aldol products.

Only recently several approaches have been developed to address these problems [89]. Denmark et al. have achieved the first catalytic, enantioselective crossed-aldol reaction of aldehydes utilizing the Lewis-base-catalyzed aldol addition of trichlorosilyl enolates [49]. More recent developments have been made in direct, catalytic crossed-aldol reactions of aldehydes using proline, although an excess of one component is needed [90].

In the Lewis base catalysis approach, the immediate aldol adduct obtained by addition of an aldehyde-derived trichlorosilyl enolate is protected as its α-chlorosilyl ether, which is less prone to further additions. The concept is illustrated in the addition of heptanal-derived enolate (Z)-**37** to benzaldehyde in the presence of phosphoramide (S,S)-**45** (Scheme 7.52). Low-temperature NMR analysis of this adduct revealed it exists in the form of the α-chlorosilyl ether **130**. Because the aldolate occurs as a chelate complex, further reactions leading to a variety of side products are prevented.

The chlorosilyl ether intermediate can be hydrolyzed to obtain either its aldehyde or acetal (Scheme 7.53). When **130** is quenched in a mixture of aqueous THF and triethylamine (basic conditions), the corresponding aldehyde is obtained in excellent yield. When dry methanol is used for quench-

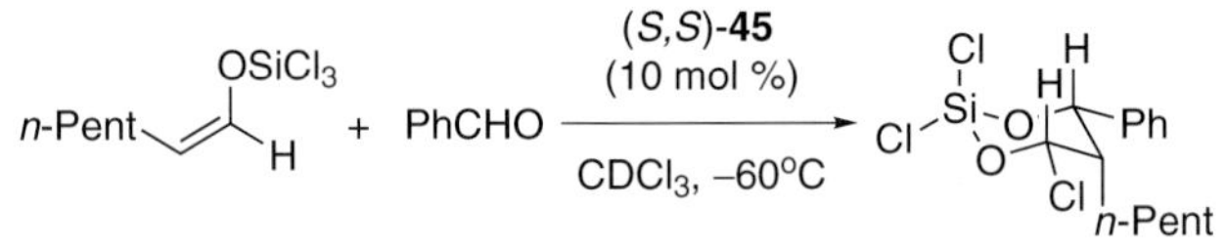

Scheme 7.52
Catalyzed aldol addition of (Z)-**37** to benzaldehyde.

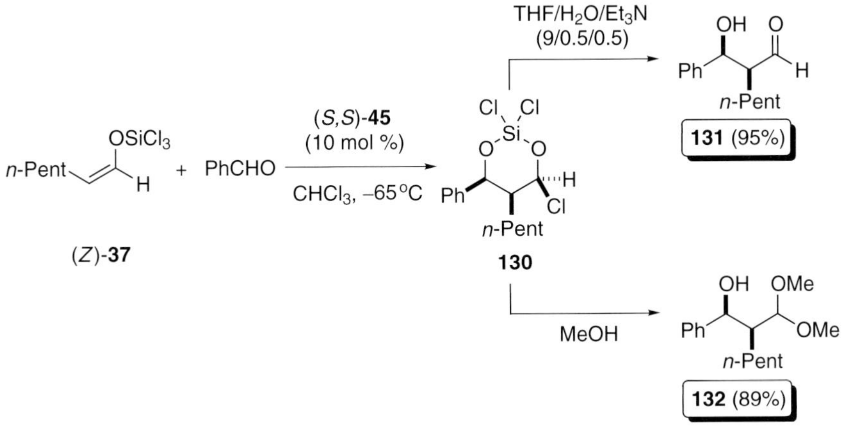

Scheme 7.53
Quenching of the chlorosilyl ether intermediate **130**.

ing, the chlorosilyl ether is converted to its dimethyl acetal **132** which can be isolated in excellent yield.

High diastereoselectivity is observed in additions of geometrically defined enolates (*Z*)- and (*E*)-**37** (Scheme 7.54). The diastereomeric composition of the aldol product strictly mirrors the *E*/*Z* ratio of the enolate, suggesting the reaction proceeds exclusively via a closed transition structure. From the correlation of *Z* enolate with *syn* diastereomer and *E* enolate with *anti* diastereomer, a chair-like transition structure can be inferred.

Although nearly perfect correlation is achieved between the *E*/*Z* ratio of the enolate and the *syn*/*anti* ratio of the aldol product, the enantiose-

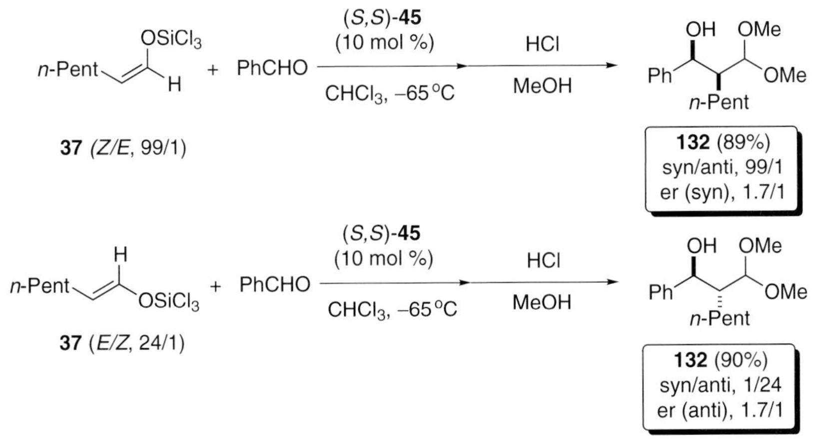

Scheme 7.54
Dependence of enolate geometry of aldol additions of **37**.

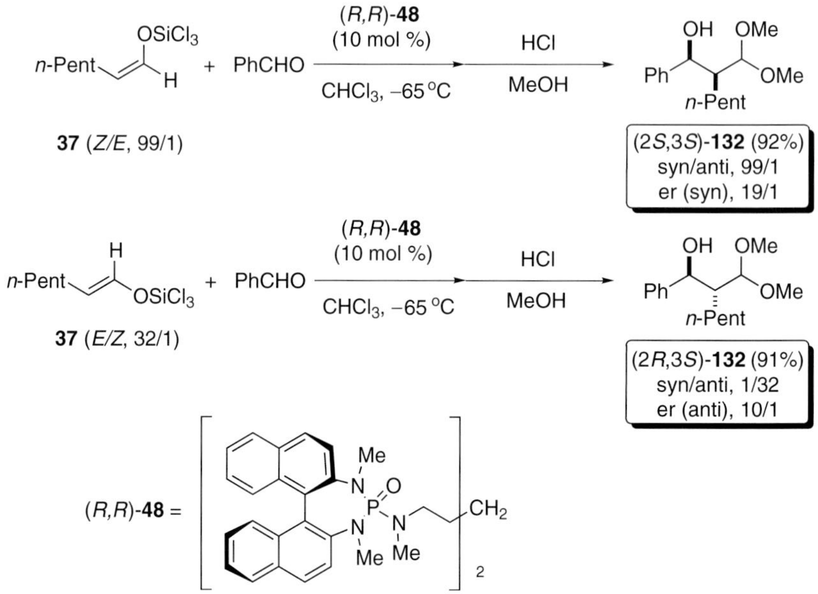

Scheme 7.55
Aldol addition of **37** to benzaldehyde catalyzed by (R,R)-**48**.

lectivity obtained by the use of **45** is rather poor. Despite an extensive catalyst survey, no significant improvement was achieved with a wide variety of monophosphoramides. A significant improvement in the enantioselectivity is achieved by using dimeric phosphoramides. Among these catalysts the binaphthyldiamine-derived dimer **48** affords the highest enantioselectivity for this transformation (Scheme 7.55).

Additions of (Z)-**12** to a variety of aldehydes in the presence of only 5 mol% (R,R)-**48** provide the corresponding aldol products in excellent yield and with exclusive *syn* selectivity (Scheme 7.56). Enantioselectivity varies significantly, good selectivity being observed for aromatic aldehydes only. There is no obvious correlation between aldehyde structure and enantioselectivity. It seems that the asymmetric induction provided by the catalyst (R,R)-**48** is most effectively transferred for benzaldehyde-like acceptors, and any structural change leads to erosion of enantioselectivity.

Aliphatic aldehydes can also be used for aldol addition of (Z)-**12**. Additions to aliphatic aldehydes are, however, very slow at −65 °C, so increased temperature (−20 °C) and longer reaction times are required for complete reaction. The absolute configuration of *syn*-**133** was assigned by conversion to the corresponding methyl ester, which was unambiguously assigned as the (2S,3S) isomer [91].

The corresponding *E* enolate also reacts with a variety of aldehydes to give *anti* β-hydroxy acetals (Scheme 7.57). This high relative diastereoselectivity

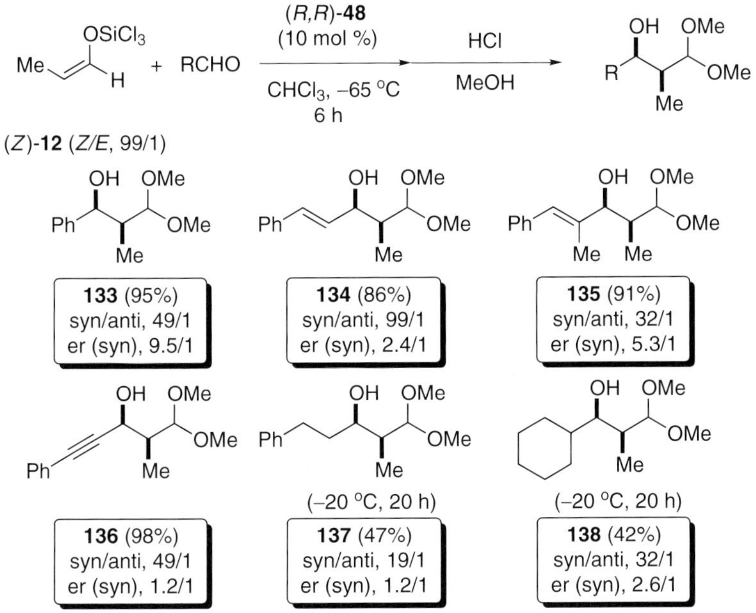

Scheme 7.56
Catalyzed addition of (Z)-**12** to different aldehydes.

contrast with the poor diastereoselectivity obtained from addition of *E* enolates derived from acyclic ketones. These trends in enantioselectivity are also different from those observed for additions of (*Z*)-**12**. The highest enantioselectivity is obtained for addition to α-methylcinnamaldehyde whereas addition to benzaldehyde provides only modest enantioselectivity. Markedly higher yields are obtained for addition of (*E*)-**12** to aliphatic aldehydes than for addition of (*Z*)-**12**, indicating that (*E*)-**12** is more reactive than (*Z*)-**12**. The absolute configuration of *anti*-**133** was established by chemically by correlation with the TBS-protected aldehyde, which has been unambiguously assigned as the (2*R*,3*S*) isomer [92].

On the basis of the individual effects of mono-substitution at the *Z* and *E* position it was envisaged that higher selectivity might be achieved by employing a disubstituted enolate. Aldol addition to benzaldehyde of α-disubstituted trichlorosilyl enolate derived from isobutyraldehyde has been examined using 10 mol% of (*R*,*R*)-**48** (Scheme 7.58) [93].

The enantiomer ratio of **140** is surprisingly low when compared with the results obtained from addition of (*E*)- and (*Z*)-**12**. For electron-rich aromatic aldehydes and electron-deficient aromatic aldehydes, however, enantioselectivity improves significantly (Scheme 7.59). Important mechanistic insights have been obtained from these observations [93]. It has been suggested that the divergence of enantioselectivity is because of the two different factors determining enantioselectivity for electron-rich and electron-

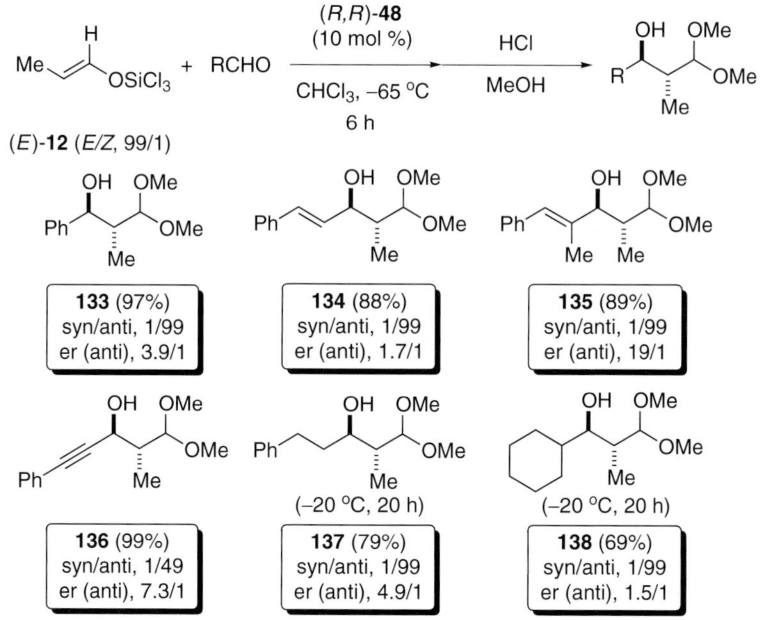

Scheme 7.57
Catalyzed addition of (*E*)-**12** to different aldehydes.

deficient aldehydes. For electron-poor aldehydes the event determining the stereochemistry is most probably the aldehyde binding process. For electron rich aldehydes, on the other hand, the stereocontrolling step is the aldolization. In both mechanistic extremes, high selectivity can be achieved. The different electronic nature of the aldehydes not only affects enantioselectivity but also reactivity.

Additions of **139** to a variety of aldehydes result in modest to good enantioselectivity, albeit with no distinct trend (Scheme 7.60). Additions to aliphatic aldehydes also proceed with good yields and moderate selectivity although elevated temperatures and long reaction times are required.

Problems associated with crossed-aldol reactions are successfully overcome by the Lewis-base-catalysis approach, and catalytic, enantioselective

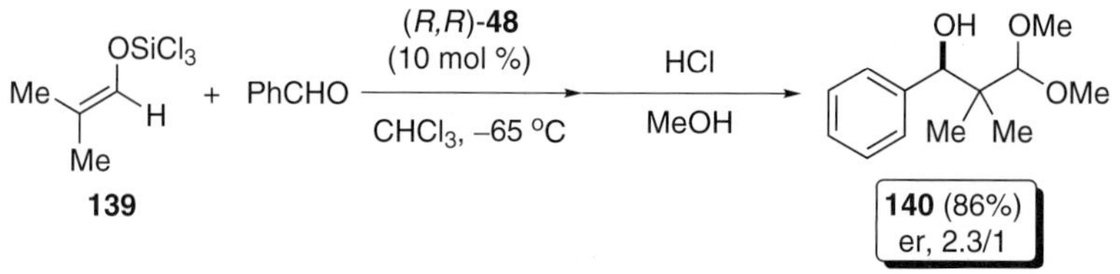

Scheme 7.58
Catalyzed aldol addition of **139** to benzaldehyde.

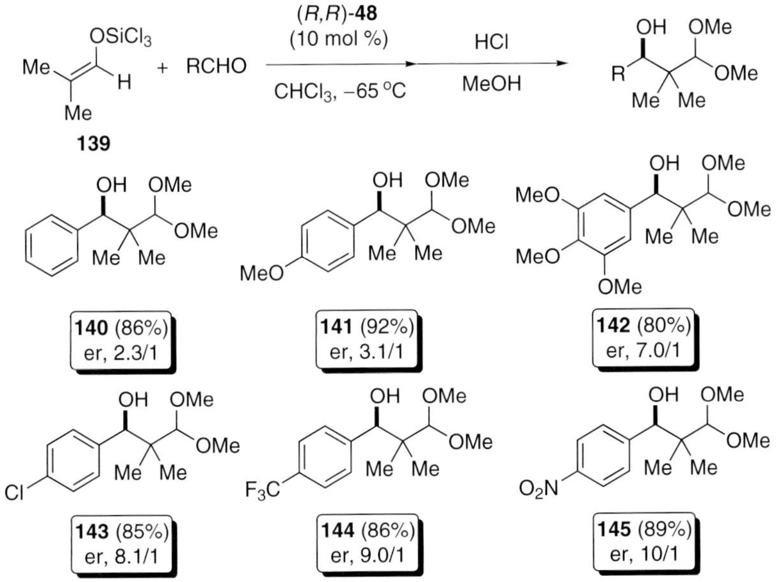

Scheme 7.59
Catalyzed addition of **139** to substituted benzaldehydes.

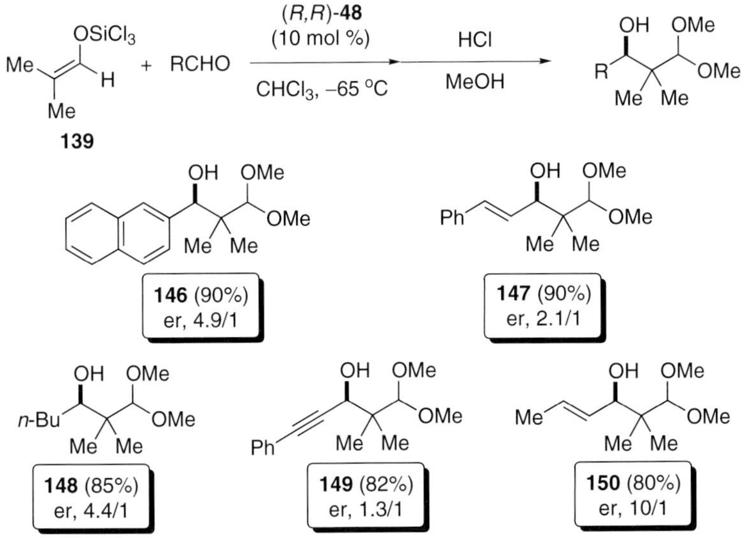

Scheme 7.60
Catalyzed aldol addition of **139** to a variety of aldehydes.

crossed-aldol reactions of aldehydes have been achieved by use of dimeric phosphoramide **48**. High diastereoselectivity can be achieved under the conditions described above by using geometrically defined trichlorosilyl enolates. When the chiral bisphosphoramide **48** is used a variety of crossed-aldol products are obtained with moderate to good enantioselectivity. The immediate aldol adduct can be recovered as the aldehyde or the acetal, depending on the quenching conditions. Thus the method discussed above will be extremely useful in enantioselective construction of a polypropionate chain.

7.7
Aldol Addition of Trichlorosilyl Ketene Acetal to Aldehydes and Ketones

For trichlorosilyl ketene acetals, enhanced nucleophilicity is expected, because of the additional oxygen substituent compared with ketone-derived enolates [32]. Trichlorosilyl ketene acetal **10** is, indeed, an extremely reactive nucleophile and reactions with a variety of aldehydes occur even at −80 °C. Aromatic, conjugated and aliphatic aldehydes all afford excellent yields of the aldol products within 30 min (Scheme 7.61). The compatibility with enolizable and sterically demanding aldehydes attests to the generality of this aldol addition.

Several chiral phosphoramides from different structural families have been examined for their capacity to induce enantioselectivity (Scheme 7.62). With 10 mol% of these phosphoramides aldol additions of **10** proceed rapidly at −78 °C to give good to excellent yields of the aldol products. Unfortunately, the enantioselectivity obtained in these reactions is poor. Modification of the reaction conditions did not significantly improve enan-

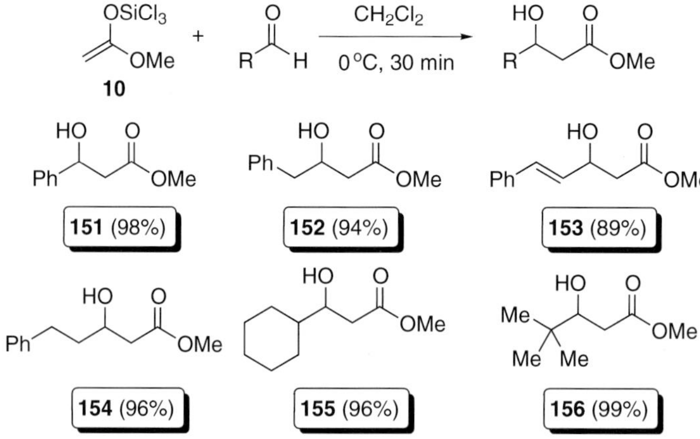

Scheme 7.61
Uncatalyzed addition of **10** to a variety of aldehydes.

Scheme 7.62
Catalyzed additions of **10** to benzaldehyde and pivaldehyde.

tioselectivity. The poor enantioselectivity observed can be explained by the competitive, rapid background reaction between **10** and aldehydes.

Although the addition of **10** to aldehydes gives modest enantioselectivity only, the extraordinary reactivity of **10** enables aldol addition to ketones. In the absence of Lewis basic promoters, ketene acetal **10** reacts with acetophenone sluggishly at 0 °C; this background reaction can, however, be completely suppressed by reducing the temperature to −50 °C. With 10 mol% HMPA the reaction gives almost quantitative yields of **157** [55]. In a survey of a variety of Lewis bases amine-N-oxides were found to be superior in promoting addition of **10** to acetophenone (Table 7.7) [55]. A variety of amine-N-oxides can promote this reaction and, among all the Lewis bases

Tab. 7.7
Survey of N-oxide promoters for addition of **10** to acetophenone.

Entry	Promoter	Temp, °C	Time, min	Conv., %
1	none	−50	240	0
2	Me_3NO	−78	240	10
3	Me_3NO	−20	50	76
4	NMO[a]	−78	70	25
5	quinuclidine N-oxide	−78	70	35
6	pyridine N-oxide	−78	70	37
7	pyridine N-oxide	−50	50	97
8	pyridine N-oxide[b]	rt	120	100

[a] N-methymorpholine-N-oxide. [b] 10 mol% of promoter was used.

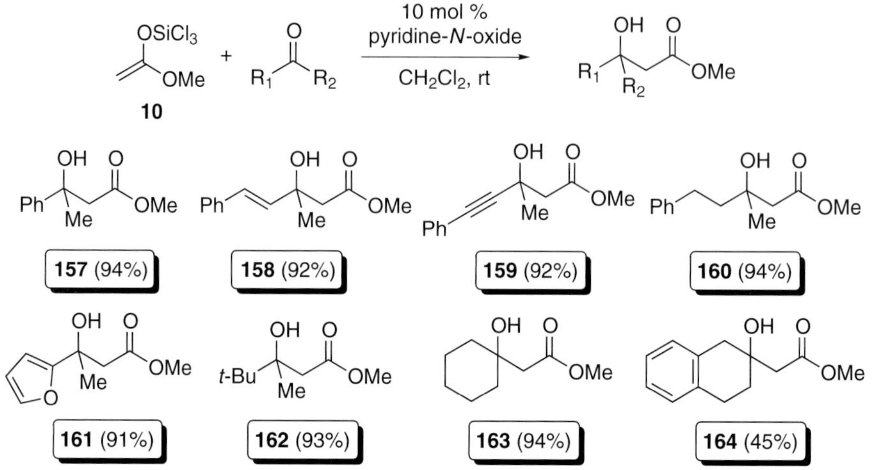

Scheme 7.63
Addition of **10** to a variety of ketones, catalyzed by pyridine-*N*-oxide.

surveyed, pyridine-*N*-oxide resulted in the highest conversion in the aldol addition. With a catalytic amount of pyridine-*N*-oxide, complete conversion can be achieved within 2 h at room temperature.

With catalytic amounts of pyridine-*N*-oxide addition of **10** to a variety of ketones has been achieved (Scheme 7.63). Excellent yields are obtained for a wide range of substrates, including highly enolizable ketones. Aldol addition to 2-tetralone is the only instance in which the reaction does not produce the expected aldol product in high yield.

To provide enantiomerically enriched aldol products the use of structurally diverse chiral N-oxides has been studied (Scheme 7.64). In this process enantioselection is clearly enhanced by use of dimeric N-oxides with 6,6′-stereogenic centers. Among these, the highest enantioselectivity is obtained by use of bis-pyridine-derived *P*-(*R*,*R*)-**46**. Interestingly, the *M*-(*R*,*R*)-**46** is equally capable of catalyzing the aldol addition, although this reaction affords the enantiomeric product with slightly attenuated enantioselectivity.

The generality of this catalyst system has been demonstrated in additions of **10** to a variety of ketones (Scheme 7.65). The enantiomeric ratio of the product ranged from modest to good, depending on the ketone structure. The crucial factor in obtaining high selectivity seems to be the size differential between the two substituents on the ketones.

Catalytic, asymmetric aldol additions to ketones have been achieved by means of the extraordinary reactivity of trichlorosilyl ketene acetal combined with Lewis-base-catalysis. The axially chiral bipyridine-*N*-oxide bearing stereogenic centers at the 6,6′-positions has excellent catalytic properties and results in synthetically useful enantioselectivity. This process provides access to enantiomerically enriched tertiary alcohols catalytically. Enantio-

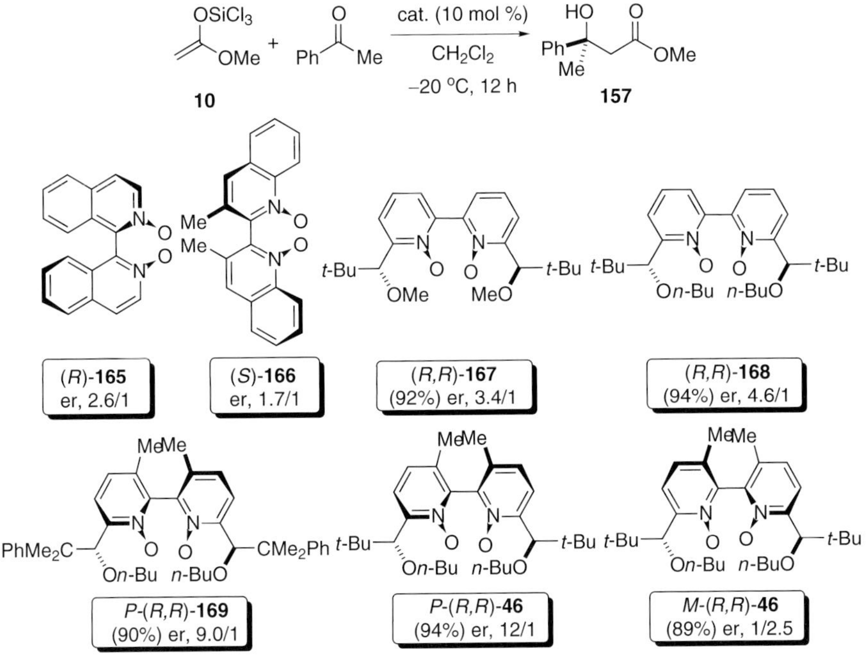

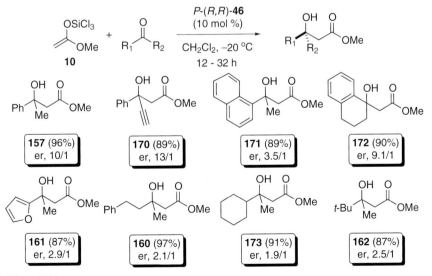

Scheme 7.64
Catalyst survey for addition of **10** to acetophenone.

Scheme 7.65
Addition of **10** to different ketones catalyzed by *P*-(*R*,*R*)-**46**.

selectivity, however, is not consistently high for different substrates, so catalyst optimization is still needed if this reaction is to be truly practical.

7.8
Lewis Base Activation of Lewis Acids – Aldol Additions of Silyl Enol Ethers to Aldehydes

The aldol reactions of enoxytrichlorosilanes described in preceding sections all involve the use of a chiral Lewis base (phosphoramide or N-oxide) to activate the nucleophile and provide the chiral environment for C–C bond-formation [53, 55, 94]. These reactions all proceed by a common mechanistic pathway that involves a cationic, hypercoordinate silicon as an organizational center for the reactants and catalyst (Section 7.9). The ability of certain Lewis bases to induce the ionization of silicon Lewis acids has intriguing potential for a new concept in Lewis-acid catalysis of organic reactions. The possibility of activating a weak Lewis acid, for example silicon tetrachloride, with a chiral Lewis base and using the resulting complex as a chiral Lewis acid for a variety of reactions has recently been demonstrated [54b].

The Lewis acidity of $SiCl_4$ is relatively weak compared with typical Lewis acids such as $TiCl_4$ or BF_3 [8]; in the presence of several different Lewis bases, however, a highly Lewis acidic silyl cation is produced (Scheme 7.66) [95]. Formation of a cationic silicon complex has been experimentally demonstrated by Bassindale and coworkers in heterolysis of halosilanes with Lewis bases [95a]. For example, ^{1}H, ^{13}C, and ^{29}Si NMR spectroscopic evidence suggested the formation of a cationic silicon complex from silyl halides and triflate in DMF. Although enhancement of Lewis acidity by Lewis bases is counter-intuitive, it can be explained by a set of empirical bond-length and charge-variation rules formulated by Gutmann [96]. Activation of silicon by ionization of a ligand has been proposed in several other systems [97]. The concept of Lewis base activation leads to an ideal opportunity for ligand-accelerated catalysis. Because the Lewis acid is active only when coordinated to the Lewis base, a stoichiometric amount of silicon tetrachloride can be used to assist rate and turnover [19].

The use of the [Lewis base–$SiCl_3$]$^+$ complex as a chiral Lewis acid was first demonstrated in the opening of *meso* epoxides to obtain enantioenriched chlorohydrins [98]. A more relevant application of this concept has

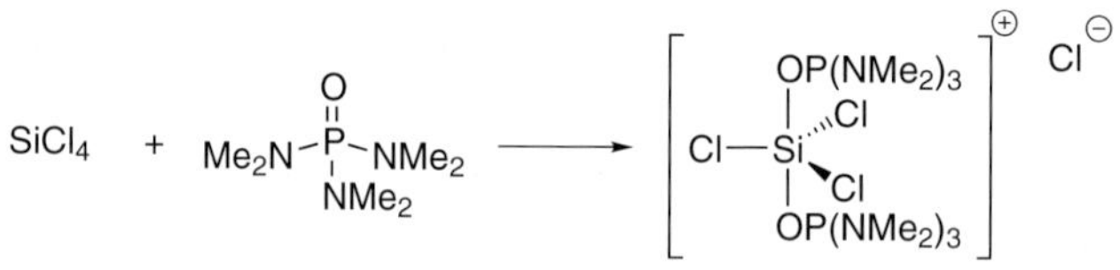

Scheme 7.66
Formation of HMPA–trichlorosilyl cation complex.

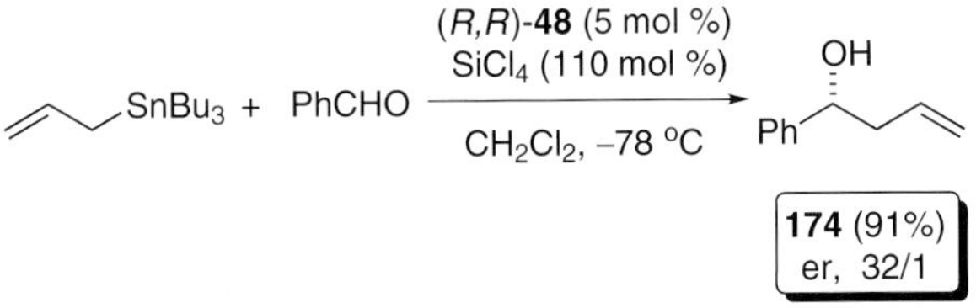

Scheme 7.67
Allylation of aldehydes using SiCl₄–bisphosphoramide complex.

been illustrated in asymmetric allylation of aldehydes (Scheme 7.67) [54b]. With allyltributyltin as an external nucleophile, the allylation of a variety of aldehydes proceeds in excellent yield; excellent enantioselectivity is obtained with the dimeric phosphoramide (R,R)-**48**.

This reaction system can be applied to aldol additions of silyl ketene acetals (Scheme 7.68) [50]. Although the aldol addition of ketene acetal **175** to aldehydes does not proceed in the absence of the Lewis base catalyst, with 5 mol% (R,R)-**48**, the aldol products are obtained in excellent yields within 15 min. This behavior is strikingly different from the addition of trichlorosilyl ketene acetal, which reacted spontaneously with aldehydes (Section 7.7). Not surprisingly, the enantioselectivity observed in these reactions is significantly better than that obtained from the reaction of trichlorosilyl ketene acetal with aldehydes, which suffers from competitive background reaction.

Excellent enantioselectivity was observed for most of the aldehydes surveyed. Sterically bulky aldehydes seem to afford lower enantioselectivity,

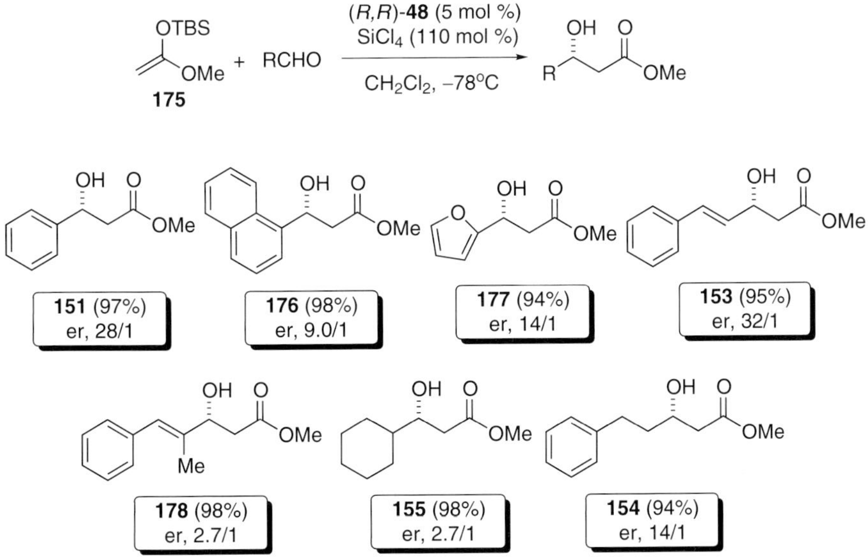

Scheme 7.68
Aldol addition of **175** to a variety of aldehydes catalyzed by SiCl₄-(R,R)-**48**.

and there is no significant electronic effect. Additions to aliphatic aldehydes are slow, but good yields of the aldol products can be obtained after 6 h with good enantioselectivity. These observations are remarkable considering that addition of trichlorosilyl nucleophiles to aliphatic aldehydes have been problematic. The absolute configuration of the benzaldehyde aldol product is *R*, which is consistent with the sense of asymmetric induction observed for the allylation.

Aldol additions with substituted ketene acetals introduce the issue of relative diastereoselection. Unlike the reactions with trichlorosilyl nucleophiles which involve closed transition structures, the mechanism of Lewis-acid-catalyzed aldol reactions usually involves an open transition structure. Under such conditions control of relative diastereoselection cannot be achieved simply by adjustment of enolate geometry.

Interestingly, the reactions of propanoate-derived ketene acetals with benzaldehyde produce the *anti* aldol products with high diastereoselectivity (Scheme 7.69). A survey of ketene acetal structures indicated that larger ester groups afford higher enantioselectivity. For the *t*-butyl propanoate-derived ketene acetal, enantioselectivity is significantly higher than for the addition of other ketene acetals. It is also important to note that the geometry of the ketene acetal does not affect the stereochemical outcome of the reaction. Starting from either the *E*- or *Z*-enriched ketene acetal **183**, the *anti* aldol product is obtained exclusively with excellent enantioselectivity. This suggests that these aldol additions do not proceed via a cyclic transition structure as alluded to above – an open transition structure can better account for the stereoconvergent aldol addition. This type of stereoconvergent *anti* aldol process is rare [16c] and the selectivity observed promises great synthetic utility for this aldol reaction.

The broad scope of this reaction has been demonstrated with a variety of aldehydes (Scheme 7.70). Aromatic and conjugated aldehydes afford high yields of the *anti* aldol product with excellent diastereoselectivity and mod-

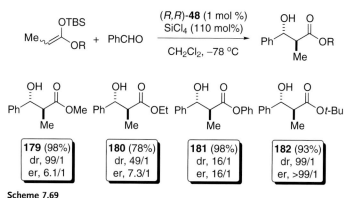

Scheme 7.69
Aldol addition of a variety of propionate-derived ketene acetals.

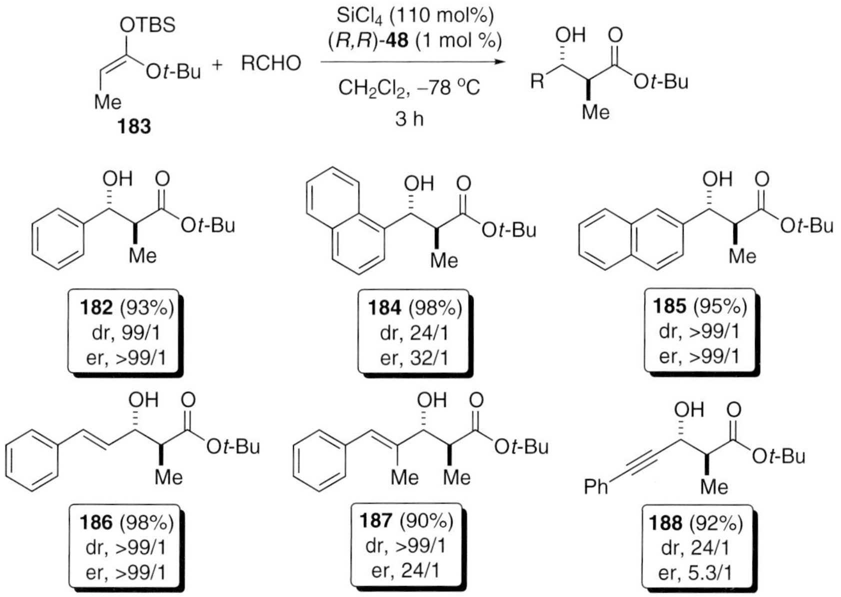

Scheme 7.70
Aldol addition of **183** to a variety of aldehydes.

est to excellent enantioselectivity. Unfortunately, aliphatic aldehydes do not react with this particular ketene acetal under these conditions.

The use of less sterically demanding ethyl ketene acetal **189** enables reactions with aliphatic aldehydes, however (Scheme 7.71). For example, combining **189** with hydrocinnamaldehyde affords **190** in 71% yield with good

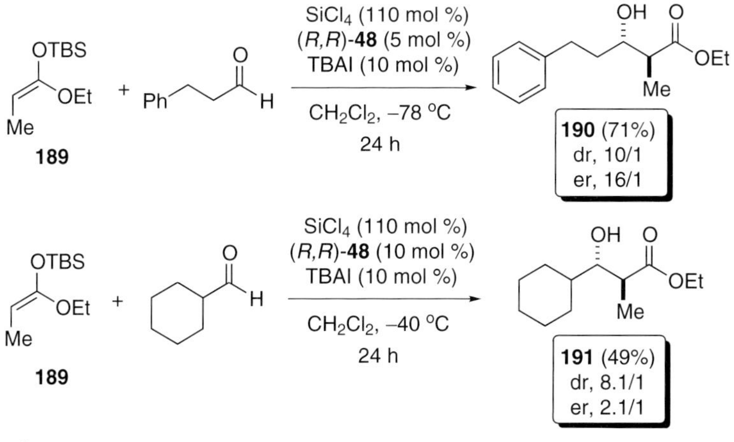

Scheme 7.71
Aldol addition of **189** to aliphatic aldehydes.

SiCl₄ (150 mol %)
(R,R)-48 (5 mol %)
TBAOTf (10 mol %)
i-Pr₂EtN (10 mol %)

OTMS
R⟍ + PhCHO → $R \overset{O}{\underset{}{}} \overset{OH}{\underset{}{}} Ph$

CH₂Cl₂, –78 °C
3 h

Me⟍ Ph	n-Bu⟍ Ph	i-Bu⟍ Ph	i-Pr⟍ Ph	Ph⟍ Ph
51 (97%) er, 49/1	**52** (99%) er, >99/1	**53** (98%) er, 99/1	**54** (95%) er, >99/1	**55** (98%) er, >99/1

Scheme 7.72
Catalyzed aldol additions of TMS enol ethers to benzaldehyde.

selectivity. The yield can be significantly improved by addition of tetrabutyl-ammonium iodide and extending the reaction time. In the addition to cyclohexanecarboxaldehyde a higher reaction temperature is also needed to achieve good conversion.

The same catalyst system can effect aldol additions of TMS enol ethers of ketones [99]. Additions of methyl ketone-derived TMS enol ethers to aldehydes proceed smoothly to afford the corresponding aldol products in excellent yield and with excellent enantioselectivity (Scheme 7.72). The use of a catalytic amount of a tetraalkylammonium salt is important for achieving complete conversion in these reactions. These studies also revealed the compatibility of the reaction system with a small amount of diisopropyl-ethylamine to scavenge adventitious HCl present in silicon tetrachloride. This modification obviates distillation of silicon tetrachloride and makes the process more practical.

The generality of this aldol addition is illustrated by the addition of 2-hexanone-derived TMS enol ether **192** to a variety of aldehydes (Scheme 7.73). Aromatic aldehydes are among the best substrates in this reaction, affording both high yields and high enantioselectivity. Heteroaromatic aldehydes are also compatible, and good yields and excellent enantioselectivity can be obtained. Sterically encumbered aldehydes react more slowly and less selectively. This effect is most evident in the reaction with α-methyl-cinnamaldehyde. Unfortunately, aliphatic aldehydes are not reactive under these conditions.

A highly regioselective vinylogous aldol reaction has also been achieved by use of the SiCl₄–bisphosphoramide system [100]. Vinylogous ketene acetals have two nucleophilic sites, i.e. the C(2)- and C(4)-positions, and reaction at the C(2)-position is usually favored, owing to the higher electron density [101]. It has been a challenge to control the reactivity of these two sites to obtain the γ-aldol adduct selectively [102]. In the presence of

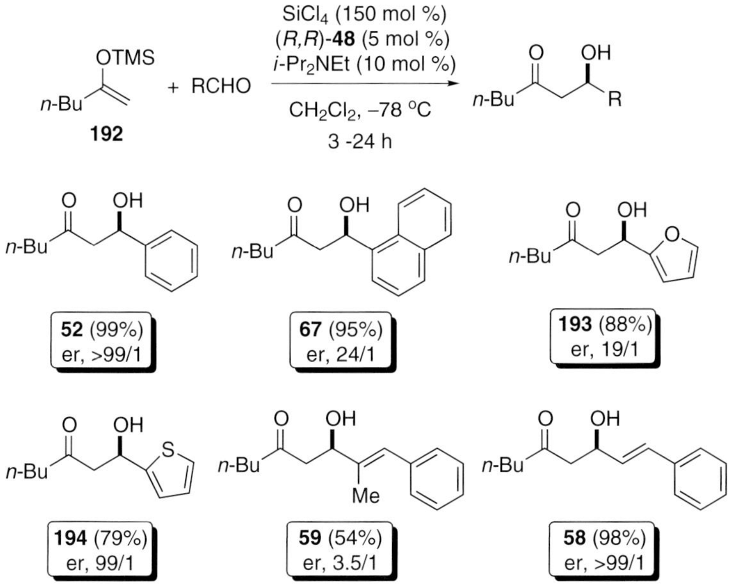

Scheme 7.73
Aldol addition of **192** to different aldehydes.

the bisphosphoramide (R,R)-**48**, crotonate-derived silyl ketene acetal **195** reacts with benzaldehyde to yield the γ-aldol adduct exclusively in good yield with excellent enantioselectivity (Scheme 7.74). The exclusive γ-selectivity is attributed to the steric differentiation between the α- and γ-positions (substituted compared with unsubstituted). In this catalyst system the reaction occurs preferentially at less sterically demanding site.

Under similar reaction conditions a variety of simple enoate-derived silyl ketene acetals undergo vinylogous aldol additions (Scheme 7.75). In the 2-pentenoate derived silyl ketene acetal a sterically bulky ester group is necessary for high regioselectivity. For example, reaction of the *t*-butyl ester-derived dienol ether yields the γ adduct **199** in good yield. The high re-

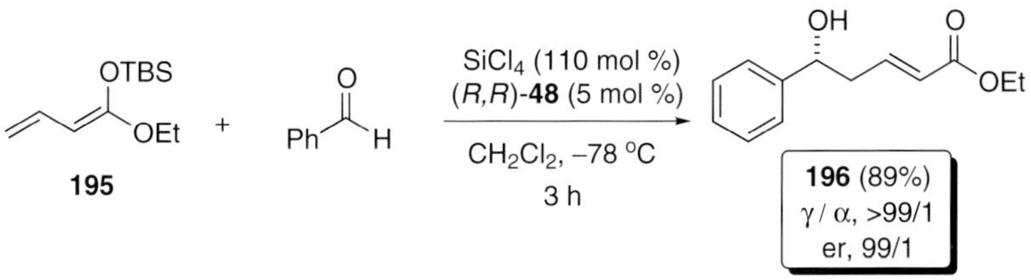

Scheme 7.74
Aldol addition of dienol silyl ether **195** to benzaldehyde.

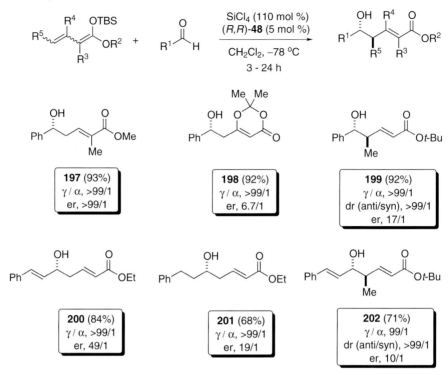

Scheme 7.75
Catalyzed aldol addition of a variety of dienol silyl ethers.

gioselectivity is complemented by high *anti* diastereoselectivity and excellent enantioselectivity. In terms of aldehyde scope, good yields and selectivity are obtained with aromatic and olefinic aldehydes, and even aliphatic aldehydes can be employed in this reaction, although longer reaction times are needed.

It is important to mention that the aldol products obtained by use of bis-phosphoramide (*R,R*)-**48** reveal the commonality of absolute configuration at the hydroxyl center (Figure 7.20). When (*R,R*)-**48** is used, nucleophiles attack the aldehyde *Re* face in **xxv**. Although stereochemical models need to be developed, the catalyst has created a highly defined environment for the aldehyde acceptor.

The bisphosphoramide–SiCl₄ complex has been successfully used as a chiral Lewis acid in highly efficient catalytic, enantioselective aldol additions of silyl ketene acetals and silyl enol ethers. Compared with aldol additions of trichlorosilyl reagents, these systems are superior in terms of preparation and handling of the nucleophiles. In particular, additions of propanoate-derived ketene acetals are one of the most stereoselective *anti* aldol additions reported to date.

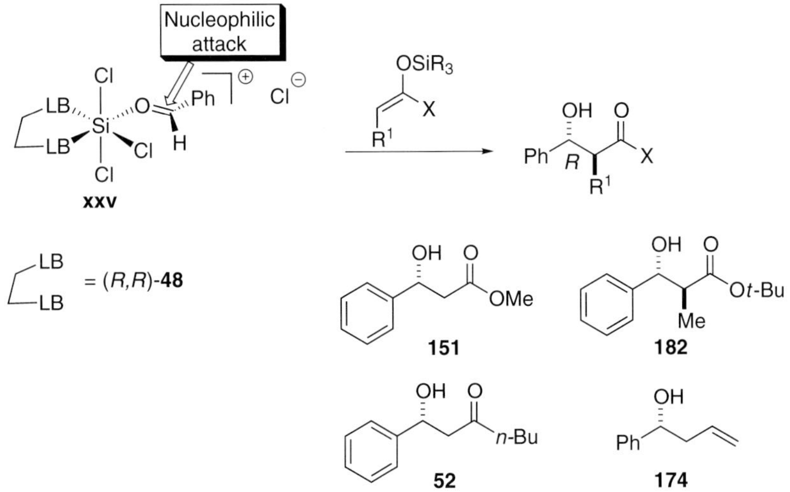

Fig. 7.20
Commonality of absolute configuration in a variety of aldol products.

7.9
Toward a Unified Mechanistic Scheme

Detailed discussion of the extensive kinetic, spectroscopic, and structural investigations that have provided the current mechanistic picture is beyond the scope of this chapter, the primary focus of which is preparative aspects of chiral Lewis base-catalyzed aldol reactions. Instead a summary of the important studies that have led to the current level of understanding will be presented, with the implications for catalyst design and reaction engineering.

As originally formulated, the foundation of Lewis base activation of the aldol addition (and subsequent stereoinduction) was based on hypothetical ternary assembly of enolate, aldehyde, and chiral catalyst in a hexacoordinate arrangement about the silicon atom (Figure 7.21) [103]. When catalysis was successfully demonstrated, the hypothesis seemed correct – i.e. that the rate acceleration arose from dual activation of the enol and the aldehyde in close proximity. Two important aspects of the reactions seemed at odds with this picture, however – rate and stereoselectivity were both difficult to rationalize. Although polarization of electron density away from the silicon atom was expected from the Gutmann analysis [96a], there was no basis for estimation of the magnitude of this effect. From analysis of simple molecular models it was, furthermore, not at all clear how single-point binding could provide the highly dissymmetric environment that induced such high facial selectivity.

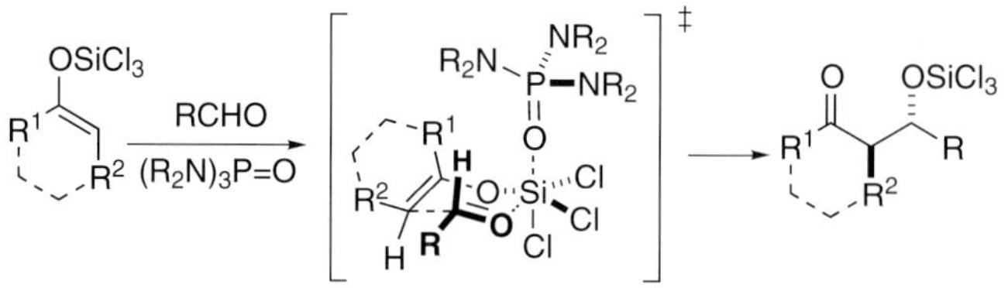

Fig. 7.21
Original hypothetical transition structure assembly.

7.9.1
Cationic Silicon Species and the Dual-pathway Hypothesis

The first experimental evidence against the simple mechanistic picture in Figure 7.21 was the observation that the diastereoselectivity of aldolization of cyclohexanone-derived enolate **20** with benzaldehyde depended on catalyst loading (Scheme 7.76). The appearance of *syn* isomers from *E*-configured enolates (at low catalyst loading) implied intervention of boat-like transition structures. Curiously, although the diastereomeric ratio changes dramatically the enantiomeric ratio of the *anti* isomer remains unchanged. This suggested that two independent pathways could be operating, one favoring the *anti* diastereomer (with high facial selectivity) and one favoring the *syn* isomer (with low facial selectivity).

Quantitative support of this hypothesis was obtained from several studies. First, the diastereoselectivity of reactions promoted by the achiral phosphoramide **203** (Figure 7.22) is dramatically dependent on catalyst loading. Figure 7.22 depicts graphically the change in *syn/anti* ratio from 1.3:1 at 200 mol% loading to 130:1 at 2 mol% loading. The excellent correlation of diastereoselectivity with inverse phosphoramide concentration provided quantitative support for the dual pathway hypothesis, namely, one phosphoramide leads to *syn* and two phosphoramides lead to *anti* [66].

The second source of quantitative evidence is the divergent behavior of chiral catalysts (*S,S*)-**45** and (*S,S*)-**81** in studies of the dependence of enantioselectivity on catalyst composition. In contrast with the highly *anti*-selective reactions promoted by (*S,S*)-**45**, diphenylphosphoramide catalyst

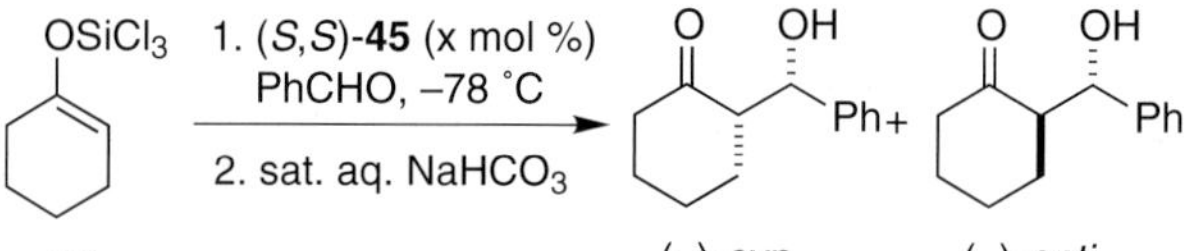

20
(−)-*syn* (−)-*anti*

10 mol % cat (94%)   syn/anti, 1/50 (er anti, 21/1)

0.5 mol % cat (53%)   syn/anti, 1/5 (er anti, 21/1)

Scheme 7.76
Catalyst loading-dependent diastereoselectivity.

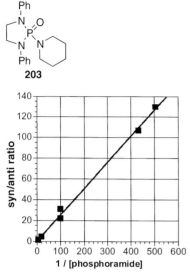

Fig. 7.22
Dependence on loading of selectivity with catalyst **203**.

(S,S)-**81** provided the *syn* aldol product in excellent diastereoselectivity (97:1), albeit with modest enantioselectivity (3.25:1 er) (Scheme 7.77) [66, 76].

With enantioselective catalysts now available for both *syn* and *anti* pathways, an important link between the steric demand of the catalyst and the resulting diastereoselectivity could be forged. According to the dual pathway hypothesis one (to *syn*) or two (to *anti*) catalyst molecules can be present in the stereochemistry determining transition structures, and that these different pathways are also stereochemically divergent. This hypothesis could be tested by making use of non-linear effects and asymmetric amplification as pioneered by Kagan [104]. The dependence of enantiomeric excess (ee) of the aldol products on the enantiomer composition of the cata-

Scheme 7.77
syn-Selective aldolization catalyzed by (S,S)-**81**.

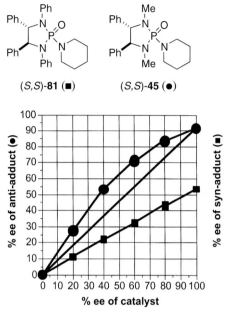

Fig. 7.23
Correlation of product and catalyst ee for (S,S)-**45** (●) and (S,S)-**81** (■).

lysts is illustrated in Figure 7.23. The linear relationship between catalyst ee and *syn*-adduct ee with phosphoramide (S,S)-**81** (Figure 7.23, ■) suggests this product arises from a transition structure involving only one chiral phosphoramide. In contrast, the obvious non-linear relationship between catalyst ee and *anti*-adduct ee with phosphoramide (S,S)-**45** (Figure 7.23, ●) suggests the participation of two phosphoramide molecules in the transition structure for aldolization [66].

The most compelling and direct evidence for the operation of dual pathways is provided by establishment of the order of the reaction in catalyst for (S,S)-**45** and (S,S)-**81** [66b]. The rate and sensitivity of these reactions required use of in-situ monitoring techniques such as ReactIR and rapid injection NMR (RINMR). First-order dependence on (S,S)-**81** ($R^2 = 1.000$) is observed for catalyzed aldol addition of **20** to benzaldehyde with typical catalyst loadings at $-35\ °C$. Importantly, the rate of reaction at very low catalyst loadings has pronounced curvature, indicative of a change in mechanism between the promoted and unpromoted pathways. RINMR analysis of the reaction catalyzed by (S,S)-**45** at $-80\ °C$ reveals aldol addition to have *second* order dependence on phosphoramide (plot of log k_{obs} against log [catalyst]; $m = 2.113$, $R^2 = 0.992$).

For Lewis-base-catalyzed aldol addition involving trichlorosilyl enolates the rate equations are rate = $k[\text{cat}][\text{enolate}][\text{aldehyde}]$ for catalyst (S,S)-**81** and rate = $k[\text{cat}]^2[\text{enolate}][\text{aldehyde}]$ for catalyst (S,S)-**45**. The experimentally determined reaction order is consistent with turnover-limiting com-

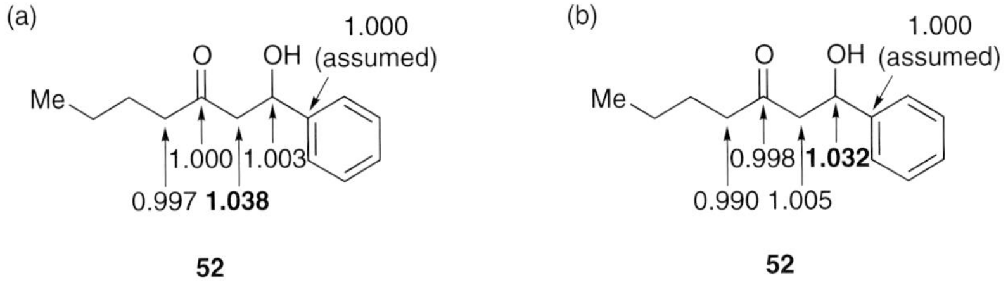

(a)

1.000
O   OH (assumed)

Me

1.000 | 1.003
0.997 **1.038**

52

(b)

1.000
O   OH (assumed)

Me

0.998 | **1.032**
0.990 1.005

52

Fig. 7.24
(a) ^{13}C KIEs (k_{12C}/k_{13C}) for a reaction taken to 5% conversion using limited aldehyde. (b) ^{13}C KIEs (k_{12C}/k_{13C}) for a reaction taken to 5% conversion using limited enol ether.

plexation or aldolization, and whereas Arrhenius activation data suggest complexation is rate-limiting they do not discount the possibility that aldolization is the turnover-limiting step. Natural abundance ^{13}C kinetic isotope effects (KIE) as pioneered by Singleton [105] provide a clear answer.

^{13}C NMR analysis of aldol product **52** from reaction of enolate **24** and benzaldehyde at 5% conversion afforded excellent results (Figure 7.24). If binding or any other pre-equilibrium process not involving the reactive centers were turnover limiting, no isotope enrichment would be expected in the aldol product. The presence of significant (1.038 and 1.032) [106] k_{12C}/k_{13C} kinetic isotope effects at the enolate carbon and the aldehyde carbonyl carbon clearly show, however, that rehybridization is occurring at both reactive centers in this transformation [107, 108]. These data, with results from the non-linear effect studies above clearly support the conclusion that the aldolization step is both stereochemistry-determining and turnover-limiting.

With evidence from a variety of sources that two phosphoramide molecules can be bound to the silicon atom of the enolate in the transition structure, formulating a picture of this assembly could be undertaken. It was reasonable to postulate that the aldehyde is also coordinated to silicon, because the stereochemical consequences of changing enolate geometry are strongly reflected in changing diastereoselectivity of the process. Thus, given the likelihood of a closed, silicon-centered transition structure, one of two possibilities arises:

- formation of a heptacoordinate silicon group; or
- ionization of a chloride, forming a cationic, hexacoordinate silicon moiety.

Support for the intermediacy of cationic silicon species is available from the effects of ionic additives on the rate and selectivity of the reaction [66a]. For reactions with catalyst (*S,S*)-**81** a clear trend emerges (Scheme 7.78). Addition of 1.2 equiv. tetrabutylammonium chloride causes marked deceleration and a diminution in enantioselectivity. Addition of 1.2 equiv. tetra-

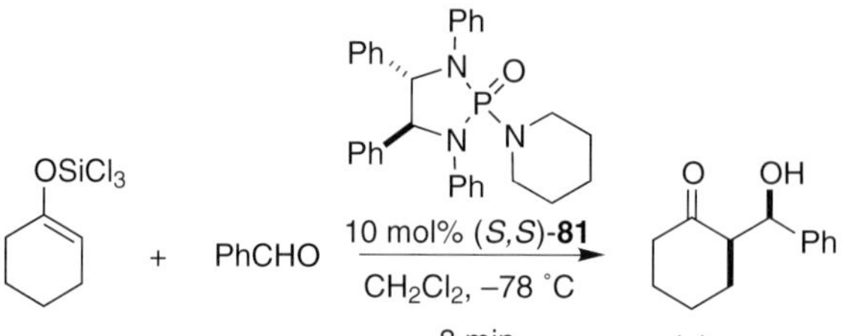

20 (+)-*syn*

no additive	44% conv.	er 3.26/1
1.2 equiv Bu_4N^+ Cl^-	8% conv.	er 1.70/1
1.2 equiv Bu_4N^+ TfO^-	92% conv.	er 3.44/1

Scheme 7.78
Effects of salts on the rate and selectivity of catalyzed aldolization.

butylammonium triflate results in moderate rate acceleration and an increase in the enantioselectivity of the overall process. The decrease in rate is consistent with a common-ion effect wherein ionization of chloride precedes the rate-determining step. The corresponding increase in rate and selectivity with tetrabutylammonium triflate, which increases the ionic strength of the medium, confirms the notion of ionization.

7.9.2
Unified Mechanistic Scheme

The available evidence from measurements of kinetics, additive effects, nonlinear studies, and stereochemical information supports a revised picture of the mechanism of phosphoramide-catalyzed aldol additions. As originally proposed, ternary association of enolate, aldehyde, and Lewis base was believed to be sufficient for activation and selectivity. Whereas unpromoted additions of trichlorosilyl enolates to aldehydes probably involve simple combination of the two reactants in a trigonal bipyramidal assembly, the catalyzed process is clearly much more complex (Figure 7.25). On binding the Lewis basic phosphoramide the trichlorosilyl enolate undergoes ionization of chloride. Depending on the size and concentration of the phosphoramide two scenarios are possible [109]. With a bulky phosphoramide, or in the limit of insufficient catalyst, aldehyde coordination and aldolization through a boat-like transition structure (with low facial selectivity) provides the *syn* aldol product (bottom pathway). Alternatively, with smaller phosphoramides or higher catalyst loading, a second molecule of catalyst can be bound to the cationic dichlorosilyl enolate to form an octahedral silicon cation [110]. On binding the aldehyde this intermediate undergoes aldolization through a chair-like transition structure organized around a hexacoordinate silicon atom (top pathway). This process occurs with a high level of facial selectivity, most probably because of the greater stereochemical influence of two chiral moieties in the assembly.

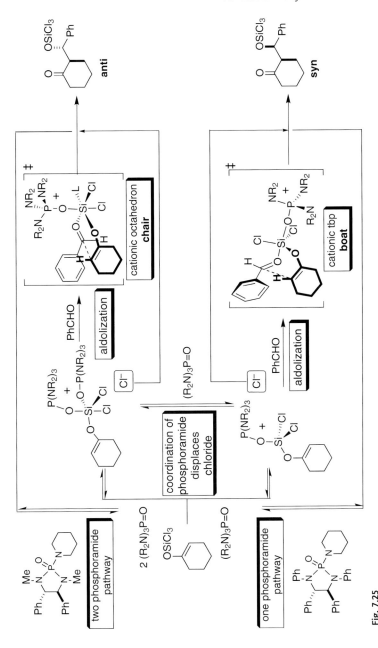

Fig. 7.25
Unified mechanistic scheme for phosphoramide-promoted aldolizations.

7.9.3
Structural Insights and Modifications

The revised mechanistic picture provides a clearer understanding of the remarkable change in diastereoselectivity with catalyst size and loading, and of the origin of rate enhancement. The reasons for the high enantioselectivity observed remain obscure, however. Insights into the stereochemical consequences of catalyst binding are provided by the solution and solid-state structures of chiral phosphoramide complexes of tin(IV) Lewis acids [70, 111].

The unified mechanistic scheme suggests a preference for 2:1 complexation with (S,S)-**45** and a preference for 1:1 complexation with (S,S)-**81**. Both scenarios are confirmed crystallographically. Single-crystal X-ray structural analysis of the 2:1 complex, ((S,S)-**45**)$_2$–SnCl$_4$ reveals interesting features (Figure 7.26):

- 2:1 complexation is confirmed,
- *cis* geometry of the complex is preferred,
- the piperidino nitrogen is planar and oriented orthogonal to the phospholidine ring

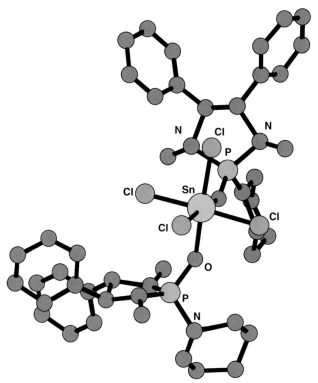

Fig. 7.26
X-ray crystal structure of ((S,S)-**45**)$_2$–SnCl$_4$.

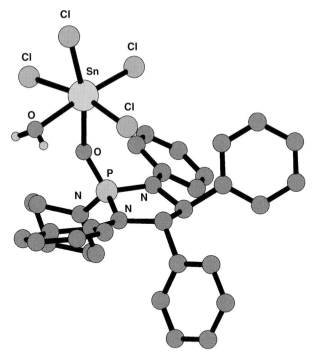

Fig. 7.27
X-ray crystal structure of (S,S)-**81**–SnCl₄–H₂O.

- the phospholidine nitrogen atoms are pyramidal, with the methyl groups disposed away from the stilbene phenyl groups, and
- the P–O–Sn unit is non-linear such that the tin moiety is oriented over the phospholidine ring.

^{119}Sn-solution NMR studies and analysis of $^1J_{P-Sn}$ coupling constants corroborate the observation of 2:1 complexes (hexacoordinate chemical shift regime) favoring the *cis* configuration.

Crystallization of (S,S)-**81** with SnCl₄ afforded a 1:1:1 complex of (S,S)-**81**–SnCl₄ with one molecule of water filling the sixth coordination site on the tin octahedron (Figure 7.27). This complexation stoichiometry is also obtained in solution, as verified by ^{119}Sn NMR studies that clearly show a doublet with ($^1J_{P-Sn}$) in the pentacoordinate chemical-shift region. The availability of an open coordination site in (S,S)-**81**–SnCl₄ suggested the possibility of incorporating a molecule of the substrate. Indeed, co-crystallization of (S,S)-**81** with SnCl₄ and benzaldehyde afforded a ternary complex, PhCHO–(S,S)-**81**–SnCl₄ (Figure 7.28). Both of these complexes had the same basic structural features as are found in the 2:1 complex ((S,S)-**45**)₂–SnCl₄.

Although these studies do indeed provide structural clues to the arrangement of groups around the central group 14 atom, there are still far too

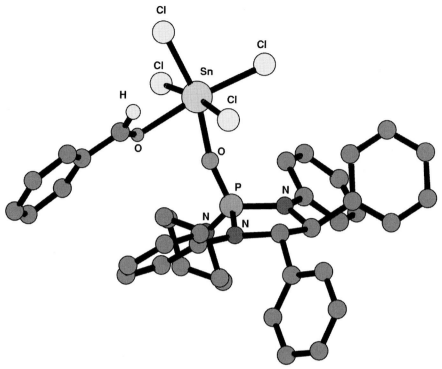

Fig. 7.28
X-ray crystal structure of PhCHO–(S,S)-**81**–SnCl₄.

many degrees of freedom to enable compelling depiction of the most favorable placement of reactive groups and alignment of combining faces. The structural insights available from these studies have, nevertheless, enabled important trends to emerge that facilitate the invention of new and better catalysts such as those that can enforce 2:1 binding by tethering and still accommodate the preferred arrangement of groups around the central atom.

Such tethered dimeric phosphoramides have been prepared from several different diamine subunits, for example those shown in Chart 7.3. In these cases the diamine subunits have been linked by aliphatic 1,*n*-diamines and have served admirably in a number reactions, for example catalytic enantioselective allylation with allylic trichlorosilanes [54c, 112] and activation of silicon tetrachloride for aldol and related reactions of trimethylsilyl enol ethers (Section 7.8). In the aldol addition of enoxytrichlorosilanes, the best results have been obtained from the use of the dimeric bis(phosphoramide) **48** for addition of aldehyde trichlorosilyl enolates (Section 7.6). A dimeric catalyst that promotes a highly enantioselective addition of trichlorosilyl enolates in general is still lacking [113].

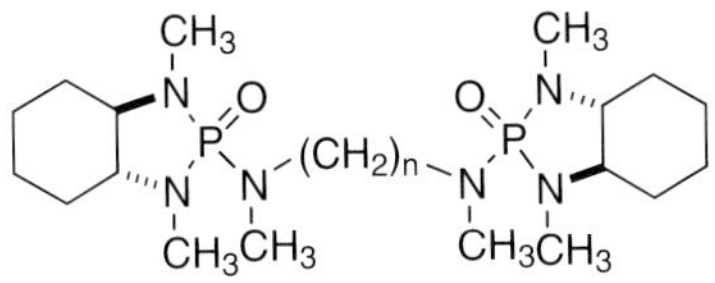

(R)-(l,l)-**204**: n = 2-6

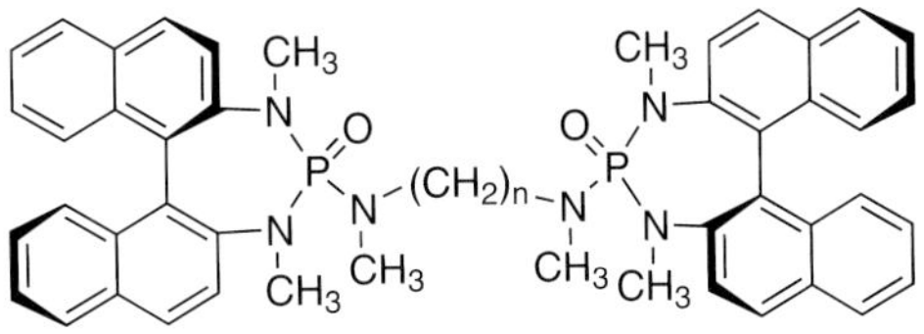

(R,R)-**205**: n = 3-6

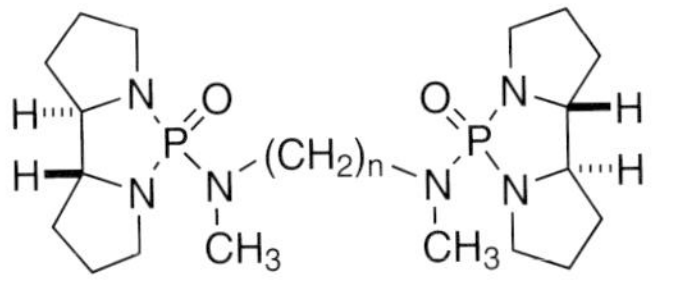

(R)-(l,l)-**206**: n = 4-6

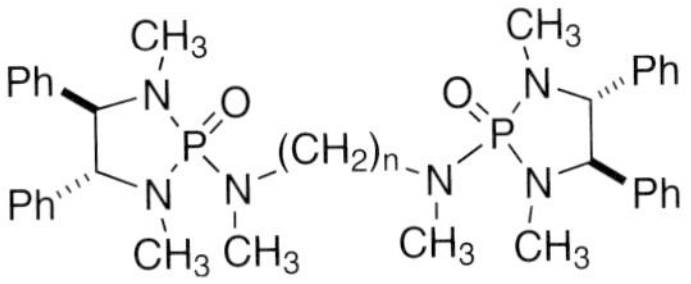

(R)-(l,l)-**207**, n = 4-8

Chart 7.3
Tethered bisphosphoramides.

7.10
Conclusions and Outlook

The phenomenon of chiral Lewis base catalysis has been successfully demonstrated for a wide variety of aldol addition reactions. This represents a fundamentally new class of reactions that embody a conceptually novel and preparatively useful addition to the growing number catalytic, enantioselective processes. Design criteria for the invention of this new variant have been formulated and documented experimentally.

Enoxytrichlorosilanes are a new class of aldolization reagents that are highly susceptible to catalysis by Lewis basic phosphoramides and N-oxides. A wide range of enolates have been prepared from simple cyclic and acyclic ketones, chiral ketones, esters, unsaturated esters, and aldehydes. Each of these classes of reagent has proven viable in aldol additions. The reactions are characterized by high yields, good functional group compatibility, excellent (and predictable) diastereoselectivity, and high enantioselectivity. There are, nevertheless, clearly identifiable limitations and shortcomings. For example, aliphatic aldehydes are a very important class of aldol partners that do not give generally acceptable results. In addition, the ability to generate substituted enolates with defined geometry (both E and Z) is still limited.

The concepts developed in this field are also applicable (and have been applied) to other reactions such as allylation [112], imine addition, Michael addition, and epoxide opening [98]. Development of chiral Lewis base activation of Lewis acids is, furthermore, a powerful extension of these concepts that has enabled a broader range of carbon–carbon bond-forming processes to be executed under the action of enantioselective catalysis (e.g. the Passerini reaction [114]). In addition, Lewis base catalysis should find application in activation of processes associated with other main group elements capable of structural changes similar to silicon.

Extensive kinetic and spectroscopic studies have revealed an unexpected mechanism involving the intermediacy of cationic silicon species. Elucidation of dual pathways proceeding via one or two-catalyst molecules has opened the door to the development of new dimeric catalysts that have proven useful in promoting faster, more selective reactions, but which have yet to find application in the aldol process specifically.

The synergistic evolution of synthetic utility and mechanistic understanding illustrates the fruitful interplay of synthesis, reactivity, and structure. These central activities constitute a chemical evergreen that will continue to address the challenges of the invention and development of new catalytic processes for years to come.

7.11
Representative Procedures

7.11.1
Preparation of Enoxytrichlorosilanes

Transition Metal Catalyzed *trans* Silylation (Section 7.2, Scheme 7.11) – Preparation of Trichloro[(1-butylethenyl)oxy]silane (24). Silicon tetrachloride (9.18 mL, 80.0 mmol, 2.0 equiv.) was added quickly to a suspension of Hg(OAc)$_2$ (127 mg, 0.40 mmol, 0.01 equiv.) in CH$_2$Cl$_2$ (40 mL). During the addition the mercury salt dissolved. Trimethyl[(1-butylethenyl)oxy]silane (6.89 g, 40.0 mmol) was then added to the solution dropwise over 10 min and the solution was stirred at room temperature for an additional 50 min. During this time the reaction mixture became somewhat cloudy once again. Removal of a sample and ^{1}H NMR analysis indicated the reaction was complete. The mixture was concentrated at reduced pressure (100 mmHg) and the resulting oil was distilled twice through a 7.5 cm Vigreux column to give 7.76 g (83%) of the trichlorosilyl enolate **24** as a clear colorless oil.

Metal Exchange via Lithium Enolate (Section 7.2, Scheme 7.11) – Preparation of (2Z,4S)-5-(*tert*-Butyl-dimethylsilyloxy)-4-methyl-3-trichlorosilyloxy-2-pentene ((Z)-35). (2Z,4S)-5-(*tert*-Butyl-dimethylsilyloxy)-4-methyl-3-trimethylsilyloxy-2-pentene (908 mg, 3.00 mmol) was dissolved in 6 mL ether at 0 °C. To this solution was slowly added MeLi (3.00 mL, 4.50 mmol, 1.5 equiv., 1.5 M in ether). The reaction mixture was stirred for 4.5 h at room temperature and then cooled to −78 °C. The reaction mixture was transferred to a cold solution of silicon tetrachloride (3.45 mL, 30.0 mmol, 10 equiv.) in 6 mL ether by use of a cannula. The reaction mixture was stirred at −78 °C for 1 h and then gradually warmed to room temp. The precipitate was left to settle at the bottom of the flask and the supernatant was transferred to another flask by means of a cannula. The volatile compounds were removed under reduced pressure and the residue was distilled by means of a Kugelrohr apparatus to afford 901 mg (2.48 mmol, 81%) (*Z*)-**35** as a clear colorless oil.

7.11.2
Aldol Addition of Ketone-derived Enoxytrichlorosilane

Aldol Addition of Achiral Enoxytrichlorosilane (Section 7.4, Scheme 7.26) – Preparation of (−)-*S*-1-Hydroxy-1-phenyl-3-heptanone (52). Trichlorosilyl enolate **24** (514 mg, 2.2 mmol, 1.1 equiv.) was added quickly to a cold (−74 °C) solution of (*S,S*)-**45** (37.1 mg, 0.1 mmol, 0.05 equiv.) in CH_2Cl_2 (2 mL). A solution of benzaldehyde (203 μL, 2.0 mmol) in CH_2Cl_2 (2 mL) was cooled to −78 °C and added quickly, via a short cannula, to the first solution. During the addition the temperature rose to −68 °C. The reaction mixture was stirred at −75 °C for 2 h then quickly poured into cold (0 °C) sat. aq. $NaHCO_3$ solution. The slurry obtained was stirred for 15 min. The two-phase mixture was filtered through Celite, the phases were separated, and the aqueous phase was extracted with CH_2Cl_2 (3 × 50 mL). The organic extracts were combined, dried over Na_2SO_4, filtered, and concentrated in vacuo. The crude product was purified by column chromatography (SiO_2, pentane–Et_2O, 4:1) to give 402.0 mg (98%) of (−)-**52** as a clear colorless oil.

Aldol Addition of in-situ-generated Enoxytrichlorosilane (Section 7.5, Scheme 7.44) – Preparation of (1*R*,4*S*)-1-Hydroxy-4-[((dimethyl)-(1,1-dimethyl)silyl)oxy]-1-phenyl-3-pentanone (*syn*-96). Trimethylsilyl enol ether **99** (548 mg, 2.0 mmol) was added dropwise over 2 min to a stirred solution of $SiCl_4$ (460 μL, 4.0 mmol, 2.0 equiv.) and $Hg(OAc)_2$ (3.1 mg, 0.010 mmol, 0.005 equiv.) in CH_2Cl_2 at room temperature. After complete addition the reaction mixture was stirred at room temperature for 1 h; volatile compounds were then removed under reduced pressure (0.3 mmHg) to give a cloudy residue. Dichloromethane (2.0 mL) was added and the mixture was cooled to −75 °C. A solution of (*R,R*)-**45** (37.0 mg, 0.1 mmol, 0.05 equiv., dried at 0.1 mmHg for 12 h) in CH_2Cl_2 was then added over 1 min via a cannula. A solution of benzaldehyde (203 μL, 2.0 mmol) in CH_2Cl_2 (1.0 mL) was then added over 1 min and the reaction mixture was stirred at −75 °C for 3 h. The reaction mixture was then rapidly poured into cold (0 °C) sat. aq. $NaHCO_3$ solution (15 mL) and the mixture was stirred for 15 min. The heterogeneous mixture was filtered through Celite, the organic phase was separated, and the aqueous phase was extracted with CH_2Cl_2 (3 × 50 mL). The organic extracts were combined, dried over Na_2SO_4, filtered, and concentrated to give a crude oil. Purification by column chromatography (SiO_2, hexane–EtOAc, 8:1) afforded 524.4 mg (85%) of a mixture of diastereomers **96** as a clear colorless oil. The diastereomeric ratio was determined by SFC analysis to be *syn/anti* 73:1.

Aldol Addition of Aldehyde-derived Enoxytrichlorosilane (Section 7.6, Scheme 7.55) – Preparation of (1*S*,2*S*)-3,3-Dimethoxy-2-pentyl-1-phenyl-1-propanol (*syn*-132). Trichlorosilyl enolate (*Z*)-**37** (496 mg, 2.0 mmol, 1.0 equiv.) was added to a cold (−78 °C) solution of the bisphosphoramide (*R*,*R*)-**48** (84 mg, 0.1 mmol, 0.05 equiv.) in CHCl$_3$–CH$_2$Cl$_2$, 4:1 (8 mL) and the mixture was stirred for 10 min. Freshly distilled benzaldehyde (0.205 mL, 2.0 mmol, 1.0 equiv.) was then added. After 6 h at −78 °C, MeOH (32 mL) was added and the mixture was stirred at that temperature for 45 min. The cold bath was removed and reaction mixture was left to warm to room temperature (total time 0.5 h), then was poured into cold (0 °C) sat. aq. NaHCO$_3$ solution and the mixture was stirred for 4 h. The reaction mixture was filtered through Celite and then washed with pentane–Et$_2$O, 1:1 (20 mL). The organic layer was separated and the aqueous layer was extracted once with pentane–Et$_2$O, 1:1 (20 mL). The combined extracts were dried over MgSO$_4$ and then concentrated in vacuo. Column chromatography (SiO$_2$, hexane–EtOAc, 85:15) then bulb-to-bulb distillation gave 491 mg (92%) *syn*-**132** as a clear, colorless, viscous liquid.

Aldol Addition of Trichlorosilyl Ketene Acetal (Section 7.7, Scheme 7.65) – Preparation of Methyl 3-Hydroxy-3-phenylbutanoate (157). Trichlorosilyl ketene acetal **10** (380 μL, 2.4 mmol, 1.2 equiv.) was added to a solution of acetophenone (240 μL, 2.0 mmol) and chiral bis-N-oxide (*P*)-(*R*,*R*)-**46** (101 mg, 0.20 mmol, 0.1 equiv.) in CH$_2$Cl$_2$ (10 mL) at −20 °C under nitrogen in a flame-dried, round-bottomed flask with magnetic stirrer. After stirring for 12 h at −20 °C the reaction mixture was transferred dropwise to a cold (0 °C) sat. aq. NaHCO$_3$ solution (20 mL) with vigorous stirring. The mixture was further stirred for 30 min at room temperature. The silicate precipitate was removed by filtration through Celite and the filtrate was extracted with CH$_2$Cl$_2$ (4 × 20 mL). The combined organic extracts were dried over MgSO$_4$ and then were concentrated under reduced pressure. The crude aldol product was separated from the catalyst by distillation and was further purified by silica gel chromatography. Analytically pure **157** (364 mg, 94%) was obtained as a colorless liquid after bulb-to-bulb distillation.

Aldol Addition of Propionate-derived Silyl Ketene Acetal (Section 7.8, Scheme 7.70) – Preparation of *tert*-Butyl (2*S*,3*R*)-3-Hydroxy-2-Methyl-3-Phenylpropanate (*anti*-182). A flame-dried, 10-mL, 2-neck flask containing a solution of bisphosphoramide (*R*,*R*)-**48** (8.4 mg, 0.01 mmol, 0.01 equiv.) in CH$_2$Cl$_2$ (5 mL) was cooled to −78 °C under nitrogen and benzaldehyde (102 μL, 1.0 mmol, 1.0 equiv.) was then added. Silicon tetrachloride (123 μL, 1.1 mmol, 1.1 equiv.) was added to the resulting solution and the reaction mixture was stirred at −78 °C for 5 min. (*E*)-1-[(*tert*-Butoxy)propenyl]-*tert*-butyldimethylsilane ((*E*)-**183**) (293 mg, 1.2 mmol, 1.2 equiv.) was then added dropwise to the reaction mixture over 5 min. The resulting mixture was stirred at −78 °C (bath temperature) for 3 h whereupon the cold reaction mixture was poured into 1:1 sat. aq. KF–1.0 M KH$_2$PO$_4$ solution (20 mL) with rapid stirring. This two-phase mixture was stirred vigorously for 1 h before filtration through Celite. The aqueous layer was washed with CH$_2$Cl$_2$ (3 × 50 mL) and the combined organic extracts were washed with brine (50 mL), dried over Na$_2$SO$_4$ (2 g), filtered, and the filtrate was concentrated in vacuo. The residue was purified by Kugelrohr distillation to yield 217 mg (93%) (2*S*,3*R*)-**182** as a colorless oil.

References

1 (a) EVANS, D. A.; NELSON, J. V.; TABER, T. R. In *Topics in Stereochemistry*; ELIEL, E. L., WILEN, S. H., Eds.; Wiley Interscience: New York, 1982; Vol. 13; pp 1–115. (b) HEATHCOCK, C. H. In *Comprehensive Carbanion Chemistry*; BUNCEL, E., DURST, T., Eds.; Elsevier: New York, 1984; Vol. 5B, p 177–237. (c) MUKAIYAMA, T. *Org. React.* **1982**, *28*, 203–331. (d) BRAUN, M. *Angew. Chem., Int. Ed. Engl.* **1987**, *26*, 24–37. (e) MUKAIYAMA, T.; KOBAYASHI, S. *Org. React.* **1994**, *46*, 1–103. (f) NORCROSS, R.; PATERSON, I. *Chem. Rev.* **1995**, *95*, 2041–2114. (g) CARREIRA, E. M. In *Modern Carbonyl Chemistry*; OTERA, J., Ed.; Wiley–VCH: Weinheim, 2000; Chapter 8. (h) PATERSON, I.; COWDEN, C. J.; WALLACE, D. J. In *Modern Carbonyl Chemistry*; OTERA, J., Ed.; Wiley–VCH: Weinheim, 2000; Chapter 9. (i) CARREIRA, E. M. In *Comprehensive Asymmetric Catalysis, Vol. I–III*; JACOBSEN, E. N.; PFALTZ, A.; YAMAMOTO, H., Eds.; Springer: Heidelberg, 1999. Chapter 29. (j) CARREIRA, E. M. In *Catalytic Asymmetric Synthesis, 2nd Ed.*; OJIMA, I., Ed.; Wiley–VCH: Weinheim, 2000; Chapter 8B2. (k) BRAUN, M. In *Stereoselective Synthesis, Methods of Organic Chemistry (Houben–Weyl)*; Edition E21; HELMCHEN, G.; HOFFMAN, R.; MULZER, J.; SCHAUMANN, E., Eds.; Thieme: Stuttgart, 1996; Vol. 3; pp 1603–1735.

2 (a) MASAMUNE, S.; CHOY, W.; PETERSON, J. S.; SITA, L. R. *Angew. Chem., Int. End. Engl.* **1985**, *24*, 1–30. (b) HELMCHEN, G. In *Stereoselective Synthesis, Methods of Organic Chemistry (Houben–Weyl)*; Edition E21; HELMCHEN, G.; HOFFMAN, R.; MULZER, J.; SCHAUMANN, E., Eds.; Thieme: Stuttgart, 1996; Vol. 1; pp 56–63. (c) COWDEN, C. J.; PATERSON, I. *Org. React.* **1997**, *51*, 1–200.

3 (a) HEATHCOCK, C. H. In *Comprehensive Organic Synthesis, Vol. 2, Additions to C–X Bonds, Part 2*; HEATHCOCK, C. H., Ed.; Pergamon Press: Oxford, 1991; Chapter 1.6. (b) LYNCH, J. E.; VOLANTE, R. P.; WATTLEY, R. V.; SHINKAI, I. *Tetrahedron Lett.* **1987**, *28*, 1385–1388. (c) OPPOLZER, W. *Pure Appl. Chem.* **1988**, *60*, 39–48.

4 (a) KIM, B. M.; WILLIAMS, S. F.; MASAMUNE, S. In *Comprehensive Organic Synthesis: Additions to C–X π-Bonds Part 2*; HEATHCOCK, C. H., Ed. Pergamon Press: Oxford, 1991; Chapt. 1.7. (b) MAHRWALD, R. *Chem. Rev.* **1999**, *99*, 1095–1120. (c) REETZ, M. T. In *Organometallics in Synthesis*;

Schlosser, Ed.; John Wiley: New York, 1994; Chapter 3. (d) Paterson, I. In *Comprehensive Organic Synthesis: Additions to C–X π-Bonds Part 2*; Heathcock, C. H., Ed.; Pergamon Press: Oxford, 1991; Chapt. 1.9.

5 Corey, E. J.; Kim, S. S. *J. Am. Chem. Soc.* **1990**, *112*, 4976–4977.

6 Duthaler, R. O.; Hafner, A. *Chem. Rev.* **1992**, *92*, 807–832.

7 (a) Myers, A. G.; Widdowson, K. L.; Kukkola, P. J. *J. Am. Chem. Soc.* **1992**, *114*, 2765–2767. (b) Myers, A. G.; Kephart, S. E.; Chen, H. *J. Am. Chem. Soc.* **1992**, *114*, 7922–7923.

8 (a) *Lewis Acids in Organic Synthesis*; Yamamoto, H., Ed.; Wiley–VCH: Weinheim, 2001; Vols. 1 and 2. (b) Santelli, M.; Pons, J.-M. *Lewis Acids and Selectivity in Organic Synthesis*; CRC: Boca Raton, 1996.

9 Kobayashi, S.; Uchiro, H.; Shiina, I.; Mukaiyama, T. *Tetrahedron* **1993**, *49*, 1761–1772.

10 (a) Ishihara, K.; Maruyama, T.; Mouri, M.; Furuta, K.; Yamamoto, H. *Bull. Chem. Soc. Jpn.* **1993**, *66*, 3483–3491. (b) Ishihara, K.; Gao, Q.; Yamamoto, H. *J. Am. Chem. Soc.* **1993**, *115*, 10412–10413.

11 (a) Parmee, E. R.; Tempkin, O.; Masamune, S.; Abiko, A. *J. Am. Chem. Soc.* **1991**, *113*, 9365–9366. (b) Kiyooka, S.-I.; Kaneko, Y.; Komura, M.; Matsuo, H.; Nakano, M. *J. Org. Chem.* **1991**, *56*, 2276–2278. (c) Corey, E. J.; Cywin, C. L.; Roper, T. D. *Tetrahedron Lett.* **1992**, *33*, 6907–6910.

12 (a) Mikami, K.; Matsukawa, S. *J. Am. Chem. Soc.* **1994**, *116*, 4077–4078. (b) Sato, M.; Sunami, S.; Sugita, Y.; Kaneko, C. *Chem. Pharm. Bull. Jpn.* **1994**, *42*, 839–845. (c) Keck, G. E.; Krishnamurthy, D. *J. Am. Chem. Soc.* **1995**, *117*, 2363–2364.

13 (a) Carreira, E. M.; Singer, R. A.; Lee, W. *J. Am. Chem. Soc.* **1994**, *116*, 8837–8838. (b) Carreira, E. M.; Singer, R. A. *J. Am. Chem. Soc.* **1995**, *117*, 12360–12361.

14 Yanagisawa, A.; Matsumoto, Y.; Nakashima, H.; Asakawa, K.; Yamamoto, H. *J. Am. Chem. Soc.* **1997**, *119*, 9319–9320.

15 (a) Evans, D. A.; Murry, J. A.; Kozlowski, M. C. *J. Am. Chem. Soc.* **1996**, *118*, 5814–5815. (b) Evans, D. A.; Kozlowski, M. C. Murgey, C. S.; MacMillan, D. W. C. *J. Am. Chem. Soc.* **1997**, *119*, 7893–7894. (c) Johnson, J. S.; Evans, D. A. *Acc. Chem. Res.* **2000**, *33*, 325–335.

16 For examples of anti-selective Mukaiyama aldol additions see: For examples of primarily anti-selective catalytic enantioselective aldol additions, see: (a) Evans, D. A.; MacMillan, D. W. C.; Campos, K. R. *J. Am. Chem. Soc.* **1997**, *119*, 10859–10860. (b) Kobayashi, S.; Horibe, M.; Hachiya, I. *Tetrahedron Lett.* **1995**, *36*, 3173–3176. (c) Yamashita, Y.; Ishitani, H.; Shimizu, H.; Kobayashi, S. *J. Am. Chem. Soc.* **2002**, *124*, 3292–3302.

17 (a) Sodeoka, M.; Tokunoh, R.; Miyazaki, F.; Hagiwara, E.; Shibasaki, M. *Synlett* **1997**, 463–466. (b) Ref. 14. (c) Krueger, J.; Carreira, E. M. *J. Am. Chem. Soc.* **1998**, *120*, 837–838. (d) Wadamoto, M.; Ozasa, N.; Yanagisawa, A.; Yamamoto, H. *J. Org. Chem.* **2003**, *68*, 5593–5601.

18 (a) Yamada, Y. M. A.; Yoshikawa, N.; Sasai, M.; Shibasaki,

M. *Angew. Chem., Int. Ed. Engl.* **1997**, *36*, 1871–1873. (b) Trost, B. M.; Ito, H.; Silcoff, E. R. *J. Am. Chem. Soc.* **2001**, *123*, 3367–3368.

19 Berrisford, D. J.; Bolm, C.; Sharpless, K. B. *Angew. Chem., Int. Ed. Engl.* **1995**, *34*, 1059–1070.

20 (a) Reichardt, C. *Solvents and Solvent Effects in Organic Chemistry*, Second Edition; VCH: Weinheim, 1988; pp 17–27. (b) Bollinger, J.-C.; Faure, R.; Yvernault, T. *Can. J. Chem.* **1983**, *61*, 328–333. (c) Gritzner, G.; Hörzenberger, F. *J. Chem. Soc. Faraday Trans.* **1995**, *91*, 3843–3850. (d) Gritzner, G. *Zeitschrift für Physikalische Chemie Neue Folge* **1988**, *158*, 99–107. (e) Gritzner, G. *Journal of Molecular Liquids* **1997**, 487–500.

21 (a) Sakurai, H. *Synlett* **1989**, 1–8. (b) Sakurai, H. In *Proceedings of the 5th International Kyoto Conference on New Aspects of Organic Chemistry*; Yoshida, Z.-I., Ohshiro, Y., Eds.; Kodansha Press: Tokyo, Japan, 1992; pp 129–157 and references cited therein. (c) Kira, M.; Zhang, L.; Kabuto, C.; Sakurai, H. *Organometallics* **1996**, *15*, 5335–5341.

22 (a) Kobayashi, S.; Nishio, K. *Tetrahedron Lett.* **1993**, *34*, 3453–3456. (b) Kobayashi, S.; Nishio, K. *Synthesis* **1994**, 457–459. (c) Kobayashi, S.; Nishio, K. *J. Org. Chem.* **1994**, *59*, 6620–6628.

23 Denmark, S. E.; Coe, D. M.; Pratt, N. E.; Griedel, B. D. *J. Org. Chem.* **1994**, *59*, 6161–6163.

24 (a) Burlachenko, G. S.; Khasapov, B. N.; Petrovskaya, L. I.; Baukov, Y. I.; Lutsenko, I. F. *J. Gen. Chem. USSR (Engl. Transl.)* **1966**, *36*, 532–537. (b) Ponomarev, S. V.; Baukov, Y. I.; Dudukina, O. V.; Petrosyan, I. V.; Petrovskaya, L. I. *Zh. Obshch. Khim.* **1967**, *37*, 2204–2207. (c) Benkeser, R. A.; Smith, W. E. *J. Am. Chem. Soc.* **1968**, *90*, 5307–5309.

25 Bassindale, A. R.; Stout, T. *Tetrahedron Lett.* **1985**, *26*, 3403–3406.

26 Denmark, S. E.; Winter, S. B. D.; Su, X.; Wong, K.-T. *J. Am. Chem. Soc.* **1996**, *118*, 7404–7405.

27 Denmark, S. E.; Stavenger, R. A. *Acc. Chem. Res.* **2000**, *32*, 432–440.

28 Mekelburger, H. B.; Wilcox, C. S. "Formation of Enolates" in *Comprehensive Organic Synthesis, Vol. 2, Additions to C–X π-Bonds, Part 2*; Heathcock, C. H., Ed.; Pergamon Press: New York, 1991; pp 99–131.

29 Petrov, A. D.; Sadykh-Zade, S. J. *Dokl. Akad. Nauk. SSSR* **1958**, *121*, 119–122.

30 (a) Chan, T.-H. "Formation and Addition Reactions of Enol Ethers in *Comprehensive Organic Synthesis, Vol. 2, Additions to C–X π-Bonds, Part 2*; Heathcock, C. H., Ed.; Pergamon Press: New York, 1991; pp. 595–628. (b) Rasmussen, J. K. *Synthesis* **1977**, 91–110. (c) Brownbridge, P. *Synthesis* **1983**, *1*, 85–104. (d) Baukov, Yu. I.; Lutsenko, I. F. *Organometallic Chemistry Review A* **1970**, *6*, 355–445.

31 (a) Mukaiyama, T.; Banno, K.; Narasaka, K. *J. Am. Chem. Soc.* **1974**, *96*, 7503–7509. (b) Ref. 1c.

32 Mayr, H.; Kempf, B.; Ofial, A. R. *Acc. Chem. Res.* **2003**, *36*, 66–77.

33 For a comprehensive review see: NELSON, S. G. *Tetrahedron Asymm.* **1998**, *9*, 357–389.

34 For examples of silicon-functionalized silyl enol ethers see (a) WALKUP, R. D. *Tetrahedron Lett.* **1987**, *28*, 511–514. (b) WALKUP, R. D.; OBEYESEKERE, N. U. *J. Org. Chem.* **1988**, *53*, 920–923. (c) WALKUP, R. D.; OBEYESEKERE, N. U.; KANE, R. R. *Chem. Lett.* **1990**, 1055–1058. (d) KAYE, P. T.; LEARMONTH, R. A.; RAVINDRAN, S. S. *Synth. Commun.* **1993**, *23*, 437–444.

35 MORI, A.; KATO, T. *Synlett* **2002**, 1167–1169.

36 YANAGISAWA, A.; NAKATSUKA, Y.; ASAKAWA, K.; KAGEYAMA, H.; YAMAMOTO, H. *Synlett* **2001**, 69–72.

37 DENMARK, S. E.; STAVENGER, R. A.; WINTERM S. B. D.; WONG, K-T.; BARSANTI, P. A. *J. Org. Chem.* **1998**, *63*, 9517–9523.

38 BURLACHENKO, G. S.; KHASAPOV, B. N.; PETROVSKAYA, L. I.; BAUKOV, Yu. I.; LUTSENKO, I. F. *J. Gen. Chem. USSR (Engl. Transl.)* **1966**, *36*, 532–537.

39 Aldol addition can occur in aqueous solvent without Lewis acid catalyst: (a) LUBINEAU, A. *J. Org. Chem.* **1986**, *51*, 2142–2144. (b) LOH, T.-P.; FENG, L.-C.; WEI, L.-L. *Tetrahedron* **2000**, *56*, 7309–7312.

40 BENKESER, R. A.; SMITH, W. E. *J. Am. Chem. Soc.* **1968**, *90*, 5307–5309.

41 PANOMAREV, S. V.; BAUKOV, Yu. I.; DUDUKINA, O. V.; PETROSYAN, I. V.; PETROVSKAYA, L. I. *J. Gen. Chem. USSR (Engl. Transl.)* **1967**, *37*, 2092–2094.

42 PEREYRE, M.; BELLEGARDS, B.; MENDELSOHN, J.; VALADE, J. J. *Organomet. Chem.* **1968**, *11*, 97–110.

43 Formation of *C*-mercurioketone: (a) HOUSE, H. O.; AUERBACH, R. A.; GALL, M.; PEET, N. P. *J. Org. Chem.* **1973**, *38*, 514–522. (b) YAMAMOTO, Y.; MARUYAMA, K. *J. Am. Chem. Soc.* **1982**, *104*, 2323–2325. (c) BLUTHE, N. MALACRIA, M.; GORE, J. *Tetrahedron* **1984**, *40*, 3277–3284. (d) DROUIN, J.; BONAVENTURA, M.-A.; CONIA, J.-M. *J. Am. Chem. Soc.* **1985**, *107*, 1726–1729.

44 Formation of *C*-(trichlorostannyl)ketones: (a) NAKAMURA, E.; KUWAJIMA, I. *Chem. Lett.* **1983**, 59–62. (b) NAKAMURA, E.; KUWAJIMA, I. *Tetrahedron Lett.* **1983**, *24*, 3347–3350. (c) ANNUNZIATA, R.; BENAGLIA, M.; CINQUINI, M.; COZZI, F.; RAINMONDI, L. *Tetrahedron* **1994**, *50*, 5821–5828. (d) KUWAJIMA, I.; NAKAMURA, E. *Acc. Chem. Res.* **1985**, *18*, 181–187.

45 DENMARK, S. E.; PHAM, S. M. *J. Org. Chem.* **2003**, *68*, 5045–5055.

46 (a) STORK, G.; HUDRLIK, P. F. *J. Am. Chem. Soc.* **1968**, *90*, 4464–4465. (b) HOUSE, H. O.; TROST, B. M. *J. Org. Chem.* **1965**, *30*, 2502–2512.

47 EVANS, D. A.; NELSON, J. V.; VOGEL, E.; TABER, T. R. *J. Am. Chem. Soc.* **1981**, *103*, 3099–3111.

48 HALL, P. L.; GILCHRIST, J. H.; COLLUM, D. B. *J. Am. Chem. Soc.* **1991**, *113*, 9571–9574.

49 DENMARK, S. E.; GHOSH, S. K. *Angew. Chem. Int. Ed.* **2001**, *40*, 4759–4762.

50 DENMARK, S. E.; WYNN, T.; BEUTNER, G. L. *J. Am. Chem. Soc.* **2002**, *124*, 13405–13407.

51 (a) Baukov, Yu. I.; Lutsenko, I. F. *Moscow Univ. Chem. Bull. (Engl. Transl.)* **1970**, *25*, 72. (b) Ref. 26.

52 Denmark, S. E.; Su, X.; Nishigaichi, Y.; Coe, D. M.; Wong, K.-T.; Winter, S. B. D.; Choi, J. Y. *J. Org. Chem.* **1999**, *64*, 1958–1967.

53 Denmark, S. E.; Stavenger, R. A.; Wong, K.-T.; Su, X. *J. Am. Chem. Soc.* **1999**, *121*, 4982–4991.

54 (a) Denmark, S. E.; Fu, J. *J. Am. Chem. Soc.* **2001**, *123*, 9488–9489. (b) Denmark, S. E.; Wynn, T. *J. Am. Chem. Soc.* **2001**, *123*, 6199–6200. (c) Denmark, S. E.; Fu, J. *Chem. Commun.* **2003**, 167–170.

55 Denmark, S. E.; Fan, Y. *J. Am. Chem. Soc.* **2002**, *124*, 4233–4235.

56 (a) Iseki, K.; Kuroki, Y.; Takahashi, M.; Kishimoto, S.; Kobayashi, Y. *Tetrahedron* **1997**, *53*, 3513–3526. (b) Peyronel, J.-F.; Samuel, O.; Fiaud, J.-C. *J. Org. Chem.* **1987**, *52*, 5320–5325.

57 Alexakis, A.; Aujard, I.; Mangeney, P. *Synlett* **1998**, 873–874.

58 Holmes, R. R.; *Chem. Rev.* **1996**, *96*, 927–950.

59 Miyano, S.; Nawa, M.; Mori, A.; Hashimoto, H. *Bull. Chem. Soc. Jpn.* **1984**, *57*, 2171–2176.

60 Lesiak, T.; Seyda, K. *J. Prakt. Chem.* **1979**, *321*, 161–163.

61 (a) Krajnik, P.; Ferguson, R. R.; Crabtree, R. H. *New. J. Chem.* **1993**, *17*, 559–566. (b) Denmark, S. E.; Fu, J.; Lawler, M. J. *Org. Synth.* **2004**, *81*, in press.

62 (a) Nakajima, M.; Saito, M.; Shiro, M.; Hashimoto, S. *J. Am. Chem. Soc.* **1998**, *120*, 6419–6420. (b) Tao, B.; Lo, M.-C.; Fu, G. C. *J. Am. Chem. Soc.* **2001**, *123*, 353–354.

63 Bolm, C.; Ewald, M.; Felder, M.; Schlingloff, G. *Chem. Ber.* **1992**, *125*, 453–458.

64 For solvent-assisted reactions of alkylsilyl enol ethers under mild conditions, see: (a) Lubineau, A.; Ange, J.; Queneau, Y. *Synthesis*, **1994**, 741–760. (b) Rajanbabu, T. V. *J. Org. Chem.* **1984**, *49*, 2083–2089.

65 Denmark, S. E.; Griedel, B. D.; Coe, D. M.; Schnute, M. E. *J. Am. Chem. Soc.* **1994**, *116*, 7026–7043.

66 (a) Denmark, S. E.; Su, X.; Nishigaich, Y. *J. Am. Chem. Soc.* **1998**, *120*, 12990–12991. (b) Denmark, S. E.; Pham, S. M. *Helv. Chim. Acta.* **2000**, *122*, 1846–1853.

67 Zimmerman, H. E.; Traxler, M. D. *J. Am. Chem. Soc.* **1957**, *79*, 1920–1923.

68 (a) Denmark, S. E.; Stavenger, R. A.; Wong, K.-T. *J. Org. Chem.* **1998**, *63*, 918–919. (b) Denmark, S. E.; Stavenger, R. A. *J. Am. Chem. Soc.* **2000**, *122*, 8837–8847.

69 Maria, P.-C.; Gal, J.-F. *J. Phys. Chem.* **1985**, *89*, 1296–1304.

70 Denmark, S. E.; Su, X. *Tetrahedron* **1999**, *55*, 8727–8738.

71 For definitions of these stereochemical terms, see: Denmark, S. E.; Almstfad, N. G. In *Modern Carbonyl Chemistry*; Otera, J., Ed.; Wiley–VCH: Weinheim, 2000; Chapter 10. pp. 300–301.

72 Enoxysilanes that react via boat-like transition structure involving trigonal bipyramidal silicon species: (a) Ref. 7(b).

(b) Gung, B. W.; Zhu, Z.; Fouch, R. A. *J. Org. Chem.* **1995**, *60*, 2860–2864.

73 Lodge, E., P.; Heathcock, C. H. *J. Am. Chem. Soc.* **1987**, *109*, 3353–3361.

74 (a) Reetz, M. T. *Angew. Chem. Int. Ed.* **1984**, *23*, 556–569. (b) Reetz, M. T. *Acc. Chem. Res.* **1993**, *26*, 462–468. (c) Reetz, M. T. *Chem. Rev.* **1999**, *99*, 1121–1162.

75 For reviews on diastereoselective aldol additions using chiral aldehydes, see: (a) Gennari, C. In *Comprehensive Organic Synthesis, Vol. 2, Additions to C–X π Bonds, Part 2*; Heathcock, C. H., Ed.; Pergamon Press: Oxford, 1991; pp. 639–647. (b) Ref. 1(k), pp 1713–1722.

76 Denmark, S. E.; Wong, K.-T.; Stavenger, R. A. *J. Am. Chem. Soc.* **1997**, *119*, 2333–2334.

77 Denmark, S. E.; Stavenger, R. A.; Wong, K.-T. *Tetrahedron* **1998**, *54*, 10389–10402.

78 Seebach, D.; Prelog, V. *Angew. Chem.* **1982**, *21*, 654–660.

79 Paterson, I.; Goodman, J. M.; Isaka, M. *Tetrahedron Lett.* **1989**, *30*, 7121–7124.

80 Denmark, S. E.; Stavenger, R. A. *J. Org. Chem.* **1998**, *63*, 9524–9527.

81 For a discussion of the coordinating abilities of various ether substituents, see: (a) Ref. 74(b). (b) Chen, X.; Hortellano, E. R.; Eliel, E. L.; Frye, S. V. *J. Am. Chem. Soc.* **1990**, *112*, 6130–6131. (c) Mori, S.; Nakamura, M.; Nakamura, E.; Koga, N.; Morokuma, K. *J. Am. Chem. Soc.* **1995**, *117*, 5055–5065.

82 Denmark, S. E.; Pham, S. M. *Org. Lett.* **2001**, *3*, 2201–2204.

83 For a discussion of chelation as a stereocontrol element in lactate-derived stannous enolate, see: Paterson, I.; Tillyer, R. *Tetrahedron Lett.* **1992**, *33*, 4233–4236.

84 (a) Martin, V. A.; Murray, D. H.; Pratt, N. E.; Zhao, Y.; Albizati, K. F. *J. Am. Chem. Soc.* **1990**, *112*, 6965–6978. (b) Paterson, I.; Oballa, R. M. *Tetrahedron Lett.* **1997**, *38*, 8241–8244. (c) Feutrill, J. T.; Lilly, M. J.; Rizzacasa, M. A. *Org. Lett.* **2002**, *4*, 525–527.

85 Denmark, S. E.; Fujimori, S. *Synlett* **2001**, 1024–1029.

86 Denmark, S. E.; Fujimori, S. *Org. Lett.* **2002**, *4*, 3473–3476.

87 (a) Paterson, I.; Gibson, K. R.; Oballa, R. M. *Tetrahedron Lett.* **1996**, *37*, 8585–8588. (b) Paterson, I.; Collett, L. A. *Tetrahedron Lett.* **2001**, *42*, 1187–1191. (c) Paterson, I.; Oballa, R. M.; Norcross, R. D. *Tetrahedron Lett.* **1996**, *37*, 8581–8584. (d) Evans, D. A.; Coleman, P. J.; Cote, B. *J. Org. Chem.* **1997**, *62*, 788–789.

88 Denmark, S. E.; Fujimori, S. *Org. Lett.* **2002**, *4*, 3477–3480.

89 Alcaide, B.; Almendros, P. *Angew. Chem. Int. Ed.* **2003**, *42*, 858–860.

90 (a) Northrup, A. B.; MacMillan, D. W. C. *J. Am. Chem. Soc.* **2002**, *124*, 6798–6799. (b) Kandasamy, S.; Notz, W.; Bui, T.; Barbas, C. F. III, *J. Am. Chem. Soc.* **2001**, *123*, 5260–5267.

91 (a) Yan, T.-H.; Tan, C.-W.; Lee, H.-C.; Lo, H.-C.; Huang, T.-Y. *J. Am. Chem. Soc.* **1993**, *115*, 2613–2621. (b) Heathcock, C. H.; White, C. T.; Morrison, J. J.; Van Derveer, D. *J. Org. Chem.* **1981**, *46*, 1296–1309.

92 PATERSON, I.; WALLACE, D. J.; COWDEN, C. J. *Synthesis* **1998**, 639–652.

93 DENMARK, S. E.; BUI, T. *Proc. Nat. Acad. Sci.* **2004**, *101*, 5439–5444.

94 DENMARK, S. E.; FU, J. *J. Am. Chem. Soc.* **2000**, *122*, 12021–12022.

95 (a) BASSINDALE, A. R.; STOUT, T. *J. Organomet. Chem.* **1982**, *238*, C41–C45. (b) BASSINDALE, A. R.; GLYNN, S. J.; TAYLOR, P. G. In *The Chemistry of Organic Silicon Compounds*; RAPPOPORT, Z.; APELOIG, Y., Eds.; Wiley: Chichester, 1998; Vol. 2, pp 495–511.

96 (a) GUTMANN, V. *The Donor–Acceptor Approach to Molecular Interactions*; Plenum Press: New York, 1978. (b) JENSEN, W. B. *The Lewis Acid–Base Concepts*; Wiley Interscience: New York, 1980; Chapter 4.

97 (a) CHOJNOWSKI, J.; CYPRYK, M.; MICHALSKI, J.; WOZNIAK, J. *J. Organomet. Chem.* **1985**, *288*, 275–282. (b) CORRIU, R. J. P.; DABOSI, G.; MARTINEAU, M. *J. Organomet. Chem.* **1980**, *186*, 25–37. (c) BASSINDALE, A. R.; LAU, J. C.-Y.; TAYLOR, P. G. *J. Organomet. Chem.* **1995**, *499*, 137–141.

98 DENMARK, S. E.; BARSANTI, P. A.; WONG, K.-T.; STAVENGER, R. A. *J. Org. Chem.* **1998**, *63*, 2428–2429.

99 DENMARK, S. E.; HEEMSTRA, J. R. JR. *Org. Lett.* **2003**, *5*, 2303–2306.

100 DENMARK, S. E.; BEUTNER, G. L. *J. Am. Chem. Soc.* **2003**, *125*, 7800–7801.

101 (a) FLEMING, I. *Frontier Orbitals and Organic Chemical Reactions*; Wiley–Interscience: New York, 1996; p. 40–47. (b) HERRMANN, J. L.; KIECZYKOWSKI, G. R.; SCHLESSINGER, R. H. *Tetrahedron Lett.* **1973**, *14*, 2433–2436.

102 For examples of stereoselective vinylogous aldol reactions see: (a) SAITO, S.; SHIOZAWA, M.; ITO, M.; YAMAMOTO, H. *J. Am. Chem. Soc.* **1998**, *120*, 813–814. (b) Ref. 13(b). (c) BLUET, G.; CAMPAGNE, J.-M. *J. Org. Chem.* **2001**, *66*, 4293–4298. (d) DE ROSA, M.; SORIENTE, A.; SCETTRI, A. *Tetrahedron Asymmetry* **2000**, *11*, 2255–2258. (e) EVANS, D. A.; KOZLOWSKI, M. C.; MURRY, J. A.; BURGEY, C. S.; CAMPOS, K. R.; CONNEL, B. T.; STAPLES, R. J. *J. Am. Chem. Soc.* **1999**, *121*, 669–685.

103 For a general discussion of this position see: DENMARK, S. E.; STAVENGER, R. A.; SU, X.; WONG, K.-T.; NISHIGAICHI, Y. *Pure & Appl. Chem.* **1998**, *70*, 1469–1476.

104 (a) GUILLANEUX, D.; ZHAO, S.-H.; SAMUEL, O.; RAINFORD, D.; KAGAN, H. B. Nonlinear Effects in Asymmetric Catalysis. *J. Am. Chem. Soc.* **1994**, *116*, 9430–9439. (b) KAGAN, H. B.; FENWICK, D. Asymmetric Amplification. *Top. Stereochem.* **1999**, *22*, 257–296. (c) AVALOS, M.; BABIANO, R.; CINTAS, P.; JIMENEZ, J. L.; PALACIOS, J. C. Nonlinear Stereochemical Effects in Asymmetric Reactions. *Tetrahedron: Asymmetry* **1997**, *8*, 2997–3017.

105 SINGLETON, D. A.; THOMAS, A. A. *J. Am. Chem. Soc.* **1995**, *117*, 9357–9358.

106 For a detailed discussion involving the determination of ^{13}C KIE from NMR integration and error analysis; Ref. 105.

107 PHAM, S. M. Ph.D. Thesis, University of Illinois, Urbana–Champaign, 2002.

108 A similar study has been completed in the addition of isobutyraldehyde trichlorosilyl enolate to benzaldehyde. Here as well, the aldolization step is shown to be rate-limiting; Ref. 93.

109 The importance of the ordering of the subsequent steps is at present unknown.

110 The configuration around the octahedral silicon cation is unknown. Moreover, given the divergent criteria for which would be more stable and which more reactive, a definitive answer must await computational analysis.

111 Attempts to identify stable complexes with silicon(IV) Lewis acids and phosphoramides have as yet been unsuccessful. However, a stable complex of bis(N-oxide) **46** with silicon tetrachloride has been analyzed crystallographically, FAN, Y. unpublished results from these laboratories.

112 DENMARK, S. E.; FU, J. *J. Am. Chem. Soc.* **2003**, *125*, 2208–2216.

113 For recent studies on linked phosphoramides in the aldol addition of ethyl ketone trichlorosilyl enolates; Ref. 45.

114 DENMARK, S. E.; FAN, Y. *J. Am. Chem. Soc.* **2003**, *125*, 7825–7827.

8
The Aldol–Tishchenko Reaction

R. Mahrwald

8.1
Introduction

The Tishchenko reaction has been known for almost 100 years [1]. The importance of catalysis in this reaction – dimerization of aldehydes to the corresponding esters and the polymerization of dialdehydes to the expected polyesters – has grown in the last 50 years. This reaction has great potential in stereoselective synthesis of defined stereocenters. Depending on the nature of substrates, reaction conditions, and catalysts, defined diastereoselective and enantioselective stereogenic centers can be created.

8.2
The Aldol–Tishchenko Reaction

The aldol–Tishchenko reaction was first studied at the beginning of the last century [2]. Although in many Tishchenko reactions the aldol–Tishchenko reaction is a competitive transformation, by use of the right reaction conditions and catalysts one can affect which pathway is taken. In recent years interest in this area has increased substantially. This reaction can be performed either with enolizable aldehydes (resulting in trimerization of aldehydes) or with ketones (resulting in formation of 1,3-diol monoesters).

8.2.1
The Aldol–Tishchenko Reaction with Enolizable Aldehydes

The classic aldol–Tishchenko reaction is used to obtain 1,3-diol monoesters by self-addition of aldehydes with at least one α-hydrogen [3]. In the first step of this reaction two molecules of aldehyde react by reversible aldol addition to give the expected aldol adduct; this is further reduced by a third molecule of aldehyde to give the 1,3-diol monoesters, **1** and **2** (Eq. (1)).

Modern Aldol Reactions. Vol. 2: Metal Catalysis. Edited by Rainer Mahrwald
Copyright © 2004 WILEY-VCH Verlag GmbH & Co. KGaA, Weinheim
ISBN: 3-527-30714-1

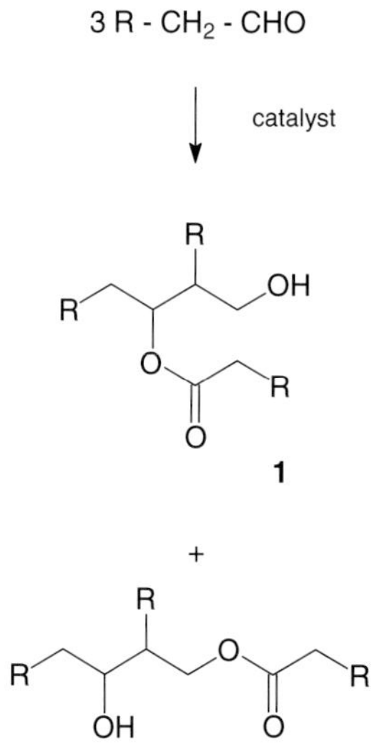

3 R - CH$_2$ - CHO

catalyst

1

+

2

Equation 1
Catalysts: magnesium-2,4,6-
trimethylphenoxide [4], Cp$_2$Sm(THF)$_2$ [5],
SmI$_2$ [6], LiO-*i*Pr [7], BINOL-Li [8],
Y$_2$O(O*i*Pr)$_{13}$ [9].

Merger et al. [10] reacted aldol adducts with aldehydes and isolated the 1,3-diol monoesters. These aldol–Tishchenko reactions were performed in the presence of metal alkoxides or without catalysts at higher temperatures. They showed that:

- the ester functionality does not come from the intermolecular combination of two aldehydes;
- hydride shift occurs in an intermediate equilibrium of hemiacetals and dioxanolen (Scheme 8.1);
- aldols are hydride acceptors – the carbonyl function will be reduced; and
- primary 1,3-diol monoesters are the thermodynamically stable products and they are formed by an acyl migration during the reaction.

Results from mechanistic study of the stereoselective aldol–Tishchenko reaction support the mechanism depicted in Scheme 8.1. First, a rapid aldol reaction occurs and by reaction with a further molecule of aldehyde the hemiacetal **4** is formed. Subsequent hydride transfer (in the intermediate equilibrium of hemiacetals and dioxanolen) yields the 1,3-diol monoester **5**.

There are only two examples of enantioselective execution of the aldol–Tishchenko reaction of aldehydes. Loog and Mäeorg used chiral binaph-

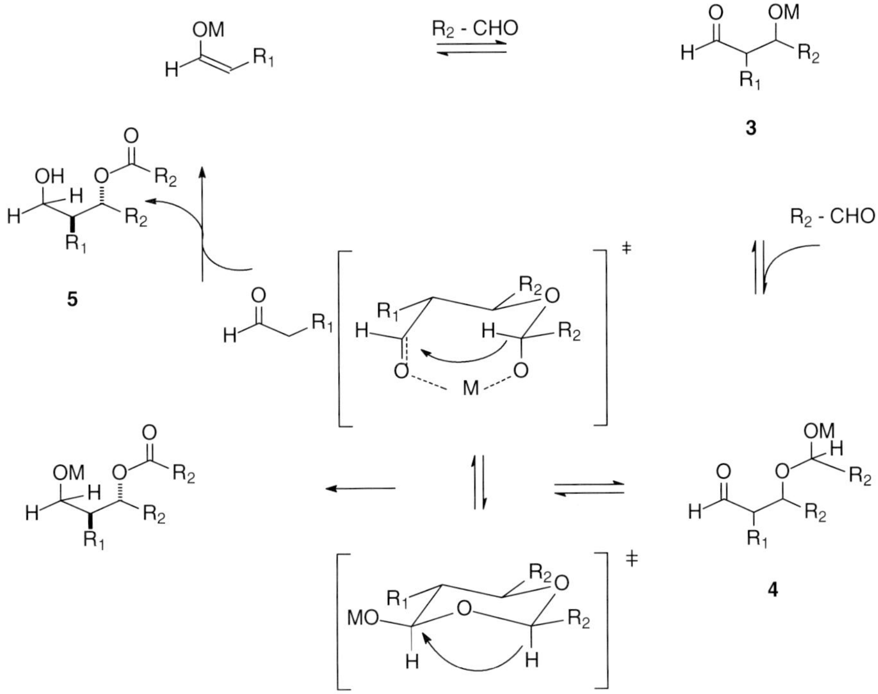

Scheme 8.1

tholate catalysts to investigate the stereochemistry of the self-addition of 2-methylpropanal [8]. 1,3-Diol monoesters were obtained with low enantio-selectivity (ee > 30%). Morken et al. recently published details of an asymmetric mixed aldol–Tishchenko reaction of aromatic aldehydes with 2-methylpropanal catalyzed by salen complexes of yttrium [9]. The results are shown in Table 8.1.

These are the first examples of enantioselective catalytic aldol–Tishchenko reactions of two different aldehydes.

8.2.2
The Aldol–Tishchenko Reaction with Ketones and Aldehydes

Ketones and aldehydes also undergo an aldol–Tishchenko reaction. Three adjacent stereogenic centers can be created by use of these reactants in the aldol–Tishchenko reaction whereas only two stereogenic centers can be formally produced by reacting enolizable aldehydes with aldehydes (Eq. (1)). This is a very effective reaction sequence in terms of chiral economy [11]. The nomenclature used in Eq. (2) and Scheme 8.2 is used throughout the following sections.

Tab. 8.1

Enantioselective aldol–Tishchenko reaction in the presence of chiral yttrium complexes.

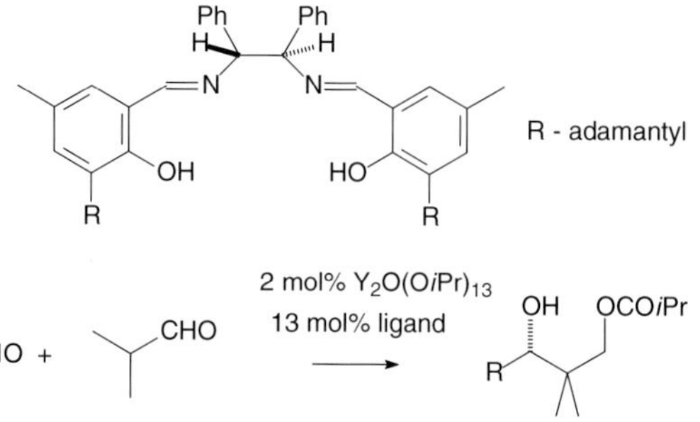

Entry	Substrate	Yield [%]	e.r. (Configuration)
1	Phenyl	70	87:13 (S)
2	4-Bromphenyl	55	85:15 (S)
3	Naphthyl	50	82:18
4	4-Methoxyphenyl	21	86:14
5	3-Phenyl-2-propenyl	50	55:45

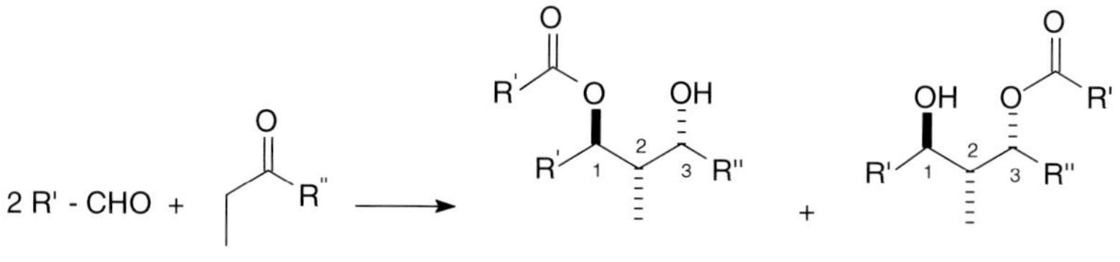

6

1,3-diol 1-monoester

7

1,3-diol 3-monoester

Equation 2

Catalysts: nickel enolates [12], SmI$_2$ [6, 13], zinc enolates [14], titanium ate complexes [15], LDA [16–18], Ti(OiPr)$_4$ [19].

The 1,3-diol 1-monoester **6** and the corresponding 3-monoester **7** were prepared with high simple stereoselectivity. Only one of the four possible diastereoisomers has been formed in all examples described in the literature. On the basis of the transition state shown in Scheme 8.1 the 1,3-diol 1-monoester **6** was formed in the 1,2-*anti*, 1,3-*anti* configuration (Scheme 8.2). The diol 3-monoester **7** was again formed by acyl migration during reaction. The extent of this migration usually depends on the steric bulkiness of the starting aldehydes.

Heathcock et al. showed that isolated nickel ketone enolates react with

$2\,R_1\text{-CHO}\,+$ [ketone structure]

6

1,2-*anti*, 1,3-*anti* 1,3-diol 1-monoester

‡

7

1,2-*anti*, 1,3-*anti* 1,3-diol 3-monoester

Entry	R_1	Diastereoselectivity
1	Ph	99 : 1
2	*t*Bu	98 : 2
3	*i*Pr	98 : 2
4	*n*Pr	97 : 3

Scheme 8.2
Reaction conditions: titanium ate complexes [15].

benzaldehyde to furnish products resulting from an aldol–Tishchenko reaction [12]. They also established the 1,2-*anti*, 1,3-*anti* configuration of
the isolated 1,3-diol monoester. These are the first examples of an aldol–
Tishchenko reaction of ketones with aldehydes. Later we found that 1,3-diol
monoesters **6** and **7** were formed with high stereoselectivity by an one-pot
aldol–Tishchenko reaction of ketones with aldehydes in the presence of
substoichiometric amounts of titanium ate complexes [15]. An instructive
example for the direction of stereochemistry during the aldol–Tishchenko
reaction is the observation that the 1,2-*anti*, 1,3-*anti* configuration of the
isolated diol monoesters **6** and **7** is independent of the configuration of the
assumed starting aldol. To demonstrate this, we have reacted the pure *syn*
aldol **8** of benzaldehyde and diethylketone with one equivalent of benzaldehyde in the presence of catalytic amounts of titanium ate complexes (Eq.

(3)). Under these conditions the 1,2-*anti*, 1,3-*anti* configured diol mono-esters **9** and **10** were formed exclusively.

Equation 3
Reaction conditions: 10 mol% BuTi(O*i*Pr)$_4$Li.

Other authors have also described achieving the same stereodirection by use of catalytic amounts of metal alkoxides [7] or LDA [16–18] (Eq. (2)). Three applications of this reaction are shown in Eqs. (4)–(6). A samarium ion-catalyzed aldol–Tishchenko reaction combined with a reductive cycliza-tion process was reported by Curran and Wolin [13]. Only one isomer was detected during this transformation (Eq. (4)). The configuration of the 1,3-diol monoester **12** (1,2-*anti*, 1,3-*anti*) was the same as that shown in Scheme 8.2 and Eq. (3).

Equation 4
SmI$_2$, Ph(CH$_2$)$_2$CHO, 81%.

13   **14**

Equation 5
LDA, PhCHO, 56%.

15   **16**   **17**

Equation 6
1. SmI_2, MeCHO; 2. K_2CO_3, MeOH, 96%.

In another example, the keto epoxide **13** was reacted with LDA and benzaldehyde to give the hydroxyester **14** as a single isomer (Eq. (5)) [16]. Because of the missing stereogenic center in the epoxide **13** only one new stereogenic center was created. In terms of the stereochemistry only this example seems to lie between the Evans–Tishchenko reduction and the aldol–Tishchenko reaction.

A samarium-catalyzed aldol–Tishchenko reaction has been used to synthesize the intermediate **17** in a highly convergent synthesis of luminacin D [20]. The diol **17** was obtained as a single isomer (Eq. (6)).

The aldol–Tishchenko reaction has also been used in the synthesis of 1'-branched chain sugar nucleosides. The 1'-hydroxymethyl group was introduced by an Sm_2-promoted aldol–Tishchenko reaction of 1'-phenylseleno-2'-ketouridine **18** with aldehydes (Eq. (7)). This is the first example of generation of an enolate by reductive cleavage of a C–Se bond by SmI_2 [21].

18   **20**

Equation 7
Synthesis of 1'-branched nucleosides.

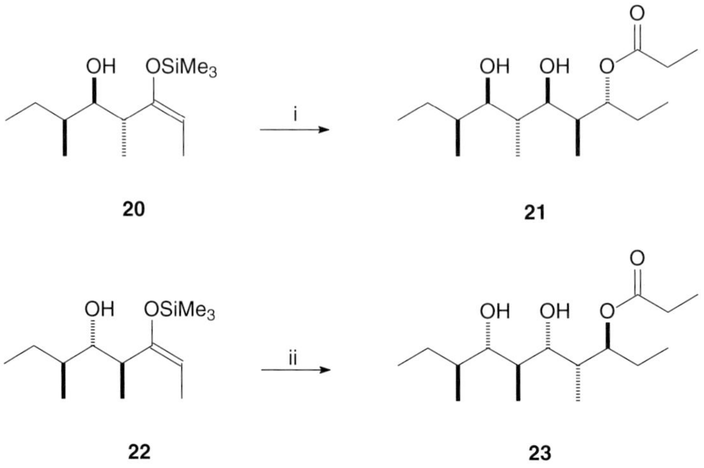

Scheme 8.3
Reaction condition: (i) 10 mol% Ti(O*i*Pr)₄,
EtCHO, 0 °C, 91%; (ii) 10 mol% Ti(O*i*Pr)₄,
EtCHO, 0 °C, 90%.

An interesting reaction was reported by Delas et al. [19]. They described aldol–Tishchenko reactions of enolsilanes (activated ketones) with aldehydes in the presence of Ti(O*i*Pr)₄ – a variation of the classic aldol–Tishchenko reaction (Scheme 8.3). They were able to obtain the stereosixtades **21** and **23** – compounds with six defined adjacent stereogenic centers – in one reaction step. The diastereoselectivity observed was very high and the stereosixtades were obtained as single isomers. This example indicates that the stereodirection of the aldol–Tishchenko reaction can be affected. Oxygen-containing functionality in the starting enolsilanes **20** and **22** has a useful stereodirecting effect on this transformation and on the configuration of the stereosixtades **21** and **23**. This paper pioneered the field of stereoselective aldol–Tishchenko reactions. The two examples given in Scheme 8.3 show the stereochemical potential of this process.

Schneider et al. recently published an enantioselective approach to chiral 1,3-*anti*-diol monoesters. Although at first glance this transformation seems to be a Tishchenko reduction of an acetate aldol (Section 8.2.3) inspection of the mechanism furnishes evidence of a retro-aldol/aldol–Tishchenko reaction. By using 10 mol% Zr(O*t*Bu)₄-TADDOL the 1,3-diol monoesters were obtained with moderate enantioselectivity (Table 8.2) [22].

8.2.3
The Evans–Tishchenko Reduction

The Evans–Tishchenko reduction is a special case of the aldol–Tishchenko reaction. The starting material is an aldol adduct, usually an acetate aldol. During the reaction the keto functionality of the starting aldol is reduced by

Tab. 8.2
Enantioselective aldol–Tishchenko reaction of aldehydes and ketones.

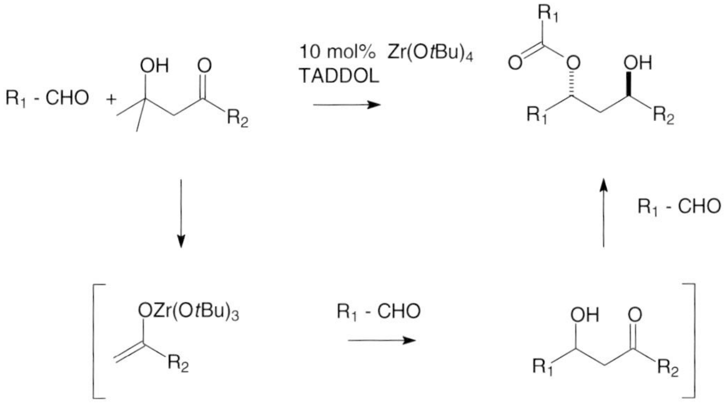

Entry	R_1	R_2	Yield [%]	ee [%]
1	tBu	nHex	88	42
2	tBu	iPr	84	57
3	tBu	cHex	75	50
4	tBu	2-ethylpropyl	69	47

reaction with an aldehyde in the presence of a Lewis acid. This reaction was first described and elaborated on by Evans and Hoyveyda [23]. The reaction was performed in the presence of substoichiometric amounts of SmI_2. The 1,3-diols were isolated in excellent yields with high *anti* stereoselectivity ($> 99:1$). The transition structure proposed for the samarium-catalyzed reduction is given in Scheme 8.4. It is very close to those described in previous sections.

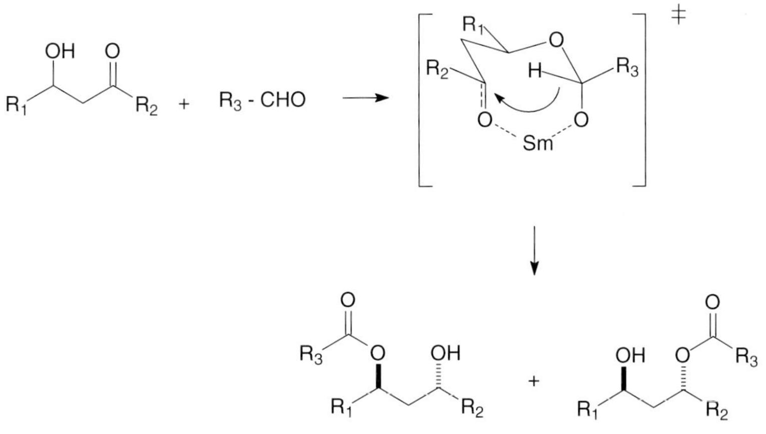

Scheme 8.4
Catalysts: SmI_2 [24–29] BuLi [30, 31], zirconocene complexes [32], $Sc(OTf)_3$ [33], ArMgBr [34].

Several other metal compounds were subsequently found to induce this reduction (Scheme 8.4). Few authors have described the acyl migration as a result of this reduction that one could expect from the reaction mechanism [30, 34]. This is an unusual result. Five applications of the SmI$_2$-mediated reduction of hydroxy ketones in natural product synthesis are given in Scheme 8.5.

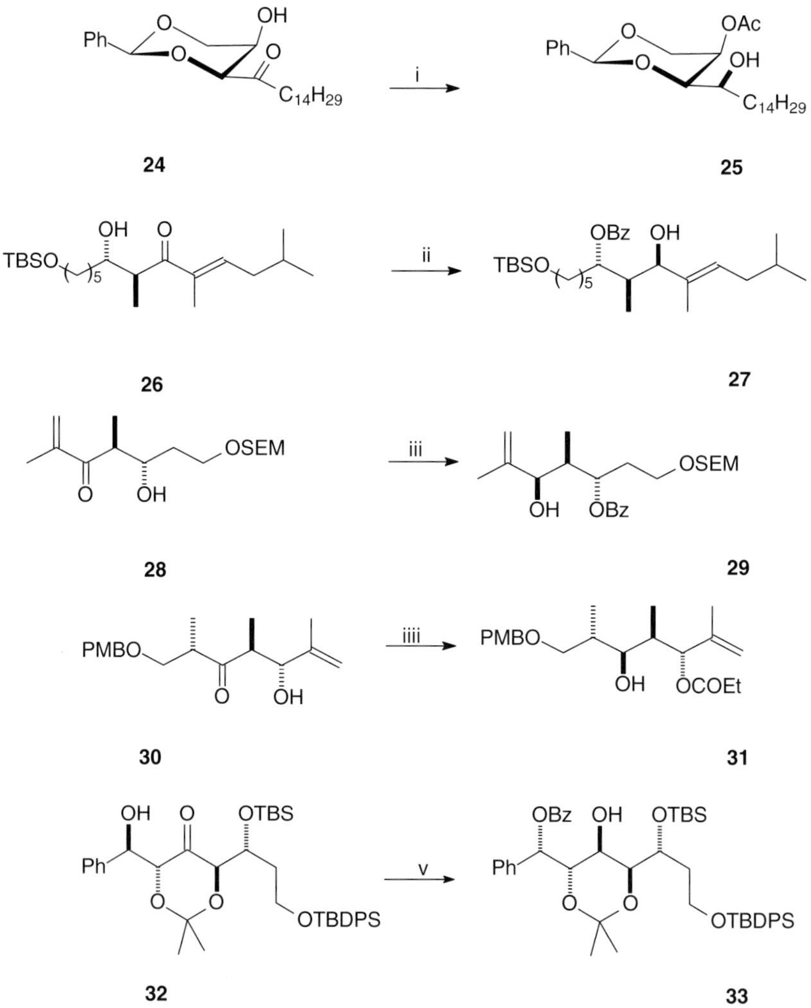

Scheme 8.5
(i) CH$_3$CHO, SmI$_2$, 80% [25]; (ii) PhCHO, SmI$_2$, 70% [26]; (iii) PhCHO, SmI$_2$, 85% [27]; (iv) EtCHO, SmI$_2$, 97% [29]; (v) SmI$_2$, PhCHO, 95% [35].

The broad variety of functional groups which can be used in this reaction are represented in these substrates. An interesting example is reduction of the hydroxyketone **32** to the hydroxybenzoate **33**. The authors obtained a single 1,2-*syn*, 1,3-*anti*-configured diastereoisomer [35]. Exclusive formation of the 1,2-*syn*, 1,3-*anti*-configured hydroxybenzoate **33** is observed, irrespective of the 1,2-*anti* configuration of the starting hydroxyketone **32**. This reaction could offer an approach to natural products containing a 1,2-*syn* configuration; these have previously been unattainable by contemporary aldol additions.

Two spectacular examples of the use of the Evans–Tishchenko reduction in natural product synthesis are given in Scheme 8.6. In 1993 Schreiber et

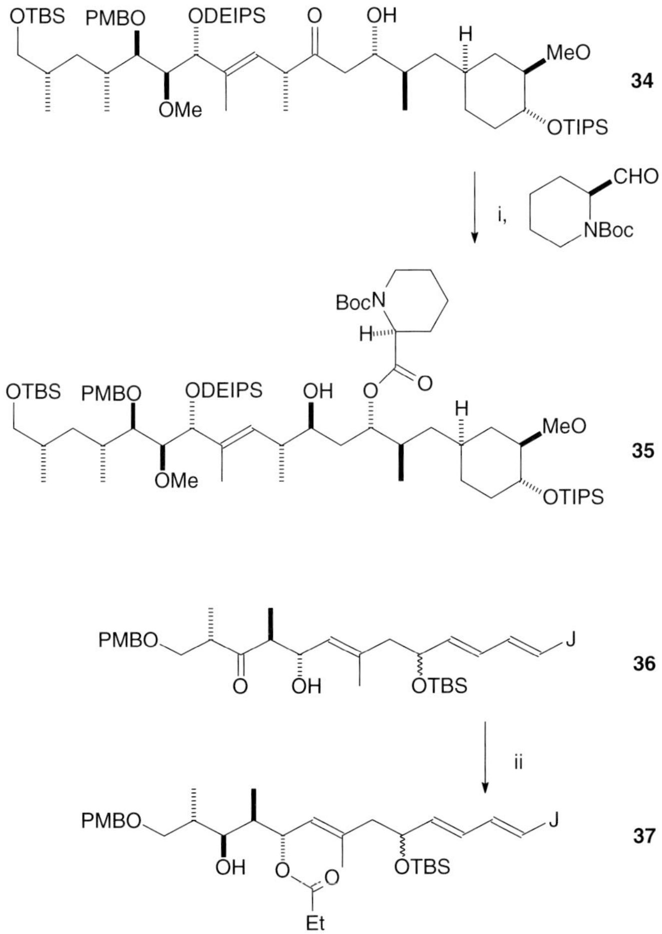

Scheme 8.6
(i) SmI$_2$–PhCHO complex, 95%; (ii) SmI$_2$, EtCHO, 92%.

al. used this reaction in the total synthesis of rapamycin to obtain the intermediate **35** [24]. This 1,3-diol monoester **35** was obtained with the correct and required stereochemistry as a single isomer by a Tishchenko reduction of ketone **34** in the presence of 30 mol% (PhCHO)SmI–SmI$_3$ (Scheme 8.6). The formation of this complex was described by Evans and Hoyveyda [23].

Paterson et al. used the Tishchenko reduction successfully for several total syntheses. In the synthesis of callipeltoside they needed the intermediate **37** for the synthesis of the aglycone [28]. Again, Tishchenko reduction of ketone **36** with propionaldehyde in the presence of SmI$_2$ yielded the diol monoester **37** with the required 1,2-*anti*, 1,3-*anti* configuration (Scheme 8.6).

Evans et al. [36] used this procedure in the total synthesis of bryostatin 2. Starting from the corresponding ketone **38** they obtained the ketone **39**, with the required configuration of the hydroxy group, by samarium-catalyzed Tishchenko reduction (Eq. (8)). These examples show the broad application of this highly stereoselective Tishchenko reduction process.

38

39

Equation 8
Reaction conditions: SmI$_2$, *p*-NO$_2$C$_6$H$_4$CHO; ds > 95:5, 76%.

The Evans–Tishchenko reduction also provides an efficient and practical solution for the oxidation of aldehydes containing sensitive electron-rich heteroatoms (e.g. aldehyde **40**, Eq. (9)). Careful selection of the sacrificial β-hydroxy ketone subsequently provides very flexible access to the desired carboxylic acid **41** (Eq. (9)). This methodology was used in total synthesis of (+)-13-deoxytedanolide [37].

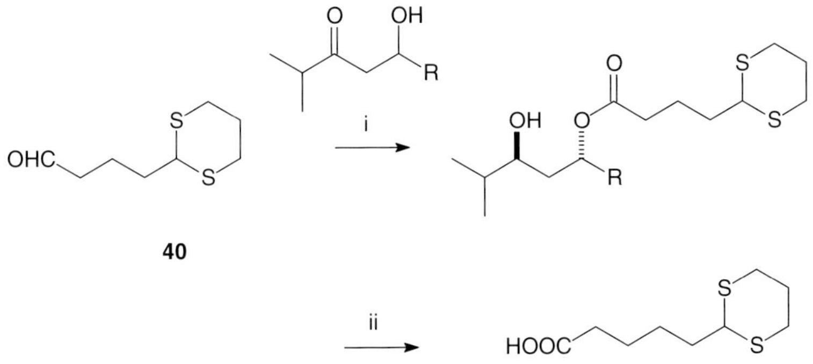

40

Equation 9
Reaction conditions: (i) 20 mol% SmI$_2$, THF, −10 °C; (ii) LiOH, aqueous MeOH.

8.2.4
Related Reactions

The samarium-mediated coupling reaction of vinyl esters with aldehydes has been described [38]. These reactions were performed with enolizable and aromatic aldehydes. Unsymmetrical diesters such as **42** were formed by means of this transformation (Scheme 8.7).

On the basis of labeling studies (reaction of vinyl acetate with PhCOD) the mechanism given in Scheme 8.7 seems plausible. An eight-membered alkoxy samarium species might be the key intermediate in this reaction. Subsequent intramolecular hydride shift, as known from the Tishchenko reaction, produces the diester.

The pinacol–Tishchenko reaction has recently been described (Eq. (10)), Figure 8.1) [39]. α-Hydroxy epoxides **43** were reacted with SmI$_2$, efficiently forming the 2-quaternary 1,3-diol monoesters **44** and **45** with high diaster-

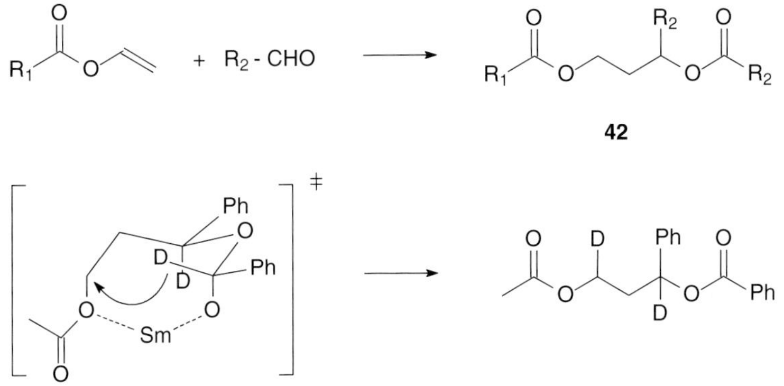

42

Scheme 8.7
Reaction conditions: 10 mol% Cp$_2$Sm(THF)$_2$.

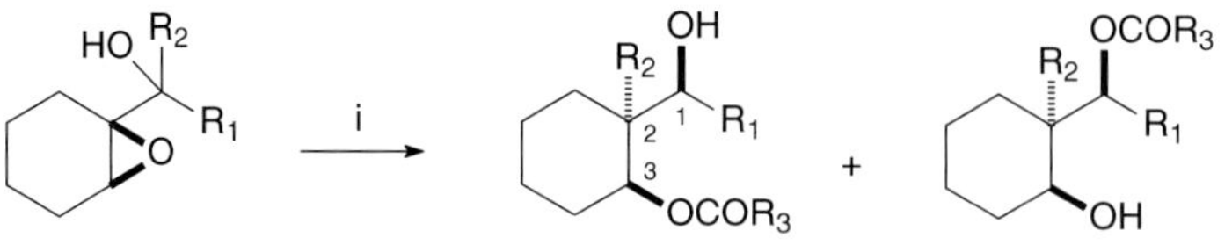

1,2-*anti*, 1.3-*anti*-diol   1,2-*anti*,1.2-*anti*-diol
1-monoester   3-monoester

43   **44**   **45**

Equation 10

Reaction conditions: (i) 10–30 mol% SmI$_2$;
R$_1$ = R$_2$ = Ph, 95%; R$_1$ = Me, R$_2$ = Ph,
96%; R$_1$ = Et, R$_2$ = Ph, 92%; R$_1$ = iPr,
R$_2$ = Ph, 92%.

eoselectivity. The reaction is similar to those already described. Rearrangement of the starting α-hydroxy epoxide **43** occurs under the reaction conditions described. Insertion into the M–O bond occurs on addition of the aldehyde R$_3$CHO, and as a consequence the hemiacetal is formed (Figure 8.1). Hydride transfer results in the formation of the 1,3-diol monoesters **44** and **45**. Only the 1,2-*anti*, 1,3-*anti* configured diol monoester is formed. This configuration is again independent of the configuration of the starting α-hydroxy epoxides. The corresponding 1,2-*syn* configured products were not detected.

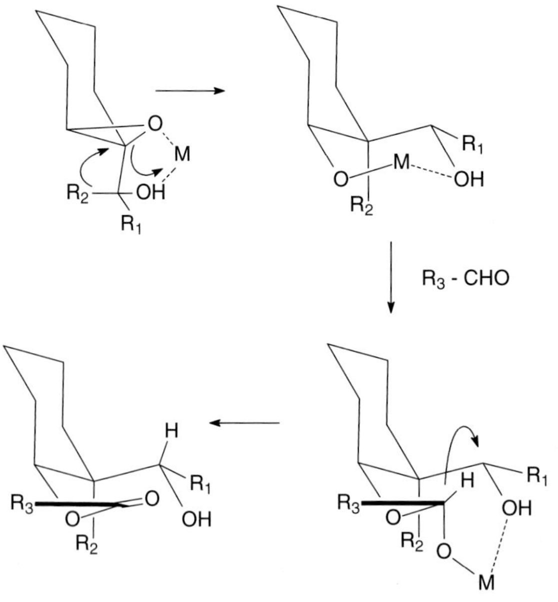

Fig. 8.1

It is clear that the Tishchenko reaction and its variations are valuable additions to the repertoire of the chemist interested in stereoselective synthesis. The exceptionally high stereoselectivity obtained in this reaction is fascinating (de > 95:5 for all the examples described). Only one diastereoisomer was formed during the reaction in all other examples.

Although much has been achieved there still is a need for control of the remaining directions of stereochemistry. This is true not only for the diastereoselectivity but also for the control of enantioselectivity.

8.3
Representative Procedures

Typical Procedure for the Aldol–Tishchenko Reaction: *rac*(1S,2S,3R)-3-Hydroxy-2-methyl-1-phenylpent-3-yl benzoate (9) and *rac*(1S,2R,3S)-1-Hydroxy-2-methyl-1-phenylpent-1-yl benzoate (10) [15]. BuLi (0.64 mL, 1.0 mmol in hexane) was carefully added, under inert conditions, to a solution of titanium(IV) *iso*-propoxide (0.32 mL, 1.0 mmol) in 1-*tert*-butoxy-2-methoxyethane (1.5 mL). After stirring for 30 min at room temperature pentan-3-one (0.5 mL, 5 mmol) and then benzaldehyde (1.0 mL, 10 mmol) were added. The solution was stirred for further 24 h at room temperature. Diethyl ether (50 mL) was added and the organic phase was extracted with water until neutral. The organic layer was isolated, dried (Na_2SO_4), filtered, and evaporated under vacuum. The pure products **9** and **10** were separated by flash chromatography with hexane–*iso*PrOH as eluent (95:5). Yield 63% (ratio of **9**/**10** = 95:5).

Typical Procedure for the Tishchenko Reduction: (7S,9R,10R)-1-Benzenesulfonyl-2,2-dimethyl-9-hydroxy-10-(*para*-methoxybenzyloxy)-7-(*para*-nitrobenzoyloxy)-undecan-3-one (39) [36]. Freshly prepared samarium diiodide (0.1 M in THF, 38 mL, 38 mmol, 0.24 equiv.) was added dropwise to a cooled (0 °C) solution of ketone **38** (7.86 g, 15.6 mmol) and *p*-nitrobenzaldehyde (23.6 g, 156 mmol. 10 equiv.) in 200 mL THF. The reaction was stirred in the dark under an argon atmosphere for 5.5 h then quenched with 150 mL sat. aqueous $NaHCO_3$. The mixture was partitioned between 250 mL EtOAc and 250 mL aqueous sat. $NaHCO_3$ and the aqueous phase was extracted with EtOAc (3 × 100 mL). The organic extracts were combined, washed with brine (1 × 75 mL), dried ($MgSO_4$), filtered, and evaporated in vacuo.

A small sample was removed and filtered through a plug of silica gel with 50% EtOAc–hexane as eluent. All the fractions containing the product were combined and assayed by HPLC (Zorbax silica gel, 0.7% EtOH–CH_2CL_2, 1.0 mL min^{-1}, 254 nm UV cut-off) to reveal the product was a 93:7 mixture of diastereoisomers (retention time major product 23.9 min; retention time minor product 32.7 min) which were shown to be from the previous aldol addition. The other diastereomer from the Tishchenko reduction could not be isolated and the only other compound visible in the HPLC chromatogram was present at <3%.

The bulk of the residue was purified by flash chromatography (8 cm × 15 cm silica gel, elution with 1.5 L of 5% EtOAc in CH_2Cl_2 and 1 L each 9, 10, 11, and 12% EtOAc in CH_2Cl_2; the mixed fractions were re-purified (8 cm × 12 cm silica gel with 9 and 10% EtOAc in CH_2Cl_2, then 6 cm × 15 cm silica gel with 10% EtOAc in CH_2Cl_2) to give 8.66 g (85% yield) of alcohol **39** as a rusty orange foam.

References

1 TISHCHENKO, W. *Chem. Zentralbl.* **1906**, *77*, 1309; TISHCHENKO, W. *Chem. Zentralbl.* **1906**, *77*, 1552.

2 (a) FRANKE, A.; KOHN, L. *Monatsh. Chem.* **1889**, *19*, 354; (b) NEUSTÄDTER, V. *Monatsh. Chem.* **1906**, *27*, 879; (c) MORGENSTERN, M. *Monatsh. Chem.* **1903**, *24*, 579; (d) KIRSCHBAUM, M. A. *Monatsh. Chem.* **1904**, *25*, 249.

3 For a review see: PETERSON, J.; TIMOTHEUS, H.; MAEORG, U. *Proc. Estonian Acad. Sci. Chem.* **1997**, *46*, 93.

4 (a) CASNATI, G.; POCHINI, A.; SALERNO, G.; UNGARO, R. *Tetrahedron Lett.* **1974**, *12*, 959; (b) POCHINI, A.; SALERNO, G.; UNGARO, R. *Synthesis* **1975**, 164; (c) CASNATI, G.; POCHINI, A.; SALERNO, G.; UNGARO, R. *J. Chem. Soc., Perkin Trans. I*, **1975**, 1527.

5 MIYANO, A.; TASHIRO, D.; KAWASAKI, Y.; SAKAGUCHI, S.; ISHII, Y. *Tetrahedron Lett.* **1998**, *39*, 6901.

6 LU, L.; CHANG, H.-Y.; FANG, J.-M. *J. Org. Chem.* **1999**, *64*, 843.

7 MASCARENHAS, C. M.; DUFFEY, M. O.; LIU, S.-Y.; MORKEN, J. P. *Org. Lett.* **1999**, *1*, 1427.

8 LOOG, O.; MAEORG, U. *Tetrahedron: Asymmetry* **1999**, *10*, 2411.

9 MASCARENHAS, C. M.; MILLER, S. P.; WHITE, P. S.; MORKEN, J. P. *Angew. Chem.* **2001**, *113*, 621, *Angew. Chem., Int. Ed. Engl.* **2001**, *40*, 601.

10 FOUQUET, G.; MERGER, F.; PLATZ, R. *Liebigs Ann. Chem.* **1979**, 1591.

11 For a discussion of "atom economy", see: a) TROST, B. *Science* **1991**, *254*, 1471; (b) TROST, B. M. *Angew. Chem., Int. Ed. Engl.* **1995**, *34*, 259.

12 BURKHARDT, E. R.; BERGMAN, R. G.; HEATHCOCK, C. H. *Organometallics*, **1990**, *9*, 30.

13 CURRAN, D. P.; WOLIN, R. L. *Synlett* **1991**, 317.

14 HORIUCHI, Y.; TANIGUCHI, M.; OSHIMA, K.; UTIMOTO, K. *Tetrahedron Lett.* **1995**, *36*, 5353.

15 MAHRWALD, R.; COSTISELLA, B. *Synthesis* **1996**, 1087.

16 BARAMEE, A.; CHAICHIT, N.; INTAWEE, P.; THEBTARANONTH, C.; THEBTARANONTH, Y. *J. Chem. Soc., Chem. Comm.* **1991**, 1016.

17 BODNAR, P. M.; SHAW, J. T.; WOERPEL, K. A. *J. Org. Chem.* **1997**, *62*, 5674.

18 ABU-HASANAYAN, F.; STREITWEISER, A. *J. Org. Chem.* **1998**, *63*, 2954.

19 DELAS, C.; BLACQUE, O.; MOISE, C. *J. Chem. Soc., Perkin Trans. 1* **2000**, 2265.

20 SHOTWELL, J. B.; KRYGOWSKI, E. S.; HINES, J.; KOH, B.; HUNTSMAN, E. W. D.; CHOI, H. W.; SCHNEEKLOTH, J. S.; WOOD, J. L.; CREWS, C. M. *Org. Lett.* **2002**, *4*, 3087.

21 KODAMA, T.; SHUTO, S.; ICHIKAWA, S.; MATSUDA, A. *J. Org. Chem.* **2002**, *67*, 7706.

22 (a) SCHNEIDER, C.; HANSCH, M. *Synlett* **2003**, 837; (b) SCHNEIDER, C.; HANSCH, M. *Chem. Comm.* **2001**, 1218.

23 EVANS, D. A.; HOYVEYDA, A. H. *J. Am. Chem. Soc.* **1990**, *112*, 6447.

24 ROMO, D.; MEYER, S. D.; JOHNSON, D. D.; SCHREIBER, S. L. *J. Am. Chem. Soc.* **1993**, *115*, 7906.

25 WILD, R.; SCHMIDT, R. R. *Tetrahedron: Asymmetry* **1994**, *5*, 2195.

26 HULME, A. N.; HOWELLS, G. E. *Tetrahedron Lett.* **1997**, *47*, 8245.

27 SCHÖNING, K.-U.; HAYASHI, R. K.; POWELL, D. R. KIRSCHNING, A. *Tetrahedron: Asymmetry* **1999**, *10*, 817.

28 PATERSON, I.; DAVIES, R. D. M.; MARQUEZ, R. *Angew. Chem.* **2001**, *113*, 623, *Angew. Chem., Int. Ed. Engl.* **2001**, *40*, 603.

29 PATERSON, I.; FLORENCE, G. D.; GELACH, K.; SCOTT, J. P.; SEREINIG, N. *J. Am. Chem. Soc.* **2001**, *123*, 9535.

30 AOKI, Y.; OSHIMA, K.; UTIMOTO, K. *Chem. Lett.* **1995**, 463.

31 IWAMOTO, K.; CHATANI, N.; MURAI, S. *J. Organomet. Chem.* **1999**, *574*, 171.

32 UMEKAWA, Y.; SAKAGUCHI, S.; NISHIYAMA, Y.; ISHII, Y. *J. Org. Chem.* **1997**, *62*, 3409.

33 GILLESPIE, K. M.; MUNSLOW, I. J.; SCOTT, P. *Tetrahedron Lett.* **1999**, *40*, 9371.

34 MELLOR, J. M.; REID, G.; EL-SAGHEER, A. H.; EL-TAMANY, E. S. H. *Tetrahedron* **2000**, *56*, 10039.

35 ENDERS, D.; INCE, S. J.; BONNEKESSEL, M.; RUNSINK, J.; RAABE, G. *Synlett*, **2002**, 962.

36 EVANS, D. A.; CARTER, P. H.; CARREIRA, E. M.; CHARETTE, A. B.; PRUNET, J. A.; LAUTENS, M. *Angew. Chem.* **1998**, *110*, 2526, *Angew. Chem. Int. Ed. Engl.* **1998**, *37*, 2354; EVANS, D. A.; CARTER, P. H.; CARREIRA, E. M.; CHARETTE, A. B.; PRUNET, J. A.; LAUTENS, M. *J. Am. Chem. Soc.* **1999**, *121*, 7540.

37 (a) SMITH, A. B. III; LEE, D.; ADAMS, C. M.; KOZLOWSKI, M. C. *Org. Lett.* **2002**, *4*, 4539; (b) SMITH, A. B. III; ADAMS, C. M.; BARBOSA, S. A. L.; DEGNAN, A. P. *J. Am. Chem. Soc.* **2003**, *125*, 350.

38 TAKENO, M.; KIKUCHI, S.; MORITA, K.-I.; NISHIYAMA, Y.; ISHII, Y. *J. Org. Chem.* **1995**, *60*, 4974.

39 (a) FAN, C. A.; WANG, B. M.; TU, Y. Q.; SONG, Z. L. *Angew. Chem.* **2001**, *113*, 3995, *Angew. Chem., Int. Ed. Engl.* **2001**, *40*, 3877; (b) TU, Y. Q.; SUN, L. D.; WANG, P. Z. *J. Org. Chem.* **1999**, *64*, 629; (c) WANG, F.; TU, Y. Q.; FAN, C. A.; WANG, S. H.; ZHANG, F. M. *Tetrahedron: Asymmetry* **2002**, *13*, 395.

Common Abbreviations Used in the Text

Ac   Acetate
Ar   Aryl

BINOL   1.1′-bi-2-Naphthol
Bn   Benzyl
Boc   *tert*-Butoxycarbonyl
BOM   Benzyloxymethyl
Bu   Butyl
Bz   Benzoate
Cp   Cyclopentadienyl
DEIPS   Diethylisopropylsilyl
Et   Ethyl
Hex   Hexyl
LDA   Lithium diisopropylamide
Me   Methyl
Ph   Phenyl
PMB   *p*-Methoxybenzyl
Pr   Propyl
SEM   2-(Trimethylsilyl)methyl
TBDPS   *tert*-Butyldiphenylsilyl
TBS   *tert*-Butyldimethylsilyl
Tf   Trifluoromethanesulfonyl
THF   Tetrahydrofuran
TIPS   Triisopropylsilyl

Index

Numbers in front of the page numbers refer to Volumes 1 and 2: e.g., 2/250 refers to page 250 in volume 2

Modern Aldol Reactions. Vol. 2: Metal Catalysis. Edited by Rainer Mahrwald
Copyright © 2004 WILEY-VCH Verlag GmbH & Co. KGaA, Weinheim
ISBN: 3-527-30714-1